D0926438

Lenk's Digital Handbook

Design and Troubleshooting

Dedication

Greetings from the Villa Buttercup!
To my wonderful wife Irene, Thank you
for being by my side all these years!
To my lovely family,
Karen, Tom, Brandon, and Justin.
And to our Lambie and Suzzie,
be happy wherever you are!
To my special readers, may good
fortune find your doorways
to good health and happy things.
Thank you for buying my books
and making me a best-seller!
This is book number 75.
Abundance!

Lenk's Digital Handbook

Design and Troubleshooting

John D. Lenk

McGraw-Hill, Inc.

New York St. Louis San Francisco Auckland Bogotá Caracas
Lisbon London Madrid Mexico City Milan Montreal New Delhi
Paris San Juan São Paulo Singapore Sydney Tokyo Toronto

Notices

IBM®	International Business Machines, Inc.
IBM PC®	
IBM PC-XT™	
IBM PC-AT™	
MS-DOS™	Microsoft Corporation
RCA®	Radio Corporation of America
Teletype®	Teletype Corporation

FIRST EDITION
FIRST PRINTING

Information contained in this work has been obtained by McGraw-Hill, Inc., from sources believed to be reliable. However, neither McGraw-Hill nor its authors guarantees the accuracy or completeness of any information published herein and neither McGraw-Hill nor its authors shall be responsible for any errors, omissions, or damages arising out of use of this information. This work is published with the understanding that McGraw-Hill and its authors are supplying information but are not attempting to render engineering or other professional services. If such services are required, the assistance of an appropriate professional should be sought.

Library of Congress Cataloging-in-Publication Data

Lenk, John D.
[Digital handbook]
Lenk's digital handbook : design and troubleshooting / by John D. Lenk.
p. cm.
Includes index.
ISBN 0-07-037516-X
1. Digital electronics—Handbooks, manuals, etc. I. Title.
TK7868.D5L43 1992
621.381—dc20 92-12064
CIP

For information about other McGraw-Hill materials, call 1-800-2-MCGRAW in the U.S. In other countries call your nearest McGraw-Hill office.

Acquisitions Editor: Dan Gonneau
Editor: B.J. Peterson
Managing Editor: Sandra Johnson-Bottomley
Director of Production: Katherine G. Brown
Book Design: Jaclyn J. Boone

EL1

Contents

Acknowledgments

Many professionals have contributed to this book. I gratefully acknowledge that the tremendous effort needed to make this book such a comprehensive work is impossible for one person and wish to thank all who contributed, both directly and indirectly.

I wish to give special thanks to the following: Dick Harmon, Ross Snyder, Nancy Teater, Mike Arnold, Owen Brown, Glen Green, Mike Ryman, Michael Slater, Barry Bronson, and Joel Salsberg of Hewlett-Packard; Bob Carlson and Martin Plude of B&K-Precision Dynascan Corporation; Joe Cagle and Rinaldo Swayne of Alpine/Luxman; Theodore Zrebiec of Sony; John Taylor and Matthew Mirapaul of Zenith; Thomas Lauterback of Quasar; Donald Woolhouse of Sanyo; J.W. Phipps of Thomson Consumer Electronics (RCA); Tom Roscoe, Dennis Yuoka, and Terrance Miller of Hitachi; Pat Wilson and Ray Krenzer of Philips Consumer Electronics.

I also wish to thank Joseph A. Labok of Los Angeles Valley College for help and encouragement throughout the years.

And a very special thanks to Daniel Gonneau, Editor-in-Chief, Steve Fitzgerald, Robert McGraw, Nancy Young, Barbara McCann, Kimberly Martin, Wayne Smith, Charles Love, Peggy Lamb, Thomas Kowalczyk, Suzanne Babeuf, Nancy Rosenblum, Kathy Greene, and Jeanne Glasser of the McGraw-Hill organization for having that much confidence in the author. I recognize that all books are a team effort and am thankful that I am working with the First Team!

And to my wife Irene, my research analyst and agent, I wish to extend my thanks. Without her help, this book could not have been written.

Preface

This book is a "crash course" in digital electronics, written with three classes of readers in mind. First are the service technicians or field-service engineers who service digital equipment. Next are the programmers/analysts who want to relate programs and software to digital hardware. Last, but not least, are the students and hobbyists who want an introduction to digital electronics, as well as information they can put to immediate use in experiments and projects.

People in the various classes of readers start from different learning points. The technicians and engineers understand electronics, but might have little knowledge of digital and programmed devices (particularly if they have been in such fields as television or communications service). Programmers might be expert in computer language and systems but know little of digital hardware, particularly microprocessor-based devices. The students or hobbyists might have only an elementary knowledge of electronics and no understanding of digital or programmed equipment.

The book brings all readers up to the same point of understanding, using the techniques found in other Lenk bestsellers. That is, the descriptions of how digital electronic devices operate are technically complete (to satisfy the technicians and engineers), but are written in simple, nontechnical terms whenever possible (for the benefit of the programmers, students, and hobbyists).

A major purpose of this book is to provide a simplified system of testing and troubleshooting for digital electronic equipment, again using basic approaches rather than specific procedures for particular equipment. Thus, the testing and troubleshooting approaches found here can be applied to any digital electronic system or equipment.

Chapter 1 provides a quick review of the number systems and alphanumeric codes used in digital electronics. Chapter 1 also describes the relationships between the number systems and electrical signals or pulses that control digital equipment.

Chapter 2 is an introduction to digital logic, which includes the basics of digital logic circuits, the logic symbols in common use, the basic principles of logic equations and logical (Boolean) algebra, and corresponding hardware functions.

Chapter 2 also provides a description of both combinational and sequential networks—including gates, decoders, encoders, code converters, data distributors and selectors, multiplexers, parity and comparison circuits, flip-flops, latches, counters, and registers. The chapter summarizes the subject of digital logic on the assumption that many readers are students who are not familiar with these basics.

Chapter 3 describes a cross-section of digital circuits, particularly ICs (integrated circuits), in common use. The chapter starts with a discussion of IC types and then proceeds with descriptions of such digital circuits as arithmetic logic units (ALUs), analog/digital (A/D) and digital/analog (D/A) converters, digital readouts and displays, and digital interface circuits.

Chapter 4 introduces the reader to microprocessors and covers such subjects as basic microprocessor functions, hardware (memory and input/output), basic system operation, timing and synchronization, and the basic digital troubleshooting approach.

Chapter 5 is devoted to test equipment used specifically to troubleshoot digital equipment. The discussions describe how features found on present-day test equipment relate to specific problems in digital troubleshooting.

Chapter 6 introduces the reader to the many techniques, procedures, and tricks that can be effective in diagnosing, isolating, and locating faults in any digital equipment. The chapter concentrates on the differences between analog and digital troubleshooting.

Chapter 7 describes some classic examples of basic digital troubleshooting. The discussions help the reader to determine if the ICs (microprocessors, gates, counters, etc.) are functioning properly before the ICs are pulled from the board.

Chapter 8 describes some additional examples of digital troubleshooting, more advanced than those discussed in Chapter 7. The examples in Chapter 8 involve more decision making and using combinations of test equipment.

Chapter 9 describes the step-by-step procedures for troubleshooting the microprocessor-based circuits of typical electronic devices. The techniques covered here are most effective when troubleshooting the system-control functions of VCRs, TVs, camcorders, CD players, CD-ROMs, tuners, frequency-synthesis circuits, etc.

Chapter 10 is devoted to troubleshooting digital equipment where the bulk of the processing circuits are contained within a few ICs. The step-by-step examples in this chapter represent typical digital video/audio equipment, such as found in digital TV, video-display terminals, and CDV (Laserdisc) special effects.

1
Digital numbers and codes

This chapter is for readers unfamiliar with the number systems and alphanumeric codes used in digital electronics. The chapter also describes the relationships between the number systems and electrical signals, or pulses, that control digital equipment.

The *decimal* (or base 10) number system is generally used in the world outside digital equipment. Base 10 means that with one digit, ten different numbers can be represented: 0, 1, 2, 3, 4, 5, 6, 7, 8, and 9. Inside digital equipment and circuits, the *binary* number system is used most often because binary numbers are compatible with electrical pulses found in computers and microprocessors (Chapter 4).

Binary numbers use only two digits, 0 and 1. The 0 can be represented by the absence of a pulse and the 1 represented by the presence of a pulse (or vice versa in some systems).

The pulses (typically about 5 V in amplitude and a few microseconds, μs, or nanoseconds, ns, in duration) can be positive or negative without affecting the binary number system (as long as only two states exist). In any event, to understand the language of digital circuits (generally referred to as *machine language*) you must examine number systems in general and the binary number system in particular.

Although digital circuits use binary numbers in the form of pulses, most computer systems use some other form of number system for assembly of computer programs. (These other systems are generally referred to as *assembly language*.) The use of assembly language is convenient because binary numbers, although compatible with pulses, are cumbersome when the values are beyond a few digits.

Shorthand number systems are used to enter and read out programs and data in a computer system. The most common shorthand number systems used are the *octal*, *hexadecimal* (or *hex*), *binary-coded decimal* (or *BCD*), and *alphanumeric* systems. The following paragraphs discuss each of these systems.

1.1 Binary numbers

Figure 1.1 compares the decimal and binary number systems. This comparison shows that the digit positions in the decimal system have weights of $10^0 = 1$, $10^1 = 10$, $10^2 = 100$, $10^3 = 1000$, and so on. In binary, the weights are $2^0 = 1$, $2^1 = 2$, $2^2 = 4$, $2^3 = 8$, and so on.

Decimal

Weight of each digit position	10^4	10^3	10^2	10^1	10^0
Example	3	8	7	0	1

$10^0 = 1$
$10^1 = 10$
$10^2 = 100$
$10^3 = 1000$
$10^4 = 10{,}000$

$$\begin{aligned} 38{,}701 &= 3 \times 10^4 + 8 \times 10^3 + 7 \times 10^2 + 0 \times 10^1 + 1 \times 10^0 \\ &= 30{,}000 + 8000 + 700 + 1 \end{aligned}$$

Binary

Weight of each digit position	2^4	2^3	2^2	2^1	2^0
Example	1	1	0	1	1

$2^0 = 1$
$2^1 = 2$
$2^2 = 4$
$2^3 = 8$
$2^4 = 16$

$$\begin{aligned} 11011 &= 1 \times 2^4 + 1 \times 2^3 + 0 \times 2^2 + 1 \times 2^1 + 1 \times 2^0 \\ &= 16 + 8 + 0 + 2 + 1 = 27 \end{aligned}$$

Fig. 1.1 Comparison of decimal and binary number systems.

Table 1.1 shows the binary equivalents of decimal numbers 0 through 31 along with the hexadecimal and octal representations.

Figure 1.2 shows a technique for converting from decimal to binary. To convert a decimal number to a binary number, perform a series of divisions by 2. In the first step, divide the decimal number by 2; use the remainder as the least significant digit, and use the quotient as the input to the next step. Continue this process until the quotient becomes 0.

When referring to a group of bits at one end of a word, the left-hand bits are called the *high-order* (or *most significant*) bits, and the right-hand bits are called *low-order* (or *least significant*) bits.

Table 1-1
Equivalents of binary, decimal, octal, and hex number systems.

Binary	Decimal	Octal	Hex
00000	0	0	0
00001	1	1	1
00010	2	2	2
00011	3	3	3
00100	4	4	4
00101	5	5	5
00110	6	6	6
00111	7	7	7
01000	8	10	8
01001	9	11	9
01010	10	12	A
01011	11	13	B
01100	12	14	C
01101	13	15	D
01110	14	16	E
01111	15	17	F
10000	16	20	10
10001	17	21	11
10010	18	22	12
10011	19	23	12
10100	20	24	14
10101	21	25	15
10110	22	26	16
10111	23	27	17
11000	24	30	18
11001	25	31	19
11010	26	32	1A
11011	27	33	1B
11100	28	34	1C
11101	29	35	1D
11110	30	36	1E
11111	31	37	1F

When referring to the bits in binary numbers, it is often necessary to refer to a particular bit or group of bits. The right-most bit is called the *least significant bit* (LSB), and the left-most bit is called the *most significant bit* (MSB).

1.2 Octal numbers

Older computers often use octal numbers because octal can provide a shorthand method for bridging the gap between decimal and binary. Present-day digital systems rarely use octal numbers directly. However, you should have an understand-

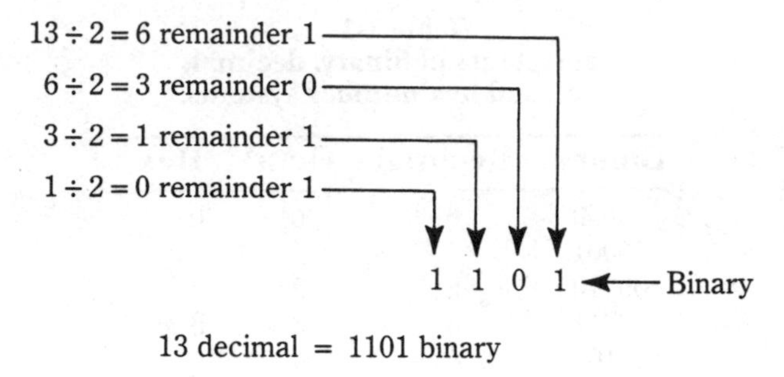

Fig. 1.2 Basic decimal-to-binary conversion.

ing of octal because 8 is a power of the binary 2, and the octal number system can be used as an aid in programming and number conversion.

Figure 1.3 shows the octal, or base 8, representation. The binary number is divided into groups of three bits, starting at the right. Each group of three is replaced by the octal equivalent. For example, the binary number 111 011 can be represented by the octal number 73.

Binary	011	111	010
Octal	3	7	2

011 111 010 binary = 372 octal

Fig. 1.3 Octal, or base 8, representation.

Note that although octal and decimal numbers look alike, they are not. (The decimal equivalent of binary 111 011 is 59.) It is much easier to convert between binary and octal than between binary and decimal.

1.3 Hexadecimal numbers

Hexadecimal (or hex) goes one step beyond octal to simplify manipulation of binary numbers. With hex, or base 16, representation of each group of four bits (or *byte*) is replaced by a single character, as shown in Fig. 1.4.

Binary	1111	0011	1010	1000
Hex	F	3	A	8

1111 0011 1010 1000 binary = F3A8 hex

Fig. 1.4 Hex representation.

Because four bits can have decimal values from 0 to 15, a way is needed to represent the decimal values 10 through 15 with a single character—the letters A through F are used. In hex, you count 1, 2, 3, . . ., 8, 9, A, B, C, D, E, F, 10, 11, . . ., as shown in Table 1.1.

When converting from binary to hex, divide the binary number into groups of four bits (bytes), and convert each group to the appropriate character. To go from hex to binary, replace each hex character by four bits. Once you are accustomed to using A for 10, B for 11, and so on, you will find hex-to-binary conversion simple.

Hex is preferred over octal because an eight-bit binary number can be represented with two hex characters, whereas it takes three octal characters for the same representation. Hex is the more compact of the two and is used most often throughout this book.

1.4 Binary codes

A number of codes are based on the binary number system. The simplest form of such coding is where decimal numbers (0 – 9) are converted into binary form using four binary digits. This four-bit system is one of the original codes used in early computers, and is still used by some systems today. With the four-bit system, generally known as *binary-coded decimal* or *BCD*, decimal 1 is represented by 0001, decimal 2 by 0010, and so on.

When the decimal number has more than one digit, four binary bits are used for each decimal digit. For example, the decimal number 3873 is represented by 16 binary bits, in groups of four, as follows:

```
   3      8      7      3   (decimal)
0011   1000   0111   0011   (BCD)
```

In a typical eight-bit microcomputer (Chapter 4), each byte can be thought of as containing two four-bit BCD numbers. With the interpretation, each byte can represent numbers in the range from 0 to 99 (decimal). This is shown in Table 1.2.

Table 1-2 Decimal equivalents of two four-bit BCD numbers (eight-bit byte).

	BCD								Decimal
Weight	2^3	2^2	2^1	2^0	2^3	2^2	2^1	2^0	
Bit number	b^7	b^6	b^5	b^4	b^3	b^2	b^1	b^0	
	0	0	0	0	0	0	0	0	= 0
	0	0	1	1	1	0	0	1	= 39
	1	0	0	1	1	0	0	1	= 99

1.5 Alphanumeric codes

Most computer systems operate not only on numbers, but also on letters and symbols (dollar signs, percents, etc.). This means that letters and symbols must be represented by binary numbers. The most common code for this, called *ASCII* (American Standard Code for Information Interchange, shortened from USASCII), is shown in Fig. 1.5.

Most significant hex digit

Least significant hex digit	0	1	2	3	4	5	6	7	Binary
0	NUL	DLE	SP	0	@	P	\	p	*0000*
1	SOH	DC1	!	1	A	Q	a	q	*0001*
2	STX	DC2	"	2	B	R	b	r	*0010*
3	ETX	DC3	#	3	C	S	c	s	*0011*
4	EOT	DC4	$	4	D	T	d	t	*0100*
5	ENQ	NAK	%	5	E	U	e	u	*0101*
6	ACK	SYN	&	6	F	V	f	v	*0110*
7	BEL	ETB	/	7	G	W	g	w	*0111*
8	BS	CAN	(	8	H	X	h	x	*1000*
9	HT	EM	)	9	I	Y	i	y	*1001*
A	LF	SUB	*	:	J	Z	j	z	*1010*
B	VT	ESC	+	;	K	[	k	{	*1011*
C	FF	FS	,	<	L	\	l	¦	*1100*
D	CR	GS	-	=	M	]	m	}	*1101*
E	SO	RS	.	>	N	↑	n	~	*1110*
F	SI	US	/	?	O	←	o	DEL	*1111*

ASCII

Fig. 1.5 ASCII versus hex conversion.

ASCII (pronounced "askey") is an eight-bit code and is thus ideally suited for hex representation. Also, because hex-binary conversion is relatively simple, both on paper and in digital circuits, ASCII can be adapted to any computer system.

Figure 1.5 shows the conversion between ASCII and hex. To convert from ASCII to hex, select the desired letter, symbol, or number, then move up vertically to find the hex MSD. Then move horizontally to the left and find the hex LSD.

For example, to find the hex code for the letter *I*, note that *I* appears in the 4 column of the hex MSD, and in the 9 column of the hex LSD. Thus, hex 49 equals the letter *I* in ASCII. Going further, the hex 49 can be converted to 0100 1001 in binary, as is generally done inside the circuits.

The process can be reversed to convert from binary to ASCII. For example, binary 0010 0100 is 24 in hex, and $ in ASCII. As an exercise, find the ASCII letters for binary 0100 1100, 0100 0101, 0100 1110, 0100 1011.

1.6 Pulses versus binary numbers

A typical digital device operates with electrical signals (generally pulses, but possibly a simple on/off signal) arranged in binary form. For example, a microprocessor (Chapter 4) performs its functions (program countings, additions, subtraction, control of motors, etc.) in response to instructions. Usually, these instructions come from a memory within the system but can also come from the outside world via a terminal. A typical microprocessor can perform many functions, with each function being determined by a specific instruction.

The instructions in a microprocessor-based computer are applied to the microprocessor as electrical pulses arranged to form a binary word. Each pulse is applied on a separate electrical line (probably a trace on a printed-circuit or *PC* board) as shown in Fig. 1.6, where a basic microprocessor has eight lines to accommodate an eight-bit binary word or byte.

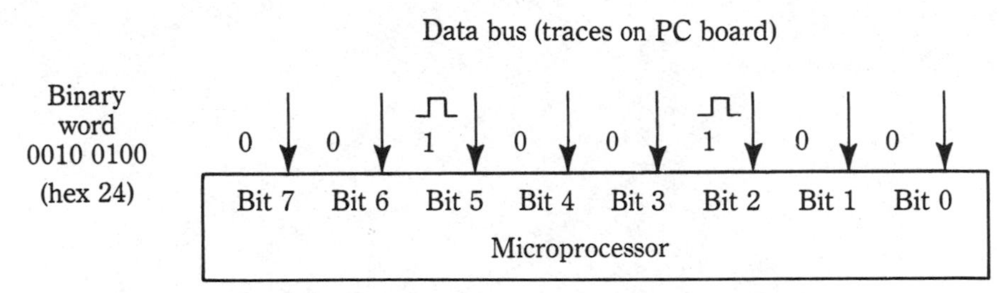

Fig. 1.6 Electrical signals versus binary numbers.

In this system, the pulses are +5 V in amplitude and a few ns in duration. The presence of a pulse indicates binary 1; the absence of a pulse (or 0 V) indicates a binary 0. So, in Fig. 1.6, the microprocessor is receiving a binary 0010 0100 that can be converted to hex 24 (or to $ in ASCII) for convenience.

For a computer-type microprocessor, this word might be an instruction to perform addition. For a control-type microprocessor, the same word might mean "start the VCR playback motor." For another microprocessor, the word might be meaningless. So, one of the first things you must do to troubleshoot microprocessor-based digital equipment is learn what functions are performed for a given set of instructions. This subject is discussed throughout the book, beginning in Chapter 4.

2
Basic digital logic

This chapter is for readers totally unfamiliar with basic digital circuits. All digital electronic equipment is based on the use of *digital logic circuits*. In turn, logic circuits are based on the use of *logic equations*. To fully understand digital electronics, you must understand logic equations and the logic circuits that are made up of *logic gates*.

As a practical matter, to understand digital electronics in its present-day form, you must also understand the use of logic circuits found in IC (integrated circuit) modules, because most digital equipment is now made up of ICs. Chapter 3 discusses typical logical circuits, including digital ICs. In this chapter, you concentrate on digital logic basics.

Note that logic equations might also be known as *logical algebra*, *Boolean algebra* (after the English mathematician George Boole), *computer algebra*, or *computer logic*. In addition to a practical working knowledge of logical algebra, you should be familiar with the symbols used in logic circuits, the binary and other digital number systems (discussed in Chapter 1), the basic logic circuit forms, and certain other problems common to all digital circuit design. These subjects are summarized in this chapter.

2.1 Logical algebra basics

Logical algebra differs from conventional algebra in two respects:

1. The symbols used in logical algebra (usually letters) do not represent numerical values.
2. Arithmetic operations are not performed in logical algebra.

Logical algebra is ideally suited to any system of intelligence based on *two opposite states*, such as on or off. Thus, logical algebra is well suited to express the opening and closing of electrical switches, the presence or absence of electrical pulses, or the polarity or amplitude relationship of pulses. Logical algebra is also quite compatible with the binary number system (and the various binary-based codes) discussed in Chapter 1.

The following is a summary of the states and quantities available in logical algebra.

1. *Logical algebra has only two discrete states.* Any pair of conditions different from each other, can be chosen. In digital electronics, the states are described as *true* or *false*, or possibly *up* or *down* (or *high* or *low*) referring to the presence or polarity of pulses. You can assign any dissimilar value or state to represent true and any other value to represent false.
2. *Logical quantities are single valued.* No quantity in logical algebra can be simultaneously both *true* and *false*. If the quantity is false, then the opposite quantity is true.
3. *A logical quantity can be either constant or variable.* Any quantity that is true is equal to any other quantity that is true. If constant, a logical quantity must remain true and false. If variable, the quantity can switch between the true and false states from time to time, but only between the two extremes.
4. *There are many ways to represent logical algebra in digital electronics.* The most common way is to represent the states by the presence or absence of pulses. Typically, the presence of a pulse represents the true state or quantity, and corresponds to a binary 1 (Fig. 1.6). With this system, the absence of a pulse represents false or binary 0. However, another system might use the exact opposite representation.

2.2 Logical algebra notation

The condition or state of a logic variable in logical algebra is generally represented by letters of the alphabet. For example, assume that the letter A represents the condition of a logic variable. In that case, the values A can assume are only true or false, because these are the only values that any logic variable may represent. (This is quite different from conventional algebra, where the letter A can represent any value from minus to plus infinity.)

The inverse, or *complement*, of the letter A is represented by the symbol $\overline{A}$ (A-bar or A-overbar), A′ (A-prime), or A* (A-asterisk or A-star). The prime and asterisk are the most popular notations in text for the complement condition, because they are available on most typewriters and word processors. However, on digital electronic diagrams, the overbar is most popular.

No matter what system of notation is used to show the complement, or inverse, condition, A and $\overline{A}$ cannot have the same value at the same time. Thus, if $\overline{A}$ is false at any given time, then A must be true, and vice versa.

If two or more logic variables are present at the same time, one might be represented by A, another by B, and so on. If at this time, B happens to be true and A happens to be true, then A = B, because a true always equals a true. Simultaneously, the complement of B (or $\overline{B}$) and the complement of A (or $\overline{A}$) are equal to each other, and are false. An instant later, the variable represented by A may change state, and A then becomes false and $\overline{A}$ (the complement) becomes true.

In most present-day logic diagrams, 1 represents true and 0 represents false. The 1 and 0 are common because they are similar to binary numbers and are easy to write. Typically, 1 means *yes*, *assertion*, *up*, *high*, *enable*, or *true*. Consequently, 0 means *no*, *negation*, *down*, *low*, *disable* (or *inhibit*), or *false*.

2.3 Logical algebra operations and symbols

This section describes the operations and symbols used when logical algebra appears on paper. Section 2.4 describes the operations and symbols used when the corresponding digital operations are shown on digital circuit diagrams (or digital logic/schematic diagrams as they might be called).

The basic operation in logical algebra (and the corresponding digital electronic circuits) are AND, OR, and NOT (or invert). Also, there are some common operations (or digital electronic functions) in logical algebra that are produced by combining the functions of these three basic operations. The most common operations are NOR, NAND, AND/OR, EXCLUSIVE OR, and EXCLUSIVE NOR.

As in conventional algebra, certain symbols (or *connectives*) are used to indicate the type of operation that is to be performed in logical algebra. Unfortunately, the use of connective symbols is not standard throughout the industry. However, the symbols shown in Fig. 2.1 are generally accepted and understood. The following paragraphs summarize the rules shown in Fig. 2.1.

Symbol or notation	Description
$\circ \cap \wedge X$	AND
$+ \cup V$	OR
$\uparrow$ $\not\mathsf{A}$	NAND
$\downarrow$ $\not\forall$	NOR
$\underline{+}$	EXCLUSIVE OR
$\overline{+}$	EXCLUSIVE NOR
$\overline{X}$ ~X X− X′	NOT X
=	equivalence
$A \cup B$	union of A and B
$A \cap B$	intersection of A and B
$A \subset B$	A belongs to B
$A \rightarrow B$ $A \supset B$	A implies B
1	True
0	False
$A = B$	A = B, unconditionally
$A \equiv B$	A is identical to B, unconditionally
$A \oplus B$	A EXCLUSIVE OR B
$\overline{A}$	NOT A
$A + B$	A OR B

Fig. 2.1 Logical algebra symbols.

The AND operation is represented by a dot between the variables (A•B). In most cases, the dot is omitted but is implied. Thus, ABC is read (A and B and C).

The OR operation is represented by a plus sign between the variables (A + B). This reads: A or B.

The NOT operation is represented by a long bar over the logical quantity to be complemented or inverted. This is similar to the inverse or complement notation. Thus, when $\overline{A}$ is subjected to NOT, the complement is removed, and when A is subjected to NOT, it becomes $\overline{A}$. Also, $\overline{A} = A$.

If the term subjected to the NOT operation contains an OR or an AND operation, in addition to variables, the OR and AND operation is also complemented (or inverted). This is, if the variables are connected by an AND, and AND is changed to an OR, in addition to changing the variables.

For example,

$$\overline{A+B} = AB \qquad \overline{AB} = A + B$$
$$\overline{A} + B = A\overline{B} \qquad \overline{AB} = A + \overline{B}$$

Thus, the complement of the AND operation is an OR, and the complement of an OR is an AND.

The equal sign (=) found in conventional algebra is carried over into logical algebra, and with the same meaning (equals, the result of, etc.) . However, the equal sign (=) with two bars is a *conditional equal* in logical algebra. That is, the conditions must be stated for the equations to be correct. For example, A = B, where A is true and B is true. The equal sign (≡) with three bars is an *unconditional equality*. For example, A ≡ B means that A is the same as B, under all contions.

2.4 Digital operations and symbols

Figure 2.2 shows the most common symbols and operations used in digital electronic circuits. These operations correspond to the logical algebra operations discussed in Sec. 2.3.

2.4.1 Truth tables

The truth table is one of the most useful tools for analyzing problems in logical algebra. A truth table consists of one vertical column for each of the logical variables involved in a given problem. The horizontal lines or rows of the truth table are filled with all possible true-false combinations that the variables can assume with respect to each other.

The vertical columns and horizontal rows of Fig. 2.2 are, in effect, truth tables for the digital electronic operations shown. The A and B columns (or inputs) represent the two variables present at the input of the digital circuit. These variables can assume four different combinations at any of four different times. That is, both true at the same time; both false at the same time; and one true and one false. There are no other possible combinations.

The remaining columns in Fig. 2.2 represent the outputs or results produced by the combination of variables at the input of each digital circuit. The output

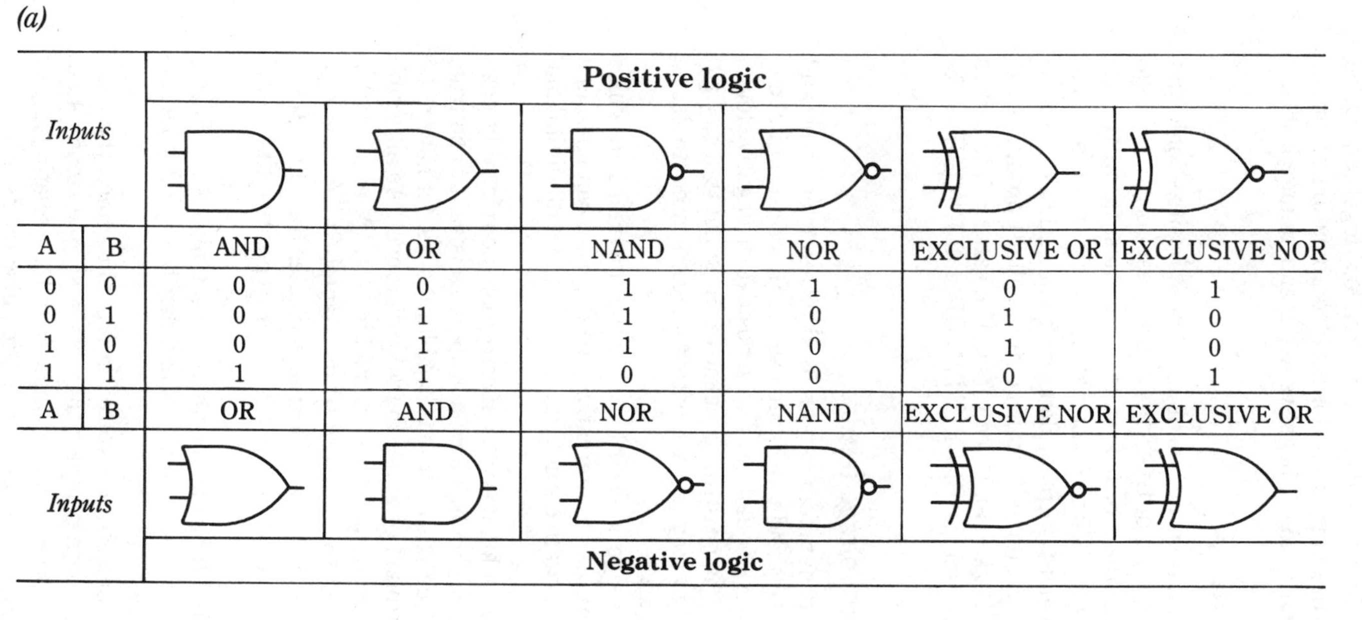

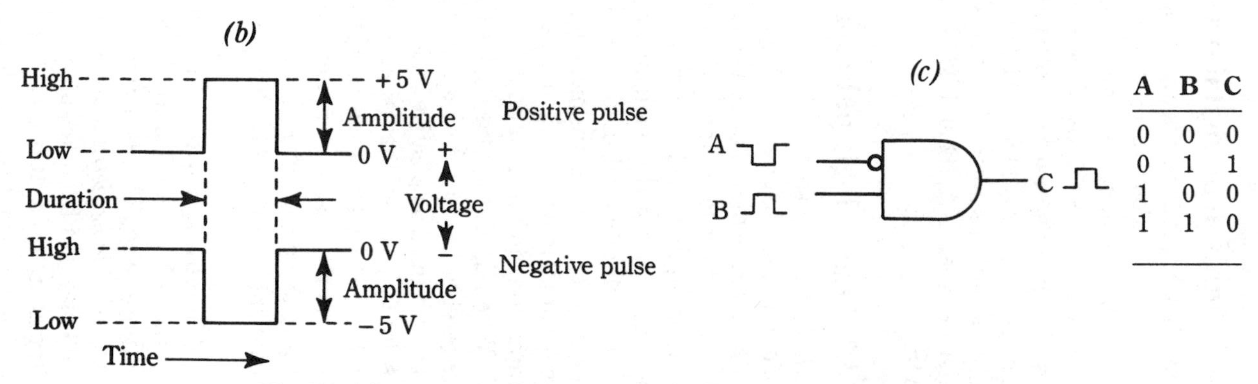

Fig. 2.2 Common symbols and operations used in digital electronics.

columns (also known as the *function* column on some digital diagrams) can contain the result, or output, when the input variables are ANDed together, ORed together, or any particular output that must occur for each combination of input.

The next section discusses positive logic, negative logic, and voltage levels.

2.4.2 Equivalent digital operations

Note that Fig. 2.2 shows digital operations or functions for both positive logic and negative logic. Positive logic defines the 1 or true state as the most positive voltage level, whereas negative logic defines the 1 or true state as the most negative voltage level. Because of the difference in definition of states, it is possible for some digital elements to have two equivalent outputs, depending on definition. As an example, a positive-logic AND circuit (or *gate* as it is often called) produces the same outputs as a negative-logic OR gate, and vice versa. The same relationships exist between NAND and NOR, as well as EXCLUSIVE-OR and EXCLUSIVE-NOR.

2.4.3 Digital-logic pulses and voltage levels

Although any system of letters and numbers can be used to represent true and false in digital diagrams, voltage levels and pulses are used in practical digital circuits. Because digital circuits can work equally well with positive or negative logic, it is necessary to define if the logic is positive or negative. In some (but not all) digital diagrams, a plus or minus sign may be used within a logic symbol to define the true state for that element or gate.

In well-prepared literature, the sign is used for all digital elements in which true and false levels are meaningful. As an alternative, the diagram can state in a note that all logic is positive true or negative true. Unfortunately, not all digital diagrams contain this notation.

When used, the plus sign within a logic symbol means that the relatively positive level of the two voltages at which the digital circuit operates is said to be true. This is defined as positive logic. Note that the true voltage level does not have to be absolutely positive (that is, above ground or above the 0 V reference). As an example, the two voltage levels at which a circuit operates could be −1.7 V and −0.9 V (which is the case for emitter-coupled logic described in Chapter 3) . A plus sign within the logic symbol (or a note specifying positive true) indicates that the −0.9 V is true and that the −1.7 V is false, because −0.9 V is closer to positive than −1.7 V.

A minus sign within the symbol (or note specifying negative true) indicates that the −1.7 V level is true (since it is more negative) and that the −0.9 V level is therefore false. This is defined as negative logic.

Digital pulse definitions. When pulses are involved in digital circuits, the true-false condition becomes somewhat more complex than when simple steady-state voltages are used. It is therefore important that you understand the relationship of positive and negative logic to pulses, as well as some basic pulse terms.

A positive pulse goes from a lower level to a higher level and back to a lower level, as shown in Fig. 2.2*b*. A *negative pulse* goes from a higher level to a lower level, and back to a higher level. Lower and higher levels represent voltage levels (just as if the pulses were steady-state voltages). Higher represents the higher voltage or *up* voltage level. Lower represents the lower voltage level or *down* voltage.

The difference between the lower and higher levels represents the *pulse amplitude* (which is typically 5 V in most present-day digital equipment).

With positive logic, the upper level of the pulses in Fig. 2.2*b* is the true condition, with the lower level the false condition.

With negative logic, the lower level of the pulses is the true condition, with the upper level the false condition.

With either system, the length of time that a positive pulse stays up or a negative pulse stays down is called the *pulse duration.*

2.4.4 The AND circuit

As shown in Fig. 2.2., the AND circuit or operation is true when all of the ANDed quantities are true, and is false when one or more of the ANDed quantities is false. In the case of a digital AND gate, the output is true (or 1) when all of the inputs are 1 and is 0 when one or more of the inputs is 0. Using positive logic, and 5 V levels, this means that the output is +5 V when all of the inputs are +5 V and is 0 V when one or more of the inputs is 0 V.

As discussed in Chapter 3, in practical digital circuits, gates are usually found in IC form. As is typical for ICs, the internal circuits are not shown on the diagrams. The symbols given in Fig. 2.2 are used instead.

In practical digital circuits, AND gates can appear in multiple-input form (usually about six inputs is maximum). Also many thousands of gates can be included in a single IC. However, no matter how many inputs are involved, all inputs to an individual AND gate must be 1 (true) for the output to be 1 (true).

Although all inputs must be true for an AND gate to produce a true output, the inputs need not necessarily be of the same polarity. That is, the internal gate circuit can be arranged to produce a 1 output when mixed 1 and 0 inputs are applied. This is shown in Fig. 2.2*c*. An inverted input is represented on the symbol by a dot (or small circle) on that input which is inverted from the normal logic on the rest of the diagram.

As an example, with positive logic, an inversion dot on an input indicates that a 0 input is required to produce a true condition. If it is assumed that the symbol of Fig. 2.2*c* uses positive logic, a 0 input is required at input A and a 1 at input B to produce a 1 output (as shown by the truth table).

2.4.5 The OR circuit

As shown in Fig. 2.2, the OR circuit of operation is true when one or more of the quantities is true, and is false only when all of the quantities are false. In practical digital circuits, OR gates can appear in multiple-input form, and many thousands of gates can be included in a single IC. However, no matter how many inputs are involved, any one true input to an OR gate produces a true output. As with AND gates the internal circuits of an OR gate can be designed so that mixed inputs produce a true output.

2.4.6 The AND-OR circuit

The AND-OR operation is not basic in the same sense as the AND and OR operation. However, the AND-OR operation is used so frequently in digital circuits (particularly in a version where the OR output is inverted) that the function can be treated as a basic operation.

Figure 2.3*a* shows the truth table and logic symbol for the AND-OR operation. Note that the AND-OR operation is made up of two AND gates at the input and one OR gate at the output.

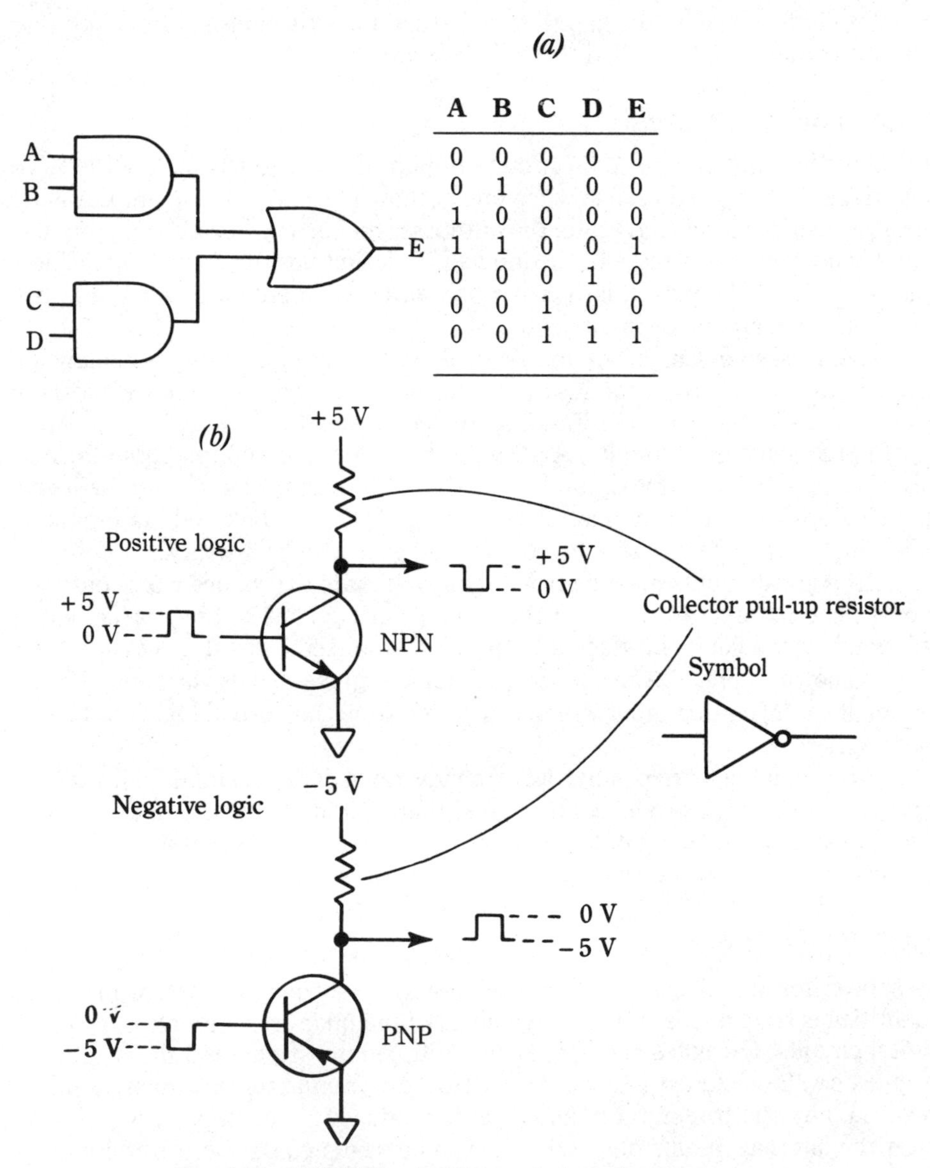

A	B	C	D	E
0	0	0	0	0
0	1	0	0	0
1	0	0	0	0
1	1	0	0	1
0	0	0	1	0
0	0	1	0	0
0	0	1	1	1

Fig. 2.3 AND-OR and NOT symbols and operations.

2.4.7 The NOT circuit

Figure 2.3*b* shows the basic NOT circuits. Note that these circuits are essentially single-stage, common-emitter transistor amplifiers.

The NOT function is often referred to as a negation, or a complementing function. The negative of a quantity can also be called the inverse, converse, or opposite. The NOT circuit is required to produce the NAND and NOR gate functions (described in Sec. 2.4.8 and 2.4.9).

The output of a NOT circuit is the inverse of the input. For example, if the input is a positive-going pulse (representing true or 1), the output will be a negative-going pulse (representing false or 0). The NOT function is often combined with amplification.

In a negative-logic system, a PNP transistor is used, because a negative voltage at the base causes the transistor to conduct, creating a ground (0-V) output that is the opposite of the negative input voltage. An NPN transistor is used with positive logic to produce the corresponding results. Note that the collector resistor is called the *pull-up* resistor with either type of logic.

The transistor circuit also can provide some amplification. The output of a diode is always less than the input, so diode gates do not have the capability of driving many other diode circuits. The transistor NOT circuit can provide pulse-amplitude restoration, and driving capability, in addition to inversion of the pulses. The driving capability of digital circuits is discussed further in Sec. 2.6.1.

2.4.8 The NAND circuit

As shown by the truth table of Fig. 2.2, the NAND output is true when one or more of the inputs is false, and it is false only when all of the inputs are true. The NAND operation is a combination of the NOT and AND operations, and typically includes an AND circuit and a NOT amplifier. The AND circuit performs the usual AND function, producing a true output only when both inputs are true. The NOT amplifier then inverts the true into a false.

With NAND, two true inputs produce a false output. If either (or both) inputs is false, the AND circuit produces a false output which is inverted to a true by the NOT circuit. The NAND circuit is generally preferred to the AND circuit for most digital applications because of the amplification provided by the NOT circuit.

2.4.9 The NOR circuit

As shown by the truth table of Fig. 2.2, the NOR output is false when one or more of the inputs is true, and it is true only when none of the inputs is true. The NOR operation is a combination of the NOT and OR operations, and it typically includes an OR circuit and a NOT amplifier. The OR circuit performs the usual OR function, producing a true output when either or both inputs are true. The NOT amplifier then inverts the true into a false.

With NOR, one or both true inputs produce a false output. If both inputs are false, the OR circuit produces a false output which is inverted to a true by the NOT circuit. The NOR circuit is generally preferred to the OR circuit for most digital applications because of the amplification provided by the NOT circuit.

2.4.10 The EXCLUSIVE-OR, EXCLUSIVE-NOR, and COINCIDENCE circuits

As shown by the truth table in Fig. 2.2, the EXCLUSIVE-OR output is true if one, but not both, of the inputs is true. The converse statement is equally accurate: the output is false if the inputs are both true or both false. The EXCLUSIVE-OR circuit is independent of polarity, and is not generally regarded as being either positive-true or negative-true. EXCLUSIVE-OR has only two inputs, and one output, unlike a multiple-input OR gate.

As shown in Fig. 2.2, the EXCLUSIVE-NOR circuit operates in the same way as the EXCLUSIVE-OR except that the output is inverted. With EXCLUSIVE-NOR, two false inputs, or two true inputs, produce a true output. One false and one true input produce a false output. For this reason, EXCLUSIVE-NOR is sometimes called the COINCIDENCE operation in certain digital literature.

2.5 Digital circuit symbols

Although there are many variations of digital elements and their circuit symbols, there are only four basic classes or groups: gates, amplifiers (or inverters), switching elements, and delay elements. In present-day digital devices, most of the elements are contained within ICs, and the symbol does not appear on the diagram. However, because individual digital elements are used in many applications (particularly between ICs), it is essential that you recognize and understand the corresponding symbols as they appear on diagrams.

This section describes the symbols that are unique to digital circuit diagrams. The symbols for other electronic parts (transistors, capacitors, resistors, etc.) are the same as the symbols used on schematic diagrams of conventional, nondigital, electronic devices.

2.5.1 Gate symbols

As discussed in Sec. 2.4, a gate is a circuit that produces an output on condition of certain rules governing input combinations. As shown in Fig. 2.4, the basic gate symbol has input lines connecting to the flat side of the symbol, and output lines connecting to the curved or pointed sides. Because inputs and outputs are easily identifiable, the symbol can be shown facing left or right or facing up or down, as necessary.

There might be two inputs to a gate. In some cases, multiple inputs can be provided by increasing the internal circuit components of the gate. In other cases, it is necessary to connect the output of two (or more) gates together, in parallel. For example, the output of three two-input AND gates can be connected in parallel, resulting in a six-input AND gate. This is sometimes known as a *wired*, *dotted*, or *implied* AND function. Digital diagrams often use the symbol shown in Fig. 2.4 when the term WIRED-AND is involved.

It is also possible for a gate to have an input other than the normal input. This is often referred to as an *extended input*. For example, a NAND gate can be made up of a diode AND gate followed by a transistor amplifier (which also inverts). An

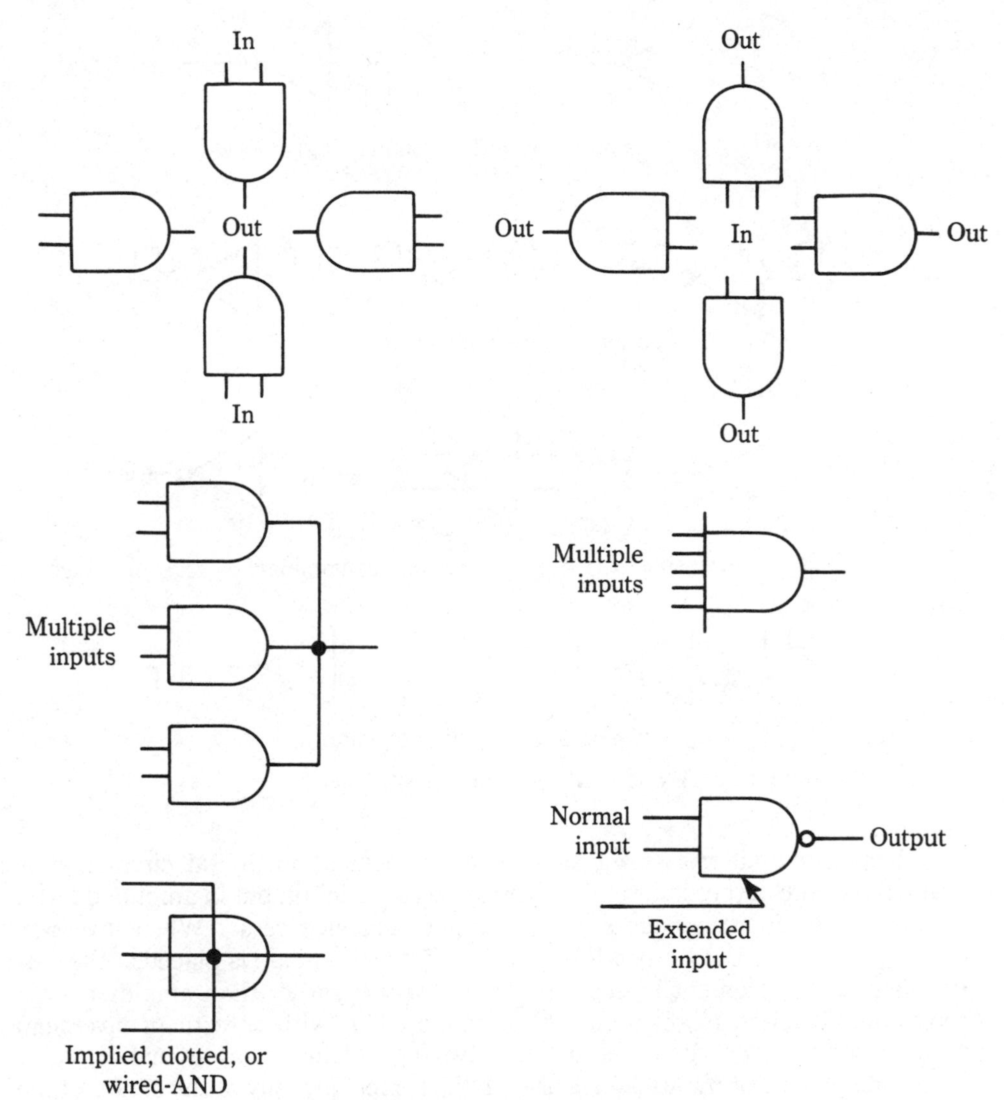

Fig. 2.4 Basic gate symbols.

input can be connected directly to the transistor base, thus bypassing the AND function. Generally, a signal at an extended input produces an output regardless of the conditions (true or false) at all of the regular inputs.

2.5.2 Amplifier symbols

As shown in Fig. 2.5, the amplifier symbol is a triangle with the input applied to the center of one side and the output taken from the opposite point of the triangle. Like gates, the amplifier can be shown in any of the four positions. When amplifiers are used in digital circuits, the driving or input signals are normally pulsed. Consequently, the output of the amplifier is an amplified form of the input pulse.

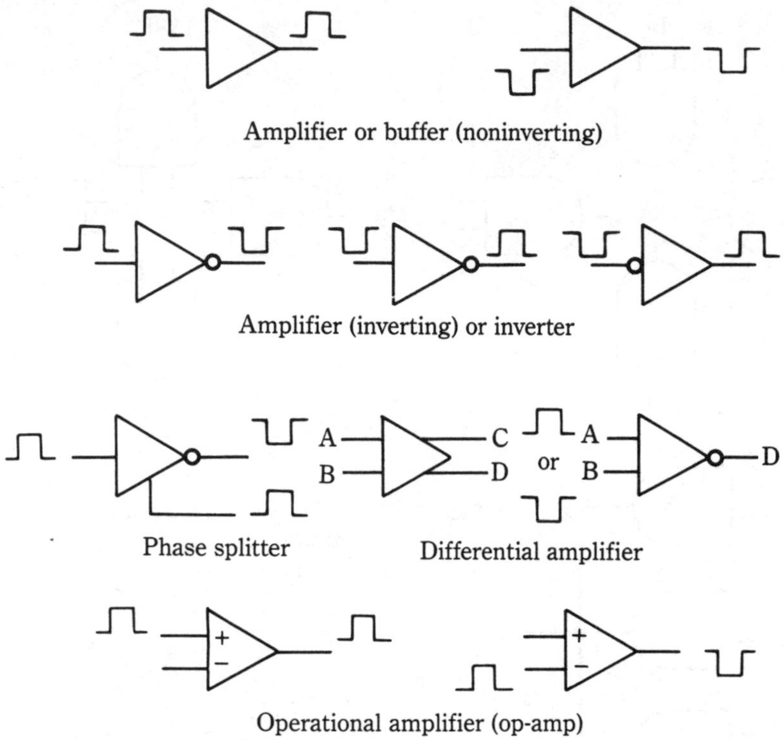

Amplifier or buffer (noninverting)

Amplifier (inverting) or inverter

Phase splitter

Differential amplifier

Operational amplifier (op-amp)

Fig. 2.5 Basic amplifier symbols.

When an amplifier is used as a separate element in digital circuits, it is assumed that the output is essentially the same as the input, but in amplified form. That is, a true input produces a true output, and vice versa. When inversion occurs, an inversion dot (or possibly an inverted pulse symbol) is placed at the output. Usually, the element is then termed an *inverter* rather than amplifier, even though amplification might occur. Also, an amplifier (with or without inversion) can be called a *buffer* when used between two logic elements or circuits.

If a single plus or minus sign is used in the symbol (usually at the center of the symbol), this usually indicates that the *input polarity* is required to turn the amplifier on. However, if the circuit is an *operational amplifier* or *op-amp*, there can be both a plus and minus sign at the inputs. An op-amp is a special type of amplifier (with two inputs) that produces an inverted output or a noninverted output, depending on which input is selected. The output is inverted if the input with a minus sign is used, and is noninverted when a signal is applied to the plus input.

One amplifier or inverter symbol often represents many amplification stages. Digital symbols, by themselves, do not necessarily imply a specific number of components, but rather relate to the overall effect. Similarly, amplifiers in digital circuits can have more than one input and output.

As an example, a *phase splitter* (Fig. 2.5) has one input and two outputs. One of the outputs is in phase with the input, and the other output is out of phase with the input. A similar case exists with *differential amplifiers*, which generally have two inputs and two outputs (some differential amplifiers have dual inputs and a single output).

Amplifier symbols can be confusing and (and are often improperly drawn on digital diagrams). Here are some rules to help unscramble the problem.

- If the symbol has one input and one output, with no inversion shown, the output and input should be of the same polarity.
- If the symbol has one input and one output, with inversion shown (dot or pulse), the output should be inverted from the input.
- If the symbol has one input and two outputs, with one output inverted, the amplifier is a phase splitter.
- If the symbol has two inputs and two outputs, the circuit is a differential amplifier. The output depends on the differential between the two input signals, and the outputs are inverted from each other.
- If the symbol has a single plus or minus sign near the center, the sign indicates the input polarity required to operate the amplifier.
- If the symbol has two inputs and one output, with plus and and minus signs at the input, the circuit is an op-amp. The output follows the input with the plus sign, and is inverted from the input with the minus sign.

2.5.3 Switching element symbols

Figure 2.6 shows the basic digital switching element symbols. Switching elements used in digital circuits are some form of multivibrator: bistable (flip-flop or FF, latch, Schmitt trigger), monostable (one-shot), and astable (free-running). According to the type of circuit, inputs cause the state of the circuit to switch, reversing the output. For example, true will switch to false and vice versa.

The FF or latch is the most common digital switching element. An FF is *bistable* in that it takes one external signal to set the FF, and another signal to reset the FF. The FF remains in a given state until switched to the opposite state by the appropriate external signal. The various types of FFs and other switching elements are discussed further in Sec. 2.8.

2.5.4 Delay-element symbols

Figure 2.7 shows the basic digital delay element symbols. A delay element provides a finite time between input and output signals. Many types of delay elements are used in digital circuits. Two frequently used delays are the *tapped delay* and delays effective only on the leading or trailing edges of pulses.

2.5.5 Modifying and identifying digital circuit symbols

Figure 2.8 shows a few modifications found on some digital diagrams. Basic digital circuit symbols are often modified to express circuit conditions. Although designers might have their own set of modifiers, together with military and industrial standards, the following modifiers are in general use (except on the diagram you are reading).

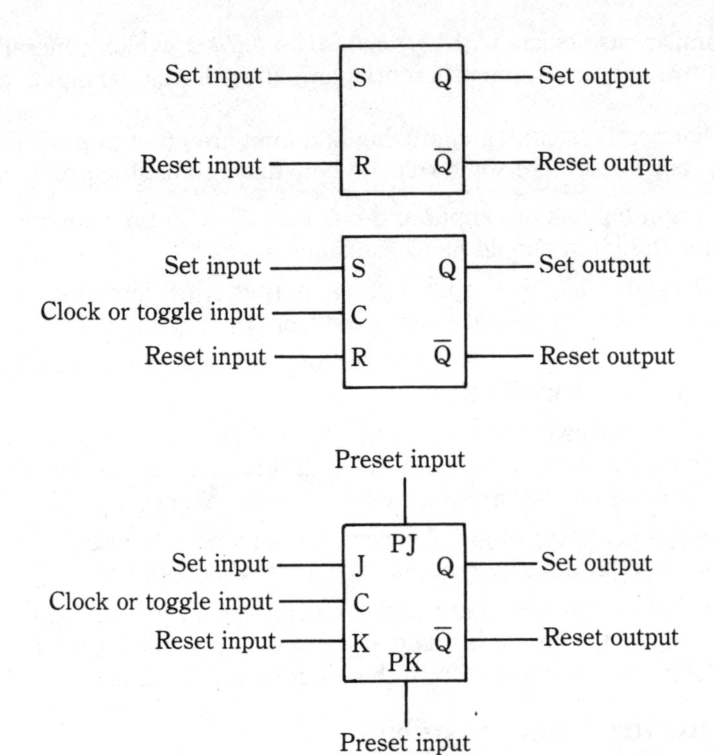

Fig. 2.6 Basic digital switching-element (FF) symbols.

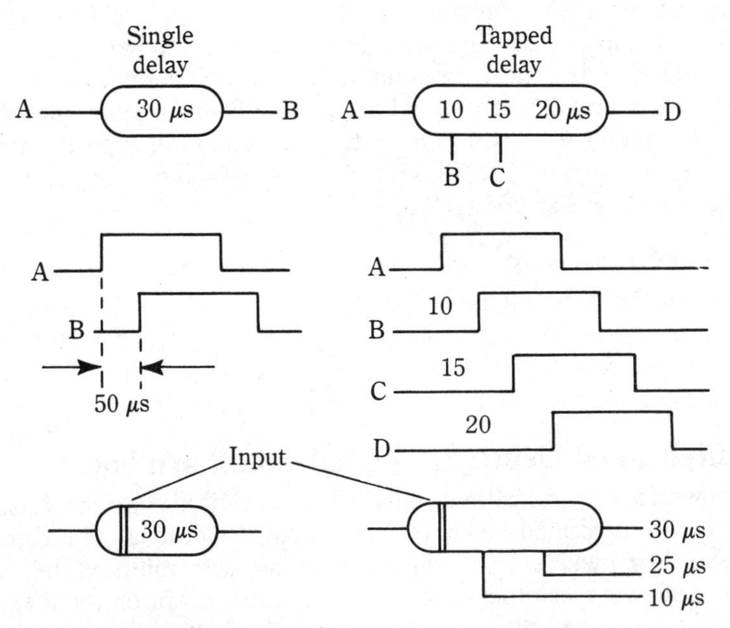

Fig. 2.7 Basic digital delay-element symbols.

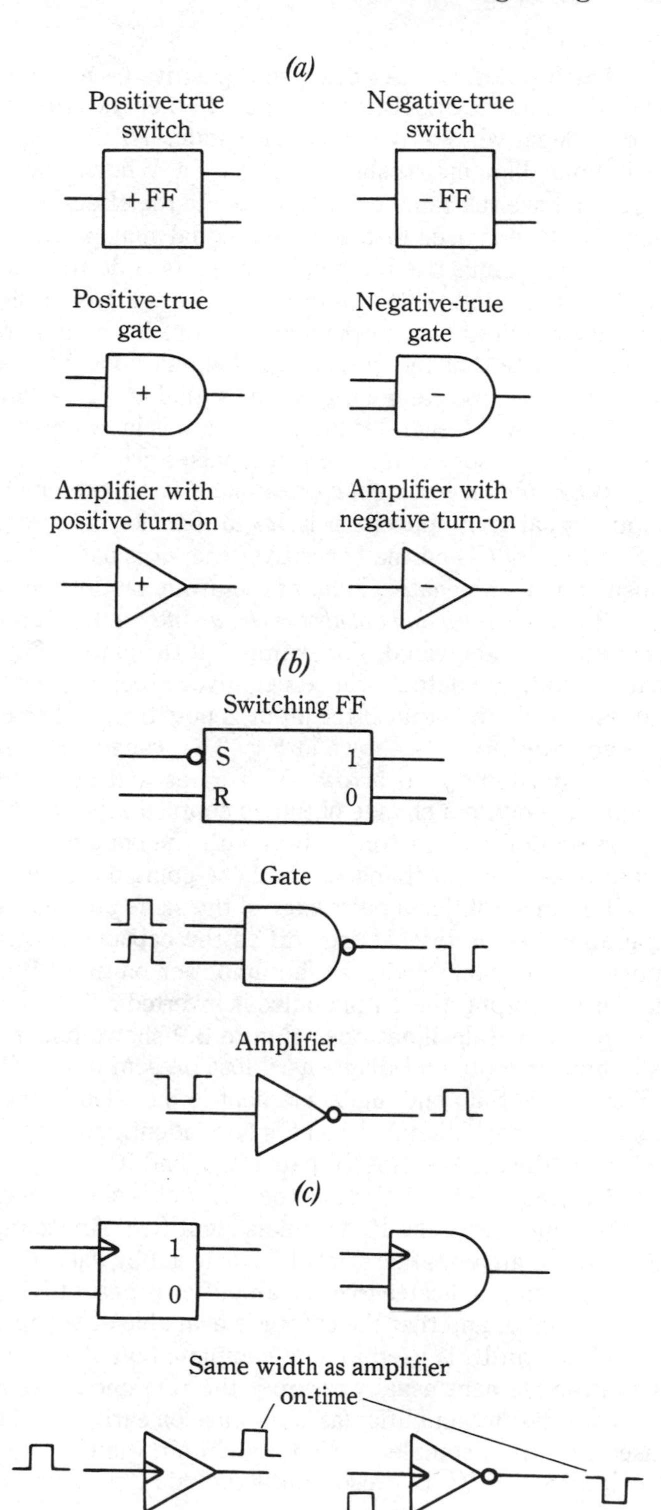

Fig. 2.8 Typical modifications found on digital diagrams.

Truth polarity As discussed, positive (+) or minus (−) indicators may be placed inside a symbol to designate whether the true state for that circuit is positive or negative, relative to the false state. This is frequently done with gates and switching elements, as shown in Fig. 2.8*a*. When all symbols on a particular digital diagram have the same polarity, a note to the effect that all logic is positive-true or negative-true is used instead of individual signs in each symbol.

Polarity signs used in amplifier symbols do not usually have any direct logic significance. Rather, the polarity signs are a troubleshooting aid, indicating the polarity required to turn the amplifier on. In the case of an op-amp, the polarity signs indicate that the output signal should follow the signal at the + input, but should be of opposite polarity to the signal at the − input.

Inversion Generally, logic inversion is indicated by an inversion dot at inputs or outputs. In some cases, inverted pulses are shown at the inputs and outputs.

When the inversion dot appears on an input, the input is effective only when the input signal is of opposite polarity to that normally required. For example, if the FF in Fig. 2.8*b* is normally positive-true, or is used on a diagram where all logic is positive-true, a negative input at the inversion dot sets the FF.

When the inversion dot appears on an output, the output is of opposite polarity to that normally delivered. For example, if the gate of Fig. 2.8*b* is used in a positive-true circuit, the output will be negative. Similarly, the amplifier of Fig. 2.8*b* produces a positive output if the input is negative, and vice versa.

ac coupling As shown in Fig. 2.8*c*, capacitor inputs to digital elements are often indicated by an arrow. With gates and switching elements, the element responds only to a change of the ac coupled input in the true-going direction. An inversion dot used in conjunction with the coupling arrow indicates that the element responds to a change in the false-going direction.

For an amplifier, a pulse edge of the same polarity as given in the symbol turns the amplifier on briefly, then off as the capacitor charges. The output is then a pulse of the same width as the amplifier on-time. With an inversion dot at the amplifier output, the output pulse is inverted.

Reference designations Figure 2.9 shows how reference designations usually appear on digital diagrams. Most present-day digital circuits are in IC form. When more than one digital element is included in the IC package, which is the usual case, each digital element is often identified with a suffix: A, B, C, and so on, which is the case with AND gates IC_{3A} and IC_{3B}.

On some digital diagrams, the element is shown enclosed by a box (or possibly dashed lines) with the IC terminals identified. An example of this is shown in Fig. 2.9, where an amplifier symbol is enclosed by dashed lines and identified as IC_{38}. This marking indicates that the amplifier is part of IC_{38}, that the input is available at terminal 3, and that the output is available at terminal 7.

When an IC is used for one complete digital element, such as an amplifier or gate, the element usually assumes the reference designation of IC rather than *G* for gate or *A* for amplifier (as is the case on early digital equipment). When an IC is used to form a complete switching circuit, the element can assume the reference designation of IC, but also should include that appropriate abbreviation (such as

FF for flip-flop, OS for one-shot, MV for multivibrator, ST for Schmitt-trigger) to identify the function.

When a digital element is composed of portions of different IC packages (not the usual case), the reference designation of both packages should be included inside the symbol, and the appropriate identifying abbreviation should be located outside the symbol (as is the case with the FF composed of IC_{3A} and IC_{7A} on Fig. 2.9).

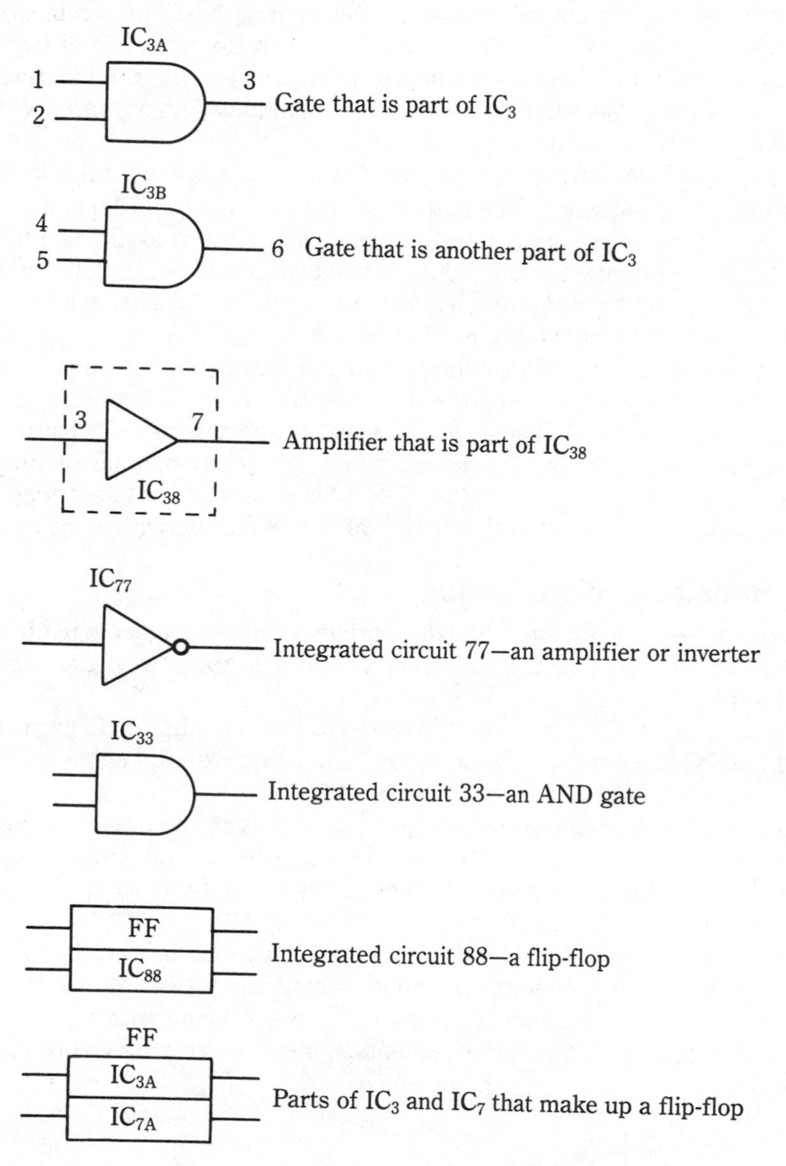

Fig. 2.9 Typical reference designations found on digital diagrams.

2.6 Basic digital circuit terms and implementation

In addition to the basic gates, amplifiers, switches, and delays described thus far, basic digital circuits can be grouped into two classifications: combinational and sequential.

Combinational digital circuits are those that have no feedback or memory (or do not depend upon feedback or memory for their operation). The combinational circuit produces an output (or outputs) in response to the presence of two or more variables (or inputs). The nature of the output depends on the combination of gates used. Adders, even and odd circuits, decoders, and encoders are examples of combinational circuits, as discussed in Sec. 2.7.

Sequential digital circuits have a memory, and possibly a feedback, and must operate in a given sequence. The outputs of sequential circuits depend on the time relation or timing of the inputs and the feedback, as well as the combination of gates or other elements. Sequential circuits usually involve flip-flops, and include such circuits as multivibrators, counters, registers, shift cells, arithmetic units, multipliers and dividers, as discussed in Sec. 2.8.

It is possible to buy most combinational and sequential networks in complete, functioning packages that require only a power source. A sampling of such IC packages is presented in Chapter 3. However, on the assumption that it might be necessary for you to design a basic digital circuit, the remaining sections of this chapter are devoted to digital circuit design. A review of these sections will also help you understand the operation of digital circuits. Start with some definitions.

2.6.1 Some basic digital terms

The terms *fan-in*, *fan-out*, *load*, and *drive* are used interchangeably in digital literature. To avoid confusion, the following definitions apply to digital circuits described in this book.

The term *fan-in* is generally applied to digital circuits in IC form. Fan-in is defined here as the number of independent inputs on a digital element. For example, a three-input NAND gate has a fan-in of 3.

Each input to a digital element represents a *load* to the circuit trying to drive the element. If there are several digital elements (say several gates) in a given circuit, the load is increased by one unit for each element. *Load*, therefore, means the number of elements being driven from a given input.

Fan-out (primarily an IC term) is defined as the number of digital elements that can be driven directly from an output without any further amplification or circuit modification. For example, the output of a NAND gate with a fan-out of three can be connected directly to three elements. The term *drive* can be used in place of fan-out.

2.6.2 Digital delay

Any solid-state element (diode, transistor, etc.) offers some delay to signals or pulses. That is, the output pulse occurs some time after the input pulse. Thus, every digital element has some delay. This delay is known as *propagation delay time*, *delay time*, or simply *delay*. Some digital literature spells out a specific time (usually in nanoseconds). Other digital literature specifies delay for a given circuit as the number of elements between a given input and a given output.

The problems of delay become obvious when it is realized that most digital circuits operate on the basis of *coincidence*. For example, an AND gate produces a true output only when both inputs are true simultaneously. If input A is true, but switches back to false before input B arrives (because of some delay of the B input), the inputs do not occur simultaneously, and the AND output is false.

The problem is compounded when delay occurs in sequential networks, which depend on timing as well as coincidence. Many of the failures that occur in digital circuits are the result of undesired delay.

2.6.3 Implementing digital functions with other digital elements

You might sometime be required to produce (or implement) a basic digital function using other digital elements. For example, you might be working with ICs that contain a number of general-purpose NAND gates. (Many manufacturers produce such ICs.) Although most of the design can be done using only NAND gates, assume that the circuit calls for one inverter, or one NOR gate, and so on. It is possible to form all of the basic digital functions discussed so far in this chapter using only NAND gates.

Figure 2.10 shows the connections required to implement all of the basic digital functions, using other digital elements. For example, to form an inverter from a NOR gate, connect one input to ground or zero. To form the same inverter with a NAND gate, connect both inputs of the NAND gate together. With either connection, the output is inverted from the input.

2.6.4 Wired-AND and wired-OR

When gate outputs are connected in parallel, an AND function or an OR function can result. These functions are often referred to as wired-OR and wired-AND. If the true condition is represented by 0 V (or ground), then the parallel outputs or gates produce an OR function. If the true condition is represented by a voltage of any value or polarity, then the function is AND (such as shown for wired-AND in Fig. 2.10). The reason is that if a gate output is 0 V (ground), all the other outputs are, in effect, shorted to ground, and the function becomes OR.

Any gates with active pull-up elements (such as the pull-up resistor in Fig. 2.3*b*) should not be connected for wired-OR. When one gate has a 0 output and the other a 1 output, the resulting output is unpredictable.

2.7 Basic combinational circuits

Many types of combinational circuits are used in digital electronics. Similarly, there are an infinite variety of combinational circuits available in IC form. The following paragraphs describe a few selected examples.

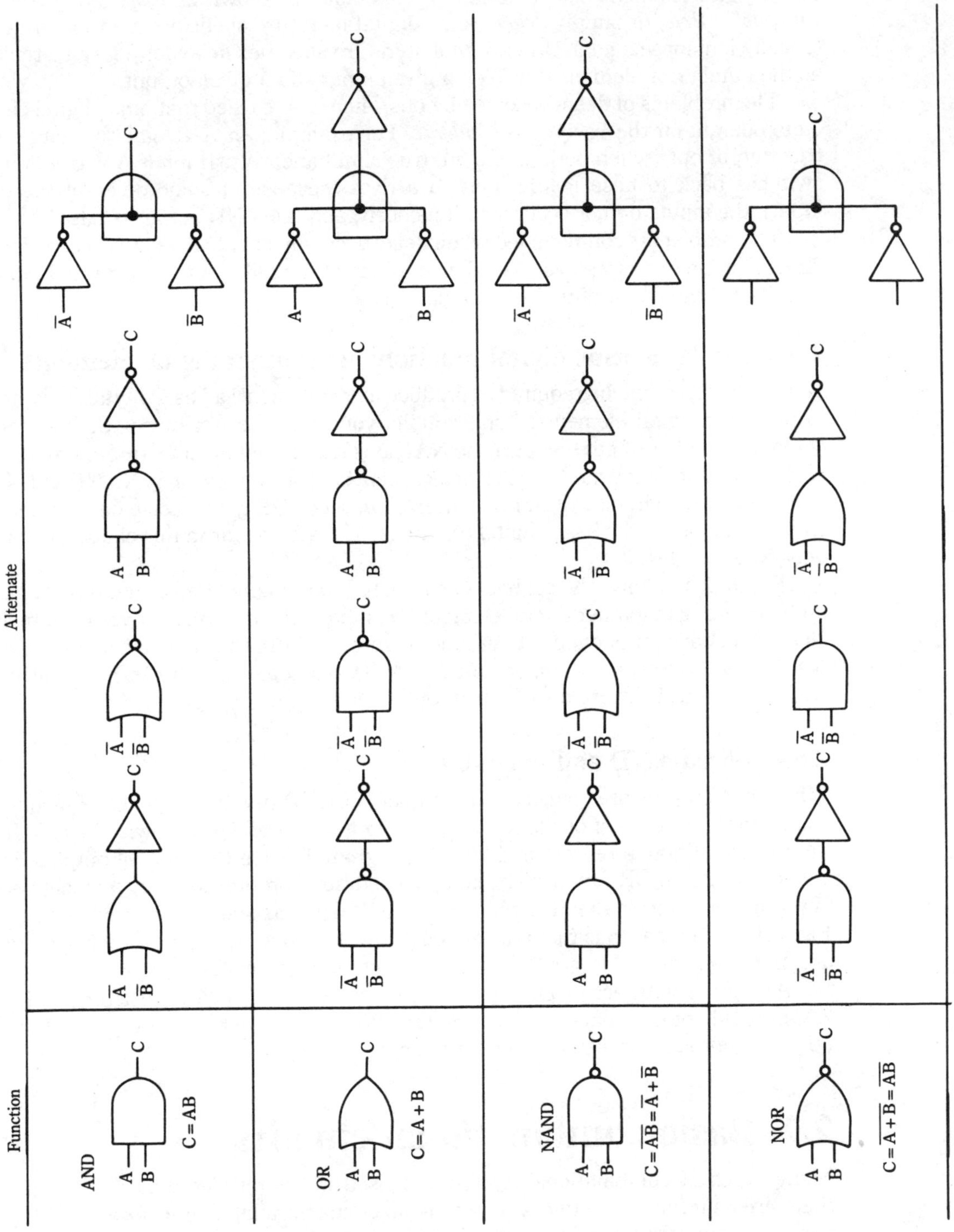

Function	Alternate
AND $C = AB$	
OR $C = A + B$	
NAND $C = \overline{AB} = \overline{A} + \overline{B}$	
NOR $C = \overline{A + B} = \overline{A}\,\overline{B}$	

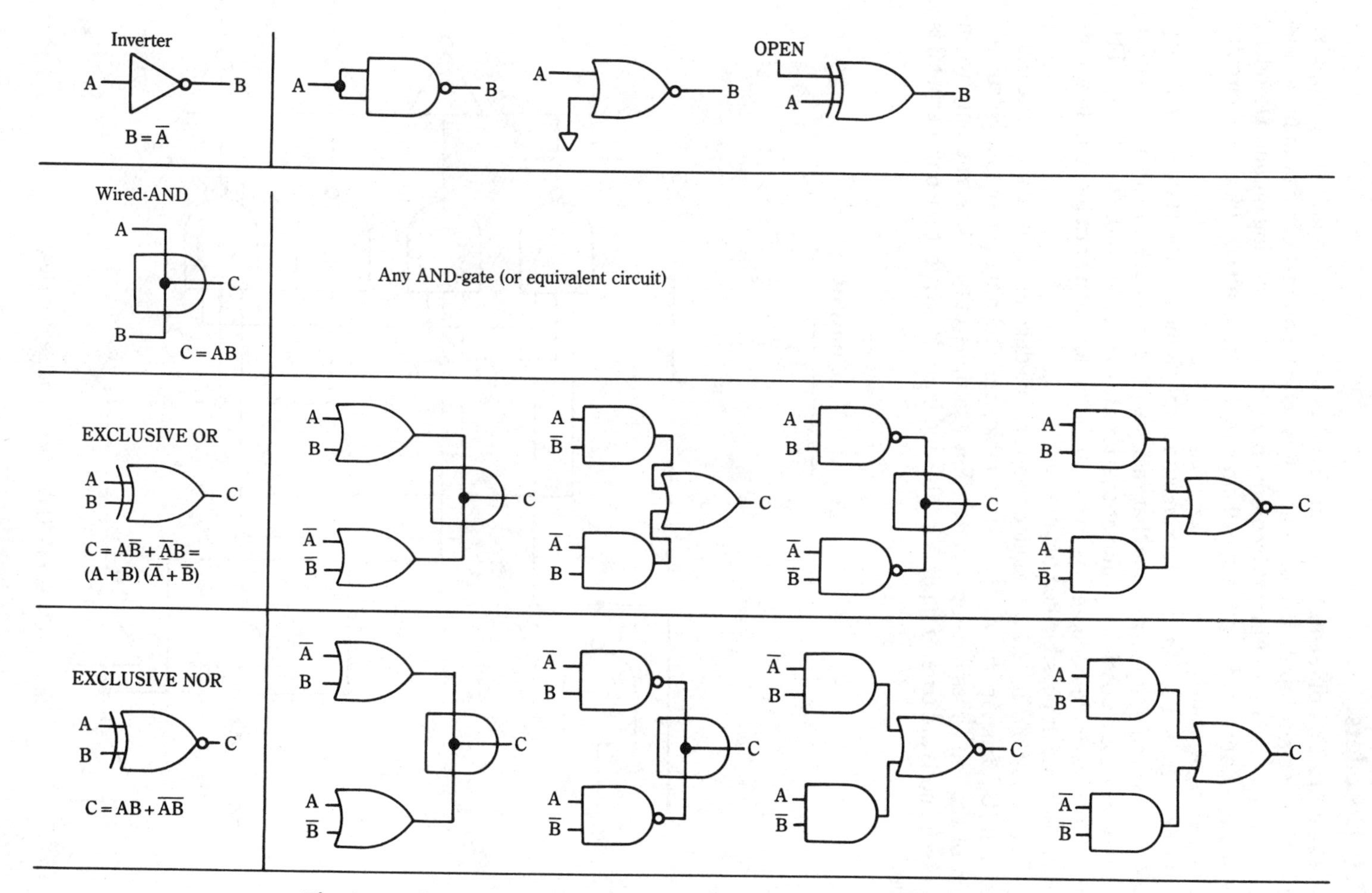

Fig. 2.10 Connections required to implement basic digital functions.

2.7.1 Decoders

The basic function of a combinational circuit is to produce an output (or outputs) only when certain inputs are present. Thus, combinational circuits indicate the presence of a given set of inputs by producing the corresponding output. Decoders are classic examples of combinational circuits. In effect, all combinational circuits are decoders of a sort.

Variable input decoders produce an output that indicates the state of input variables. For example, if you assume that each variable can have only two states (0 or 1), there are four possible combinations for two variables (00, 01, 10, and 11). Thus, a two-variable decoder has two inputs (one for each variable) and four outputs (one for each possible combination).

The circuit of Fig. 2.11 is a gated two-variable decoder. Such circuits are often found in IC form. As shown in the truth table, one and only one output is true for each of the four possible input states. For example, output 4 is true only when inputs A and B are true. If input A is false and input B is true, then only output 3 is true.

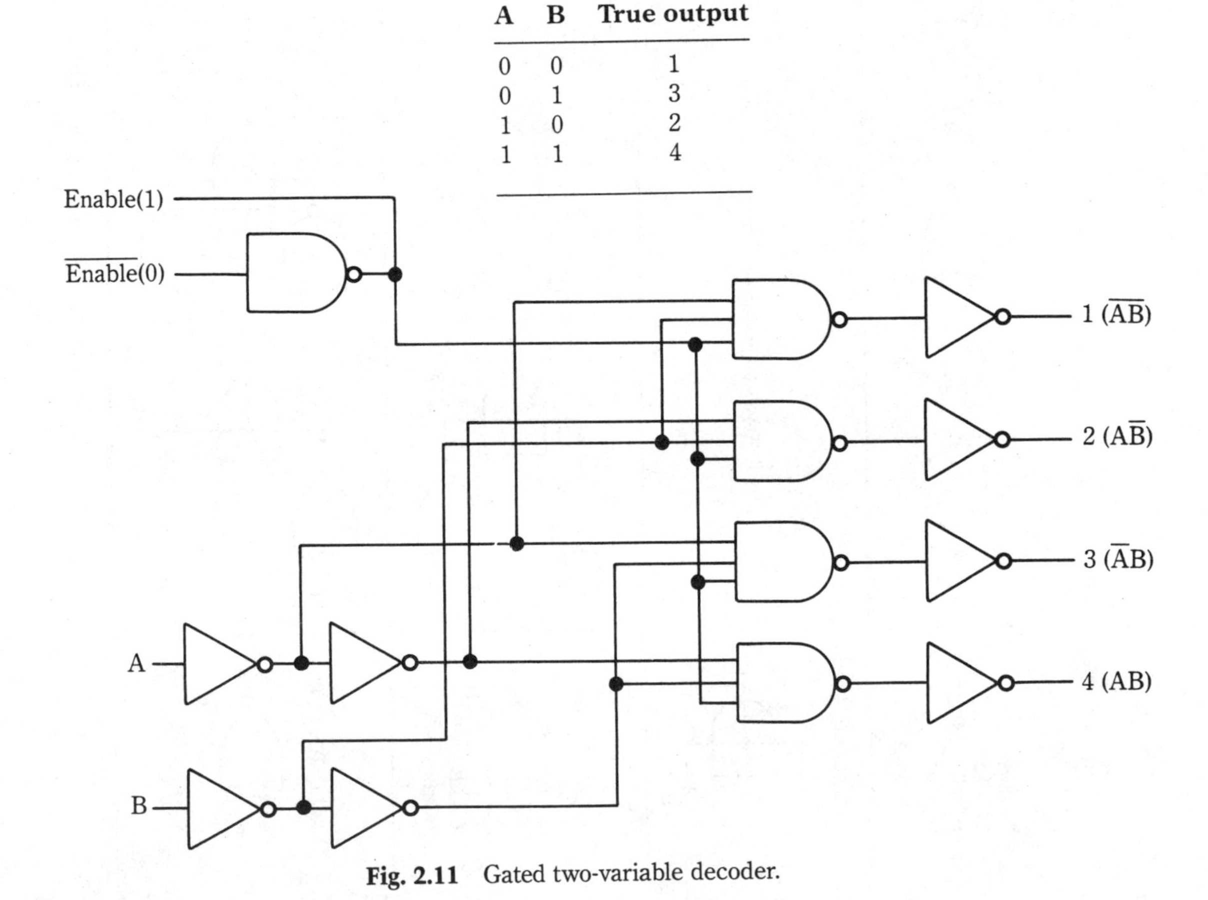

A	B	True output
0	0	1
0	1	3
1	0	2
1	1	4

Fig. 2.11 Gated two-variable decoder.

Note that all four outputs are available only when an *enable, strobe, gate,* or *clock* signal is present. Such turn-on or gate signals can be in pulse form or can be a fixed dc voltage. Note that the enable signal for this particular digital decoder can be true or false (complemented).

Even and odd decoders indicate the number of variables that are true (or false) on an even or odd basis. For example, an odd decoder with three variables produces a true output only when an odd number of variables is true.

Majority and minority decoders are used to indicate the majority (or minority) state of the input variables. For example, if a majority decoder is used with three variables, the single output is true only when two or three inputs are true. With a three-input minority decoder, the output is true if any one of the inputs does not agree with the other two inputs. If all three inputs agree, either true or false, the output is false.

Code converters are a very common type of digital decoder found in IC form. Such decoders convert one type of digital code to another. For example, a binary-to-decimal decoder converts a four-bit binary number into a decimal equivalent.

Figure 2.12 shows an *address-decoder* which is a form of code converter. In this circuit, the four binary-coded address inputs A, B, C, and D are decoded to select (or *address*) only one of 16 mutually exclusive outputs. Note that Fig. 2.12 shows the truth table and block diagram of the IC rather than the complete circuit. This block representation is typical of the diagrams found in digital electronics.

With the circuit of Fig. 2.12, all 16 outputs are at 1 unless both enable and data inputs (pins 18 and 19) are at 0. When enable and data inputs are 0, the addressed output is at 0, with all other outputs at 1. For example, if A, B, and C are at 0 and D is at 1 (binary 1000 or decimal 8), then the address output is 8 (8 is at 0, all other outputs are at 1).

Note the circuit of Fig. 2.12 is housed in a 24-pin IC. Power is applied to pins 12 and 24. The remaining pins are for active signals.

The device of Fig. 2.12 can also be used as a 1- to 16-line *data distributor* by setting enable at 0 and connecting binary information to the data input. (Data distributors are discussed further in Sec. 2.7.4.) This information is routed, unchanged, to the output selected by the address inputs. For example, if A and B are at 1, C and D are at 0, and enable is at 0, whatever data pulses appear on the data input (pin 19) will be distributed to output 3. All other outputs remain at 1.

2.7.3 Encoder circuits

The term *encoder* is used here to indicate any combinational circuit that provides the opposite function of a decoder. For example, there are many decoders that convert four-bit BCD (binary-coded decimal) numbers to 10-bit decimal numbers. An encoder, in this context, converts a 10-bit decimal number to the BCD equivalent. However, both encoders and decoders are forms of combinational circuits in that they both produce an output (or outputs) only when certain inputs are present.

The circuit of Fig. 2.13 is a decimal-to-BCD encoder. Operation of the circuit is as follows. Assume that the 3 decimal input is true, and that all other inputs are false. Under these conditions, the output of inverter gate 3 is false while the out-

Inputs						Outputs															
Enable	**Data**	**D**	**C**	**B**	**A**	**0**	**1**	**2**	**3**	**4**	**5**	**6**	**7**	**8**	**9**	**10**	**11**	**12**	**13**	**14**	**15**
0	0	0	0	0	0	0															
0	0	0	0	0	1		0														
0	0	0	0	1	0			0													
0	0	0	0	1	1				0												
0	0	0	1	0	0					0											
0	0	0	1	0	1						0										
0	0	0	1	1	0							0									
0	0	0	1	1	1								0								
0	0	1	0	0	0									0							
0	0	1	0	0	1										0						
0	0	1	0	1	0											0					
0	0	1	0	1	1												0				
0	0	1	1	0	0													0			
0	0	1	1	0	1														0		
0	0	1	1	1	0															0	
0	0	1	1	1	1																0
0	1	X	X	X	X																
1	0	X	X	X	X																
1	1	X	X	X	X																

X = 1 or 0 All outputs = 1 unless otherwise specified

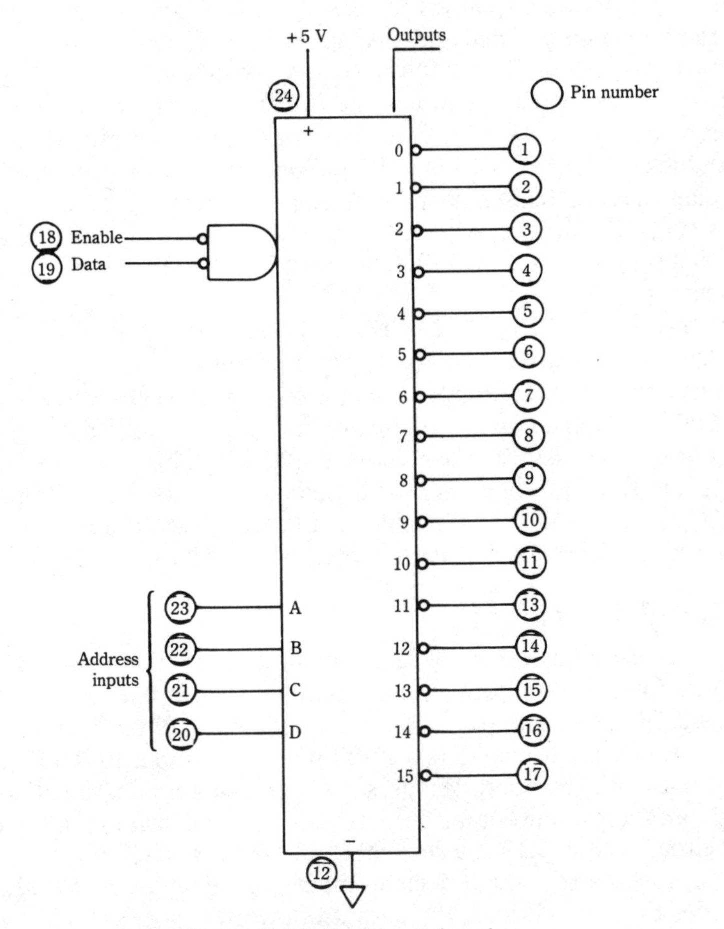

Fig. 2.12 Typical address decoder.

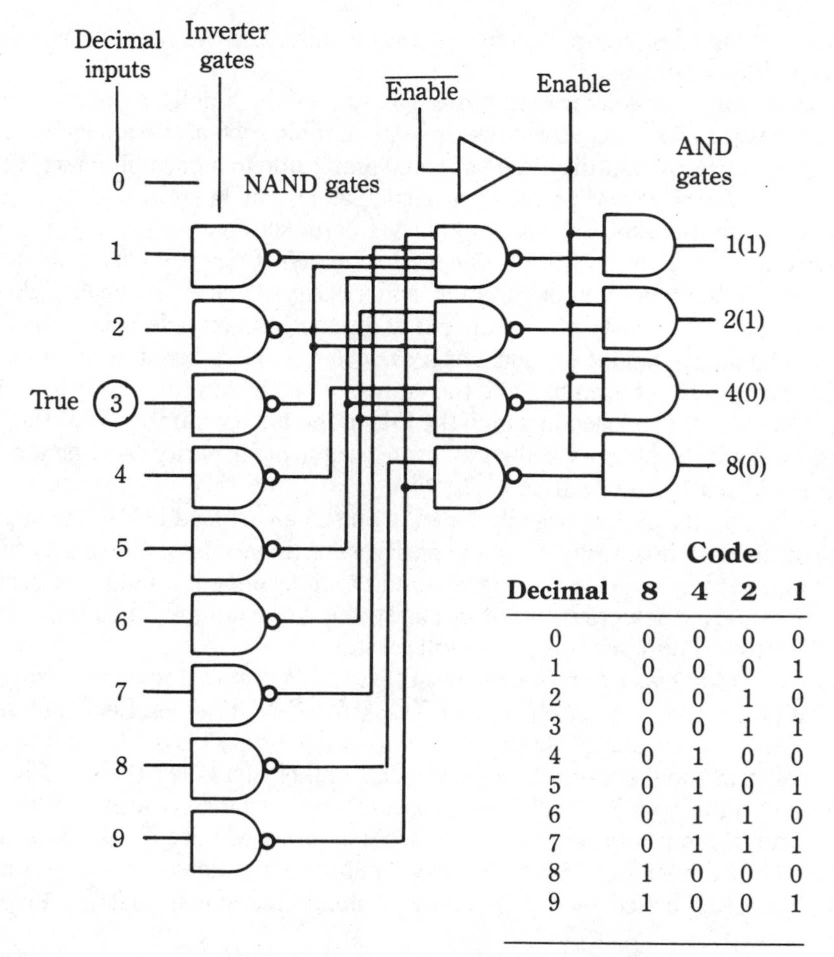

	Code			
Decimal	**8**	**4**	**2**	**1**
0	0	0	0	0
1	0	0	0	1
2	0	0	1	0
3	0	0	1	1
4	0	1	0	0
5	0	1	0	1
6	0	1	1	0
7	0	1	1	1
8	1	0	0	0
9	1	0	0	1

Fig. 2.13 Decimal-to-BCD encoder.

puts of all other inverter gates are true. Inverter gate 3 is connected to output NAND gates 1 and 2. Any false input to a NAND gate produces a true output. Thus, output of NAND gates 1 and 2 are true. The inputs to output NAND gates 4 and 8 are all true. All true inputs to a NAND gate produces a false output. Thus, the outputs of NAND gates 4 and 8 are false. The binary output from the NAND gates is then 0011, or decimal 3. This number is applied to the IC output through the AND gates when the enable signal is present.

Note that the 0 decimal input line is not connected. When the decimal input is 0, lines 1 through 9 are all false, the outputs of the inverter gates are all true, and the output NAND gates are all false, producing a binary 0000, or decimal 0.

2.7.3 Parity and comparator circuits

Special codes have been designed for detecting errors that might occur in digital counting. Any complex digital network that operates on a binary counting system is subject to counting errors because of circuit failure, noise on the transmission line, or similar occurrence.

As an example, a defective amplifier can reduce the amplitude of a pulse (representing a binary 1) so that the pulse appears as a binary 0 at the following circuit. Similarly, noise in the circuit can be of equal amplitude to a normal binary-1 pulse. If the noise occurs at a time when a particular circuit is supposed to receive a binary 0, the circuit can react as if a binary 1 is present.

Parity check A *parity check* is one method of detecting errors in digital circuits. *Parity* refers to the quality of being equal, and a parity check is an equality-checking code. The coding consists of introducing additional bits or pulses into the binary number. The additional bit is known as a *parity bit*, and can be either a 0 or a 1.

The parity bit is chosen to make the number of all bits in the binary group even or odd. If a system is chosen in which the bits in the binary number, plus the parity bit, are even, the system is known as even parity. Even parity is in general use, although odd parity can be used.

Once the parity bit has been included (a pulse has been added on the appropriate line of the data bus, Chapter 4), the parity word (binary bits, plus parity bit) can be examined at any point in the digital circuit to determine if a failure or error has occurred. A parity detection circuit or parity checker examines the parity word to see if the desired odd or even parity still exists.

Figure 2.14 shows a very simple parity-checking circuit known as a *parity tree*. The 8-bit parity tree consists of seven EXCLUSIVE-NOR gates. Each gate has a 0 output if and only if one of its two inputs is at a 1 level. Thus, the output of the parity tree is in the 0 state if there is an odd number of 1s over the eight inputs.

The circuit of Fig. 2.14 detects the presence of a single error in a word. If two errors occur, the output does not indicate that an error has occurred. Thus, simple parity systems detect an odd number of errors but cannot detect an even number of errors. More advanced parity-checking systems recognize that an error has

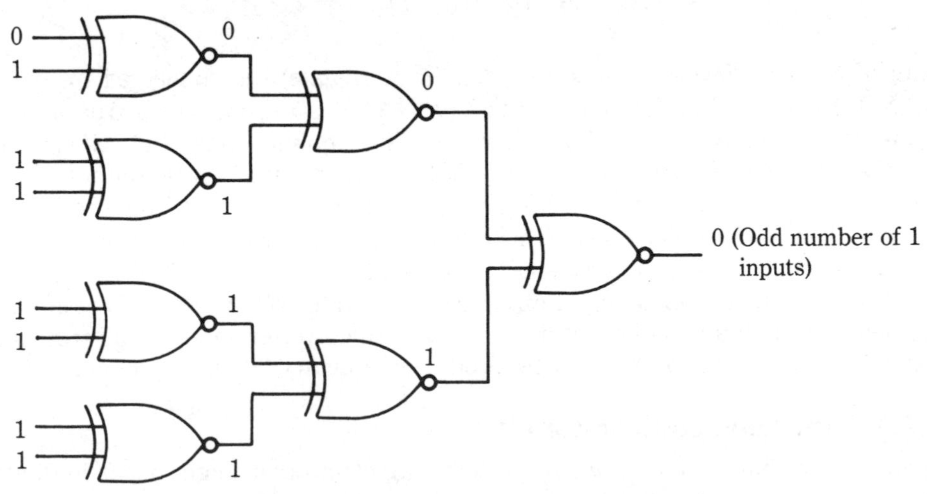

Fig. 2.14 Simple parity-checking circuit (parity tree).

occurred and detect which bit is in error. Several extra bits must be added to the system word to make these parity checks.

Parity circuits are often used with peripheral equipment such as floppy disks and CD-ROMs discussed in Chapter 4. However, most complex digital equipment uses CRC (cyclic redundancy check) circuits to check the accuracy of bits, also discussed in Chapter 4.

2.7.4 Data distributors and selectors

Data distributor and selector circuits, also known as *demultiplexers* (DEMUX) and *multiplexers* (MUX), are very similar to decoders. The two basic circuits described in this section are universal in that the circuits can be rearranged to meet almost any digital system need where data distribution or selection is involved.

Data selector or MUX Figure 2.15*a* shows a four-input *data selector*, or *MUX*. This circuit selects data on one or more input lines and applies the data to a single output channel, in accordance with a binary code applied to control lines. Information present on input lines X0 through X3 is transferred to output Q0, in accordance with the state of control inputs A and B, as shown by the truth table.

Data selection from more than four locations (or inputs) can be implemented with a multiple data-selector circuit as shown in Fig. 2.15*b*. Here, one bit is selected from one of 16 inputs, and transferred to the output in accordance with the truth table.

The basic data selector also can be used to implement a network that provides multiple outputs from multiple inputs as when in Fig. 2.15*c*. Here, each output is selected from one of its own group of four different inputs. For example, with control lines A and B both at 1, data lines X3, X7, and XM are enabled, and all other data lines are turned off.

Data distributor or DEMUX Figure 2.16*a* shows both a two-channel and four-channel *data distributor*, or *DEMUX*. This circuit distributes a single channel of input data to any number of output lines, in accordance with a binary code applied to control lines.

Information applied to input X is distributed to outputs Z0 through Z3, in accordance with the binary numbers applied to control inputs A and B (for the states 0 and 1 of inputs A and B). Information applied to input Y is distributed to outputs Z4 and Z5 in accordance with the state of control input Y.

The size of the distribution function can be increased by increasing the number of basic distributor circuits, as shown in Fig. 2.16*b*, where one bit of information is distributed to one of 16 locations. It is also possible to distribute more than one bit of information using the circuit of Fig. 2.16*c*. Here, N bits of information are distributed to one of eight locations. Information of inputs X through M is distributed in accordance with control variables A and B.

2.8 Basic sequential circuits

There are many types of sequential circuits used in digital electronics. Similarly, there are an infinite variety of sequential circuits available in IC form. The following paragraphs describe a few selected examples.

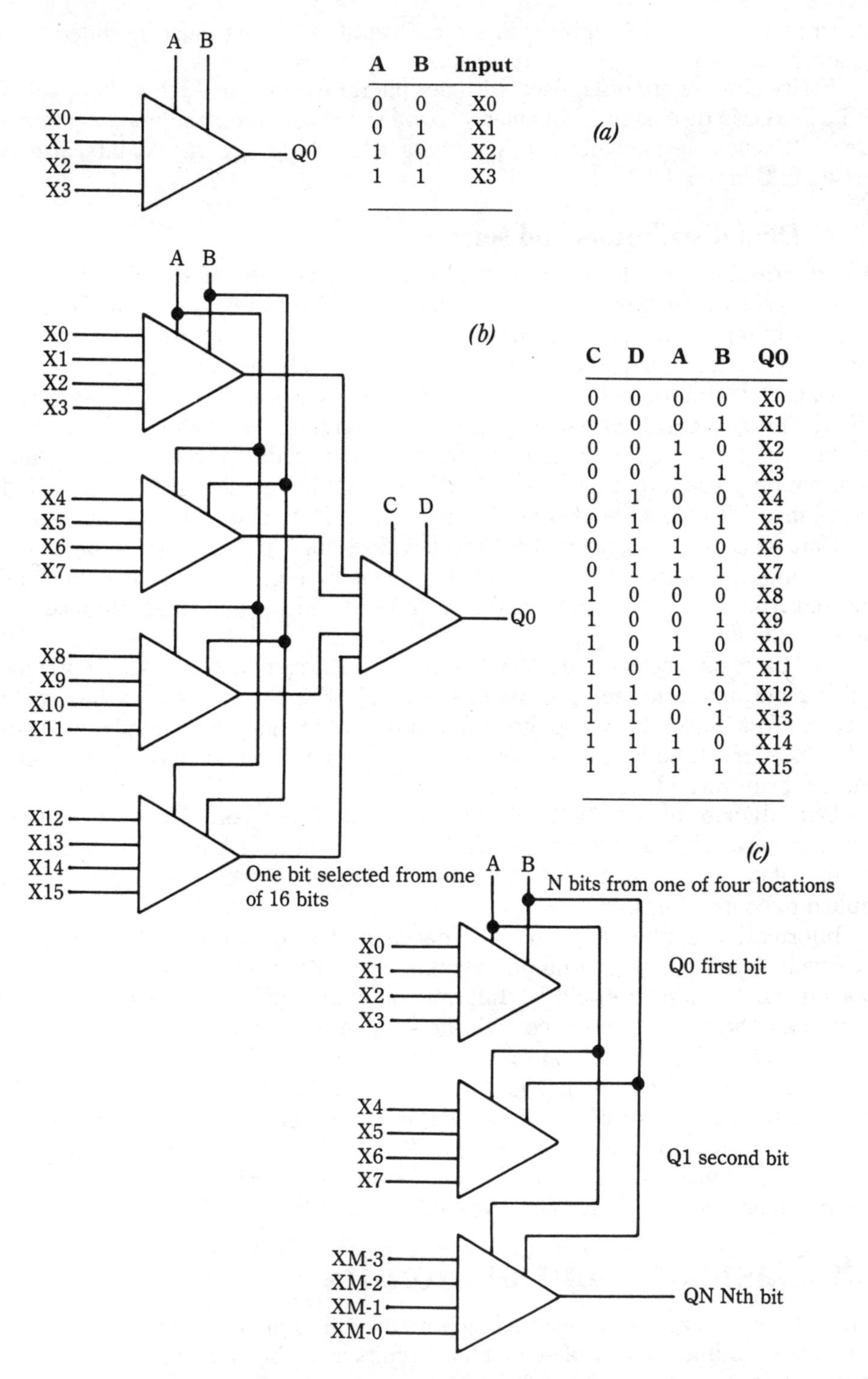

A	B	Input
0	0	X0
0	1	X1
1	0	X2
1	1	X3

C	D	A	B	Q0
0	0	0	0	X0
0	0	0	1	X1
0	0	1	0	X2
0	0	1	1	X3
0	1	0	0	X4
0	1	0	1	X5
0	1	1	0	X6
0	1	1	1	X7
1	0	0	0	X8
1	0	0	1	X9
1	0	1	0	X10
1	0	1	1	X11
1	1	0	0	X12
1	1	0	1	X13
1	1	1	0	X14
1	1	1	1	X15

Fig. 2.15 Four-input data selector or MUX.

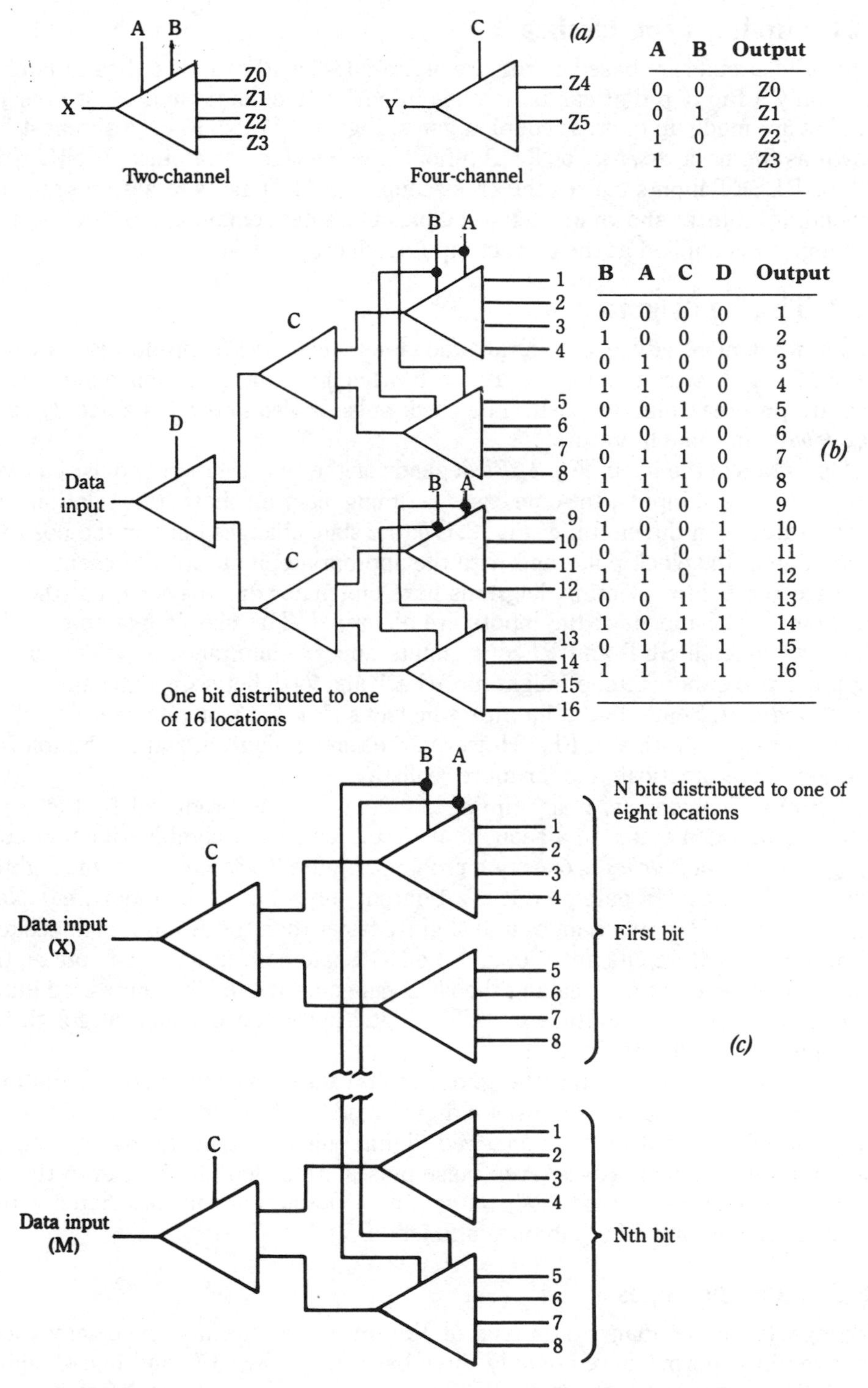

A	B	Output
0	0	Z0
0	1	Z1
1	0	Z2
1	1	Z3

B	A	C	D	Output
0	0	0	0	1
1	0	0	0	2
0	1	0	0	3
1	1	0	0	4
0	0	1	0	5
1	0	1	0	6
0	1	1	0	7
1	1	1	0	8
0	0	0	1	9
1	0	0	1	10
0	1	0	1	11
1	1	0	1	12
0	0	1	1	13
1	0	1	1	14
0	1	1	1	15
1	1	1	1	16

Fig. 2.16 Two-channel and four-channel data distributor or DEMUX.

2.8.1 Flip-flops and latches

Sequential circuits are based on the use of flip-flops (or FFs) and latches. A latch is essentially a flip-flop that can be latched in one state or the other. Most present-day FFs are made up of cross-coupled gates. Figure 2.17*a* shows the simplest FF, known as the basic *reset-set* or RS flip-flop. The presence of a pulse at either the SST or RESET inputs causes the cross-coupled NAND gates to assume the corresponding state, as shown in the truth table. The gates remain latched in the state until a pulse is applied at the correct input to change states.

2.8.2 Timing diagrams

A somewhat more advanced circuit, known as a *clocked-RS* flip-flop is shown in Fig. 2.17*b*. This circuit changes states only when both the input pulse and a *clock pulse* are present simultaneously. The clock pulse is also known as a *gate pulse* or *trigger pulse* in some literature.

Operation of the FF in Fig. 2.17*b* depends on the time relation of pulses as well as the presence of inputs. In some cases, a timing diagram shows this relationship. As an example, in the circuit of Fig. 2.17*b*, the state changes only on the positive-going edge of the clock pulse and with the appropriate input pulse present.

Race condition Timing diagrams have one major drawback in that they do not show what happens if the inputs are abnormal. The classic example of this occurs when both SET and RESET inputs appear simultaneously. The circuit might move to either state, or might move back and forth between states (known as a *race condition*. Some digital literature includes *flowcharts* or *flow tables* to show the theoretical operation of FFs. However, the timing diagram and truth table are generally more practical, and far more realistic.

Glitches Obviously, many timing problems can be produced by the rapid switching between states in a sequential circuit, or even a combinational circuit. The generation of *glitches* is one such problem. Figure 2.17*c* shows a typical glitch situation where a NOR gate produces a 1 output only when both inputs are 0. Now assume that the B input changes to 0 slightly faster than the A input changes to a 1. For a time, both inputs are at 0, and the NOR gate produces a 1. Of course, the A input soon becomes a 1, making the NOR gate output a 0. This undesired glitch pulse is of the same amplitude as other digital pulses on the line (although the glitch pulse is shorter in duration).

Depending on the circuit, the glitch could cause a serious malfunction or a minor problem such as flickering of a digital readout. Note that the term *glitch* is now generally applied to any undesired digital pulse or rapid change of voltage level on a line. In some cases, even noise pulses are called glitches, even though the noise is generated by external sources and is not necessarily associated with a race condition or any other circuit malfunction.

2.8.3 Basic FF types

Although there are many variations of FFs in digital circuits, especially those packaged in IC form, there are only three basic types: RS, JK, and master-slave. Figure 2.18 summarizes the basic FF types.

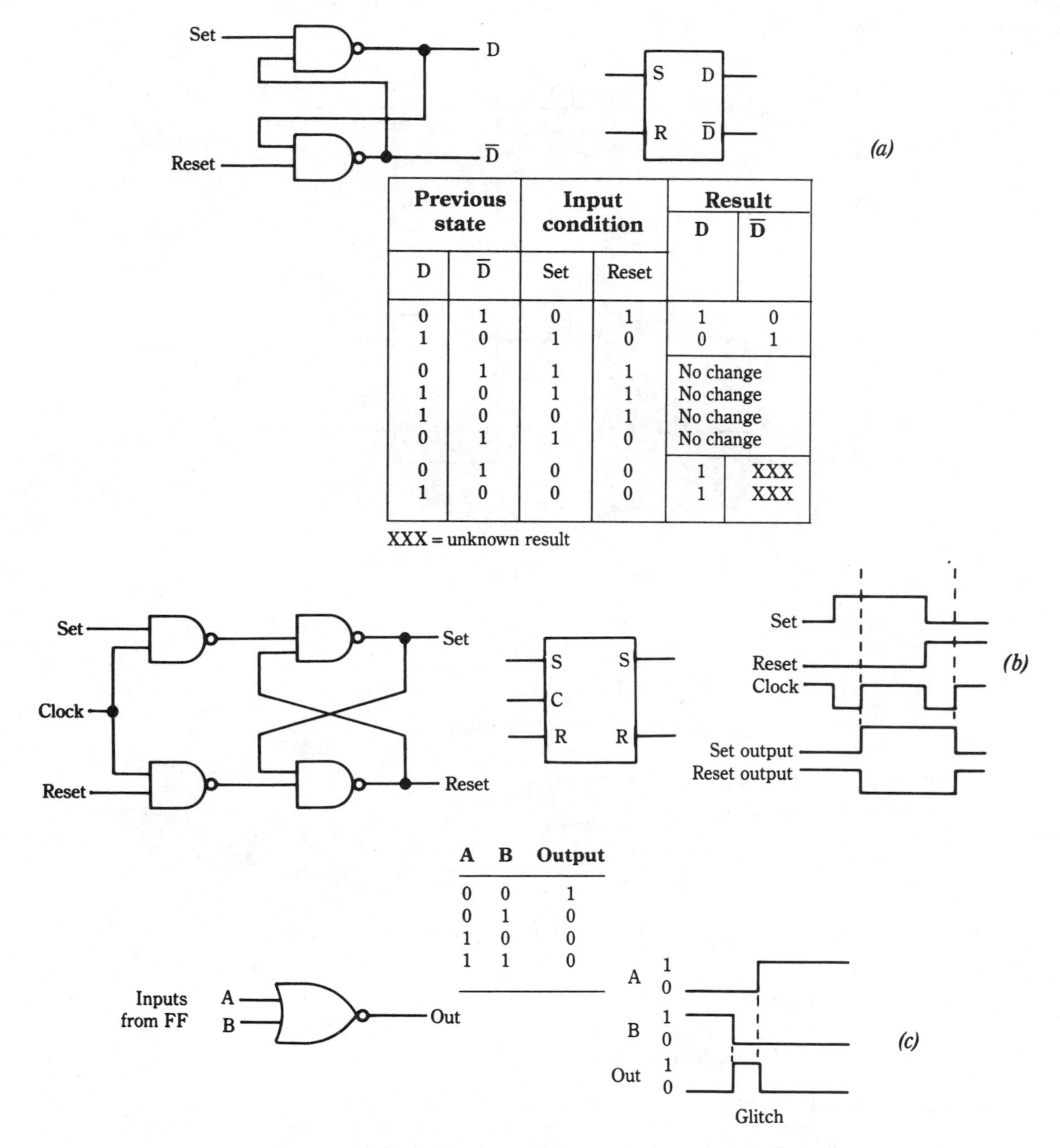

Previous state		Input condition		Result	
D	$\overline{D}$	Set	Reset	D	$\overline{D}$
0	1	0	1	1	0
1	0	1	0	0	1
0	1	1	1	No change	
1	0	1	1	No change	
1	0	0	1	No change	
0	1	1	0	No change	
0	1	0	0	1	XXX
1	0	0	0	1	XXX

A	B	Output
0	0	1
0	1	0
1	0	0
1	1	0

Fig. 2.17 Basic FF circuits, truth table, and timing diagrams.

Gated RS The gated RS FF shown in Fig. 2.18*a* is an improved version of the basic RS FF of Fig. 2.17. To set a gated RS FF, it is necessary to have two positive pulses appearing simultaneously at the inputs of the SET gate. The reset condition requires two pulses at the RESET gate.

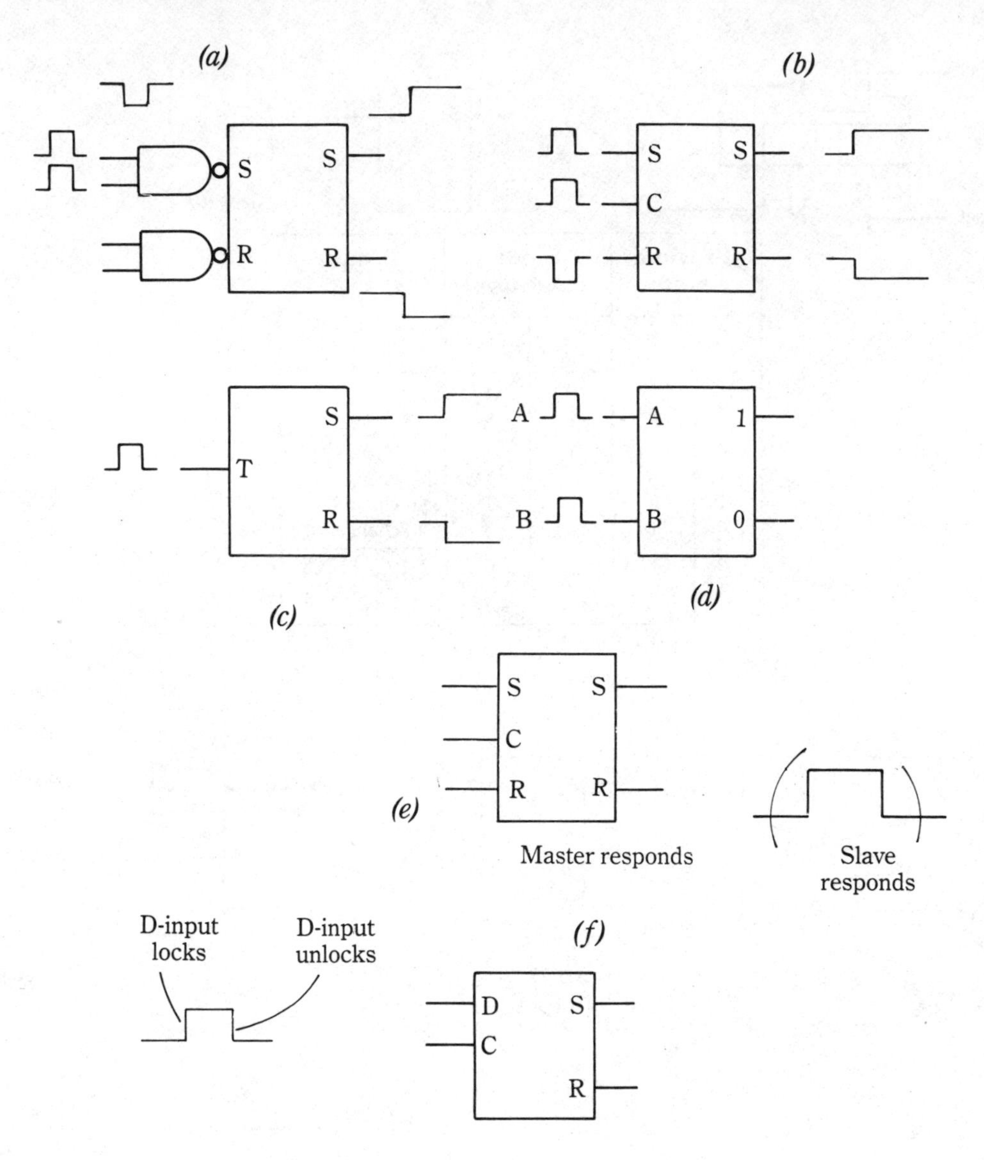

Conditions

Data input	New Clock	Old output S	Old output R	New outputs S	New outputs R	
—	0	1	0	1	0	May not be
—	0	0	1	0	1	changed
1	1	—	—	1	0	Set
0	1	—	—	0	1	Reset

Fig. 2.18 Summary of basic FF types.

(g)

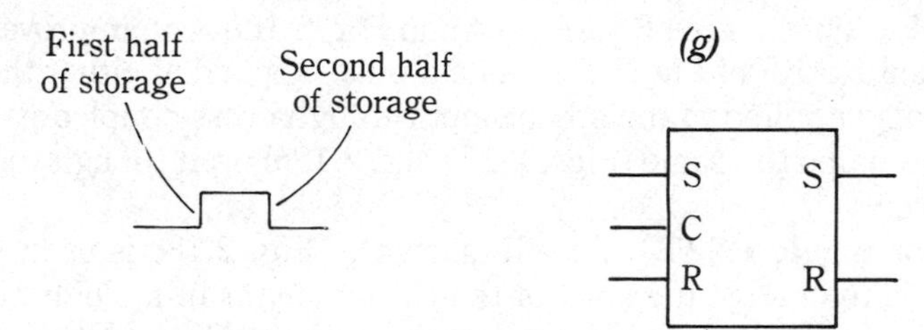

Clock	Input R	Input S	Master	Slave
1	1	1	Same as input	Previous state of slave
1	1	0	Same as input	Previous state of slave
1	0	1	Same as input	Previous state of slave
1	0	0	Previous state of master	Previous state of slave
0	Any		Previous state of master	Same as master

(h)

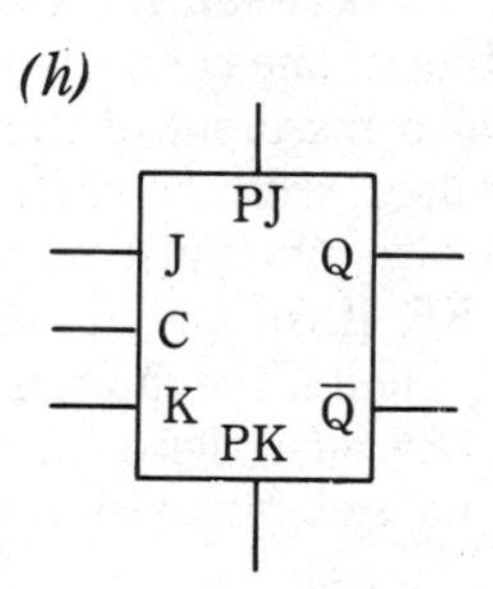

Synchronous			Asynchronous		
J	**K**	**Q**	**PJ**	**PK**	**Q**
0	0	Q	0	0	Q
0	1	0	0	1	0
1	0	1	1	0	1
1	1	$\overline{Q}$	1	1	*

* See text.

Steered RS FF The steered RS FF shown in Fig. 2.18*b* is an improved version of the basic clocked RS FF of Fig. 2.17*b*. A data pulse applied at either the SET or RESET input is also applied to the other input through cross-coupled gates. Note that the symbol remains the same (Figs. 2.17*b* and 2.18*b*) even though the internal circuits are different.

Self-steered or toggle RS FF The FF shown in Fig. 2.18*c* is used when the primary purpose is to change the state of the FF by means of a single trigger (or clock or timing) pulse. Each clock input pulse causes the SET and RESET outputs to toggle between high and low, or between 1 and 0.

This basic toggle operation has one major drawback. As soon as the FF changes states, the conditions at the input gates reverse and, if the timing pulse is still present (this is very possible because the timing-pulse duration is usually longer than the FF delay time), it is quite possible that a second pulse (or glitch) can appear at the output of the input gates. This changes the state of the FF once more.

The condition can repeat several times (a race condition can occur) until the timing pulse is no longer present. Under these conditions, it is very difficult to predict the end result. That is, the FF can end up being SET or RESET when the timing pulse is absent. For this reason, the basic toggle FF is used only when the toggle function is of primary importance.

Cascaded or JK FF To solve the problem of the basic toggle FF (not knowing the end result of SET or RESET between clock pulses), two gated-RS FFs (Fig. 2.18*a*) can be *cascaded* (the output of one connected to the input of another, and the output of the second returned to the input of the first). The basic cascaded FF requires two timing or clock pulses, with the common inputs of the two FFs separated, as shown in Fig. 2.18*d*. Also, one pulse must be delayed from the other.

With the basic cascaded FF, if the output FF is set, the input FF will be RESET when pulse A arrives. Then if the input is RESET, the output FF will be RESET when pulse B arrives. Because there is no feedback action after the first change for each FF, there are no ambiguous situations, and the end result is predictable.

The approach of Fig. 2.18*d* was used in early digital circuits and is sometimes used (in modified form) today. Early designs of FFs with this capability were known as JK FFs, in contrast to RS FFs. However, the term *JK* is now generally applied to any FF made up of cascaded elements in which the outputs are connected back to the input.

Most present-day JK FFs require only one clock input, together with a SET and RESET input, to change states. The FFs can be *preset* to either state. The term *JK* is also applied to the master-slave FF, described next.

Master-slave FF The master-slave configuration (Fig. 2.18*e*) is used extensively in IC FFs. The circuit is similar to the basic cascaded FF, with the slave FF being the output FF. However, with master-slave, the outputs of the slave FF are not tied back (internally, within the IC) to the inputs of the master FF. This is done for greater flexibility. For example, if you want to use only the master or slave portion of the IC, you can do so. In other circuits, both master and slave are used, but you must tie the slave outputs back to the master. Note that the master-slave symbol is the same as that for any clocked FF.

The other major difference is that the clock inputs to both master and slave FFs are tied together. This eliminates the need for two clock or timing pulses. Typically, the master FF is triggered by the positive-going edge, and the slave FF by the negative-going edge of the same pulse, as shown in Fig. 2.18*e*.

Note that in the FF circuits discussed thus far, the clock line could be labeled *strobe, trigger, timer, gate* or some similar term. No matter what term is used, when the clock line is zero, the FF output cannot be changed by an input at either SET or RESET. Also, the circuits described thus far are *double-rail* because they provide for two inputs (in addition to the clock input).

Single-rail FFs Figure 2.18*f* shows the symbol and truth table for single-rail or D-type FFs. Such FFs require only one data (D) input, in addition to the clock (C) input. Within the IC, a single data line (from terminal D) is connected to one gate input. An inverter is connected between the D input and the opposite gate input. This ensures that the SET and RESET levels cannot be high (or at 1) at the same time.

Single-rail or D-type FFs are especially suited to counters and registers (Sec. 2.8.4 and 2.8.5). Note that the D-type FF shown in Fig. 2.18*f* triggers on the leading edge (or positive-going edge if positive logic is assumed) of the clock pulse. Once the clock line goes to 1, all input changes affect the FF. Thus, the FF is locked to the input. The D input is not unlocked until the clock input trailing edge (negative-going voltage) falls below the threshold of the input gates.

Double-rank FF The double-rank FF shown in Fig. 2.18*g* is similar to the cascaded FF (Fig. 2.18*d*) and the master-slave (Fig. 2.18*e*). The term *double-rank* is used because the circuit is essentially two complete FFs connected together, as in the case of a master-slave. However, unlike master-slave, where the clock inputs of both master and slave are connected directly to the clock input, the double-rank has an inverter connected between the clock input and the slave. Note that the double-rank FF is often called a JK FF, even though the FF is not truly JK-wired (outputs returned to inputs).

Because of the inverted input to the slave circuit, the double-rank FF operates on a principle called *pulse dodging*. This term means that half the storage cycle occurs on the 0-to-1 transition of the clock line, and the other half occurs on the 1-to-0 transition.

When the clock line is at 0, the output of the master FF is unchangeable, while the output of the slave FF is determined by the state of the master, as shown in the truth table of Fig. 2.18*g*. When the clock is at 0, the output of slave (and thus the output of the complete FF) is locked to the previous stage of the master (last input). On the other hand, when the clock is at 1, the slave FF is locked and the master responds to the input. The double-rank FF cannot be put into a race condition easily.

Again, note that the symbol is the same for double-rank, master-slave and basic steered FFs. Only the truth table (or possibly the timing diagram) can be used to tell the differences in operation.

Preset FF The FF shown in Fig. 2.18*h* is capable of being preset (to SET or

RESET) using dc voltages at the PJ and PK inputs. This can be done without regard to the clock (asynchronously) or any other input. The FF can also be switched synchronously using the J and K inputs together with a clock pulse.

When switched asynchronously, the FF acts like an RS FF (or basic latch), and can be used for *jam transfer* of data (transfer of data into the FF with no regard for the previous state of the FF). With asynchronous or preset operation, the master and slave are coupled together, and the outputs are set immediately.

When switched synchronously, the rising clock pulse cuts the slave off from the master. As the clock line rises still higher, the pulses or voltage levels at the J and K inputs are set into the master. Then, when the clock returns to 0, the state of the master is transferred to the slave, which in turn sets the output levels.

The synchronous outputs follow the conventional definitions of a JK FF, as shown by the truth table of Fig. 2.18*h*. If both J and K inputs are at 0, the Q output remains in the previous state. If both J and K are at 1, the Q output changes states.

The preset inputs are also shown by a truth table. Note that the preset (PJ and PK) truth table is identical to the synchronous input (J and K), with one exception. When both PJ and PK inputs are at 1, there is no immediate effect on the outputs. The PJ or PK input that drops to 0 last controls the final state of the FF.

2.8.4 Counters

Three basic types of counters are used in digital electronics. All are available in IC form. The three types are *serial* or *ripple counters*, *synchronous counters*, and *shift counters*. The basic differences among the three is in the method of counting. Counters can be made to count either serially or in parallel. In all three cases, the counters are made up of FFs connected in series and/or parallel.

Serial counters Figure 2.19*a* shows the basic serial counter, implemented by connecting the clock input of each stage or FF to the output of the previous stage. (The FF shown in Fig. 2.18*h* can be used in a serial counter.) In serial counters (also known as ripple counters), the FFs operate in the toggle mode (changing state with each clock pulse) . The output of each FF drives the clock input of the following stage.

Serial counters are able to operate at higher speeds than most other types of counters, require little or no gating, and few interconnections, and present only one clock input to the driving or input line. A limitation of all serial counters is that a change of state may be required to ripple through the entire length of the counter. With serial, each FF changes on each 1-to-0 (or 0-to-1) transition of the previous stage.

Serial counters fall into two groups: *straight binary* and *feedback*. Straight binary counters divide the input by 2^N, where N is the number of counting elements (FFs). With the straight binary counter of Fig. 2.19*a*, the clock input of the LSB FF receives the count-line input, and the Q output of each FF is connected to the clock input of the following FF.

The count-sequence table of Fig. 2.19*a* shows the state of each FF for all counts from 0 to 7. When several pulses are applied to the clock line, the A, B, and C FFs are all at 1. This results in a 111, or a binary 7. In this circuit, all FFs can be reset simultaneously. When a 1 is applied to the RESET line (PK inputs), all FFs are reset (the Q output is 0, and the output is 1).

Each FF also can be preset on an individual basis. This permits a number to be entered in the counter before a count is taken. For example, if it is desired to preset a count of decimal 3 into the counter before the first count, FFs A and B are set to 1, and FF C is set to 0 (binary 011 or decimal 3). The first count then changes FFs A and B to 0 and FF C to 1, resulting in binary 100 or decimal 4.

Feedback serial counters also count in binary code, but a number of higher-value states are eliminated by the feedback. The basic feedback counter divides the input frequency by $2^{X-1} + 1$, where X is the number of FFs. A division by any arbitrary number can be done using various combinations of straight and feedback counters.

Figure 2.19*b* shows a feedback counter. The corresponding count-sequence table shows the state of each FF for all counts from decimal 0 through 9 in a BCD format. This counter also has a BCD-to-decimal decoder, providing both a decoded output (indicating the count in decimal form), and the outputs (true and complementary) of each FF stage. Each FF has a RESET function tied to a common RESET line, permitting all stages (FFs) to be RESET (or CLEARED) to 0 simultaneously. The counter is not provided with any preset function. The BCD-to-decimal decoder is similar to those discussed in Sec. 2.7.

Synchronous counters A synchronous or *clocked* counter is one in which the next state depends on the present state, and all state changes occur simultaneously with a clock pulse. Because all FFs change simultaneously, the output (or count) can be taken from synchronous counters in *parallel form* (in contrast to *serial form* for ripple counters). Synchronous counters can be designed to count up, down, or do both, in which case they are often called *bidirectional counters.*

Shift counters Shift counters are a specialized form of clocked counter. The name is derived from the fact that operation is similar to a shift register (Sec. 2.8.5). In general, the shift counter produces outputs that are easy to decode, and normally requires no gating between stages (thus permitting high-speed operation). Generally, IC shift counters are found in two forms: *decade shift counters* and *ring shift counters.*

The decade shift counter shown in Fig. 2.19*c* is sometimes called a *switch-tail ring counter, Johnson counter,* or simply a *shift counter.* FFs A through E provide a decade output, but require decoding as shown by the output decoding table.

For example, if you want a count of 8, you monitor the simultaneous condition of the FF D Q output and the FF C Q output. FF A represents the least significant bit, FF E the most significant bit. The output of FF E is fed back into the FF A clock input as shown. This feedback presets the counter to 0 before the counting can begin.

The ring shift counter shown in Fig. 2.19*d* is a typical ring counter with error correction. In some applications, it is important to eliminate (or minimize) the decoding logic that follows a counter. A ring counter is useful in such applications. Ring counters use one FF for each count (a 10-count device or decade counter requires 10 FFs). Only one FF is true at any time, so no decoding is required.

In the circuit of Fig. 2.19*d*, the K input of the last FF (FF M) is connected to a

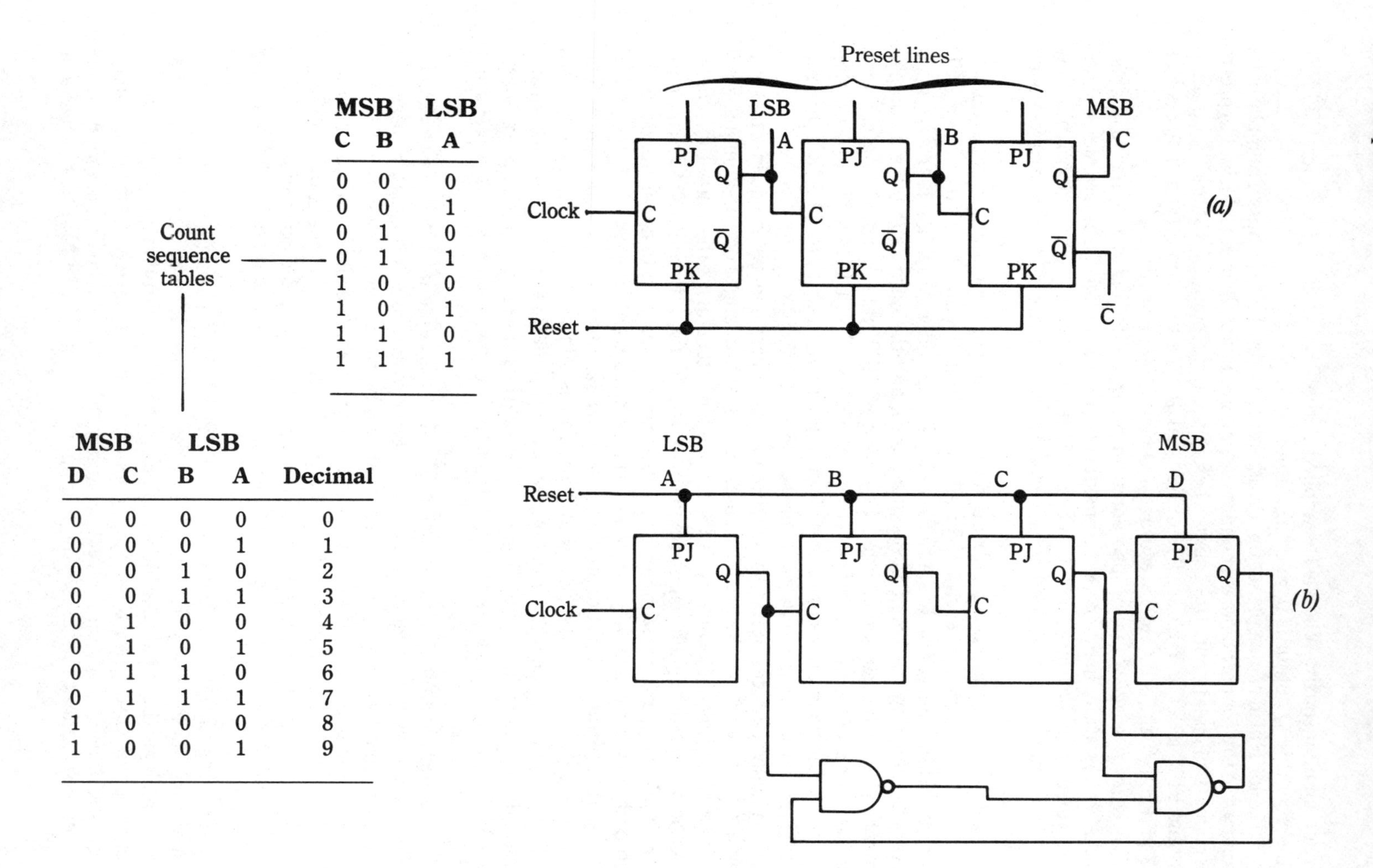

MSB C	B	LSB A
0	0	0
0	0	1
0	1	0
0	1	1
1	0	0
1	0	1
1	1	0
1	1	1

MSB D	C	B	LSB A	Decimal
0	0	0	0	0
0	0	0	1	1
0	0	1	0	2
0	0	1	1	3
0	1	0	0	4
0	1	0	1	5
0	1	1	0	6
0	1	1	1	7
1	0	0	0	8
1	0	0	1	9

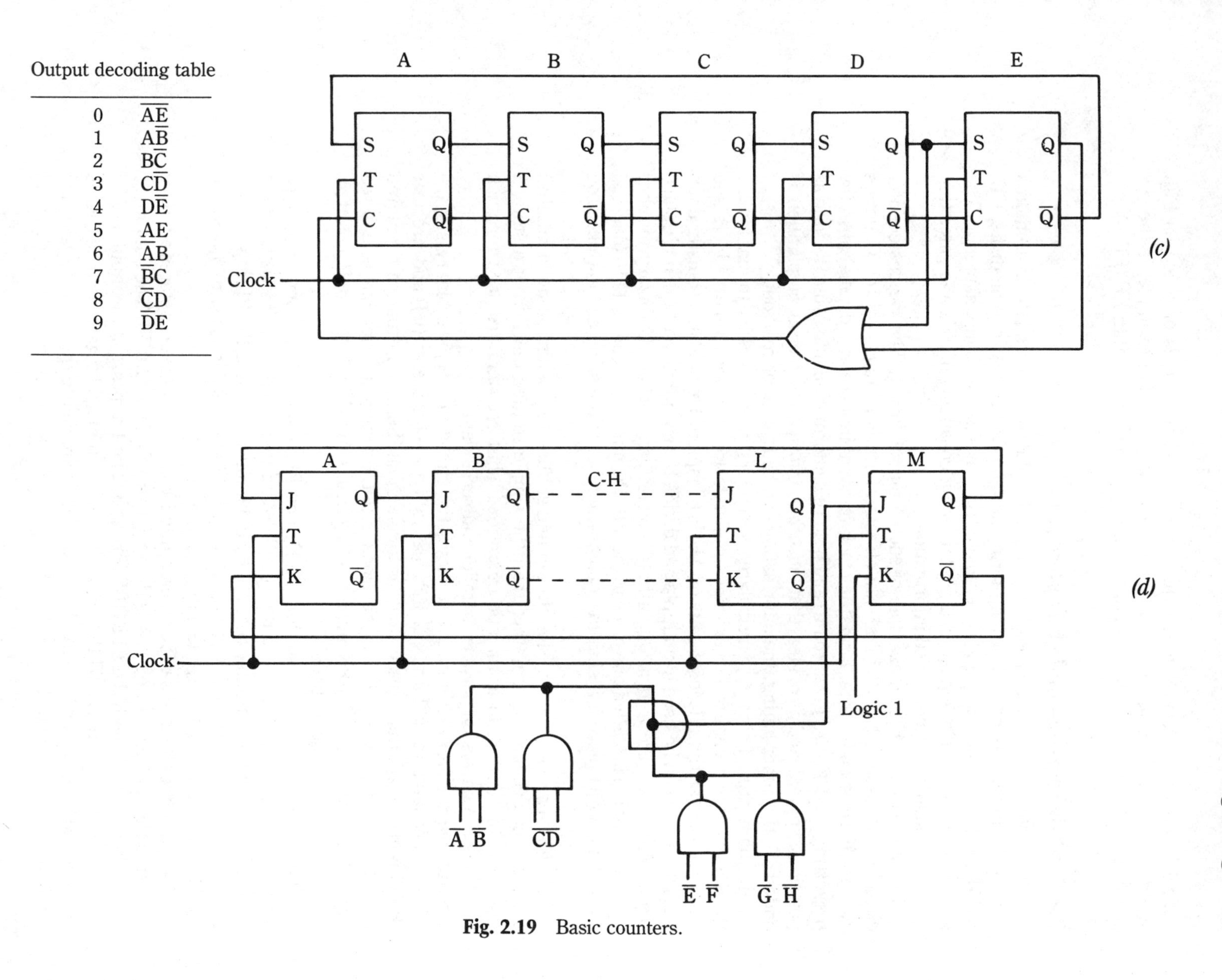

Output decoding table

0	$\overline{AE}$
1	$A\overline{B}$
2	$B\overline{C}$
3	$C\overline{D}$
4	$D\overline{E}$
5	AE
6	$\overline{A}B$
7	$\overline{B}C$
8	$\overline{C}D$
9	$\overline{D}E$

Fig. 2.19 Basic counters.

logic 1 level at all times. Note that no output decoding table is provided on Fig. 2.19*d* as is the case for a decade shift counter, Fig. 2.19*d*. For example, in the circuit of Fig. 2.19*d*, if you want a count of 8, you monitor the eighth FF (FF L).

2.8.5 Shift registers and shift elements

The term *register* can be applied to any digital circuit that stores information on a temporary basis. Permanent or long-term storage is generally done on disks, CD-ROMs, magnetic devices, and other memories, as discussed in Chapter 3. The term *register* is usually applied to a digital circuit consisting of FFs and gates that can store binary or other coded information.

A storage register is an example of such a digital circuit. The various counter circuits discussed in Sec. 2.8.4 are, in effect, a form of storage register. Counters accept information in serial form (the clock or count input) and a parallel form (by presetting the FFs to a given count) and hold this information (the count) as long as power is applied and provided that no other information (serial or parallel) is added. With proper gating, the information can be read in or read out.

In many digital circuits, particularly computers, it is necessary to manipulate the data held in registers. For example, registers are used with binary adders and subtractors in arithmetic logic units or ALUs (Chapter 3) to hold and transfer data. Registers also can be used to multiply and divide binary numbers. Both multiplication and division are a form of shifting. For example, when the binary number 0111 (decimal 7) is shifted one place to the left, the number becomes 1110 (decimal 14). Thus, a one-place left shift for a binary number is the same as multiplication by two.

A shift register is a circuit for storing and shifting (or manipulating) a number of binary or decimal digits (rather than the simple storage function of a storage register). In addition to arithmetic operations, shift registers are used for such functions as conversion between parallel and serial data.

Storage registers Figure 2.20*a* shows the circuit of a typical binary storage register available in IC form. All FFs are JK type with preset (PJ) and preclear (PK) inputs, in addition to the clock or toggle input. Serial information is entered into the register by means of the clock input. Parallel information is entered through the gates and PJ/PK inputs. The circuits are, in effect, ripple counters with gates for parallel entry of data.

The gates are arranged so that the PJ and PK inputs receive opposite states when information is applied to the parallel data lines, and the STROBE line is enabled. For example, assume that the PJ input must be 0 (and the PK input 1) for the Q input to be 1. Under these conditions, the STROBE line is set to 0, and the data line is set to 1, when the Q output is to be 1.

Now assume that a 1 is to be entered into FF A. This presets or stores a binary 0001 into the register, since FF A is the LSB (with FF D the MSB). The data-input line is then set to 1, with the STROBE line at 0. The 1 appears at the input of gate B, along with a 1 from the STROBE line (inverted from a 0 by gate A). The two 1s produce a 0 output from gate B. This 0 output is applied to the PJ input and sets the Q output of FF A to 1.

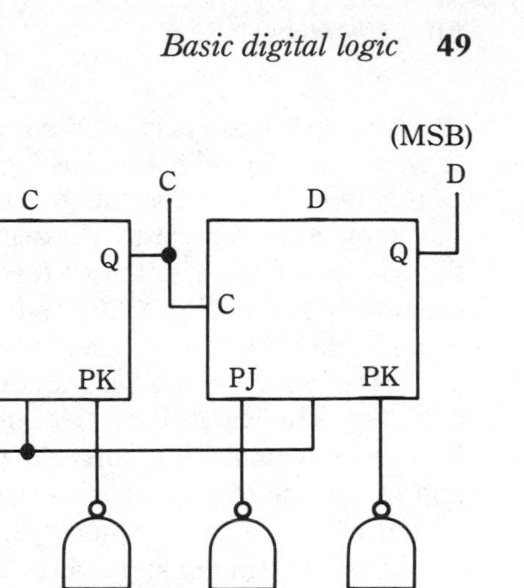

(a)

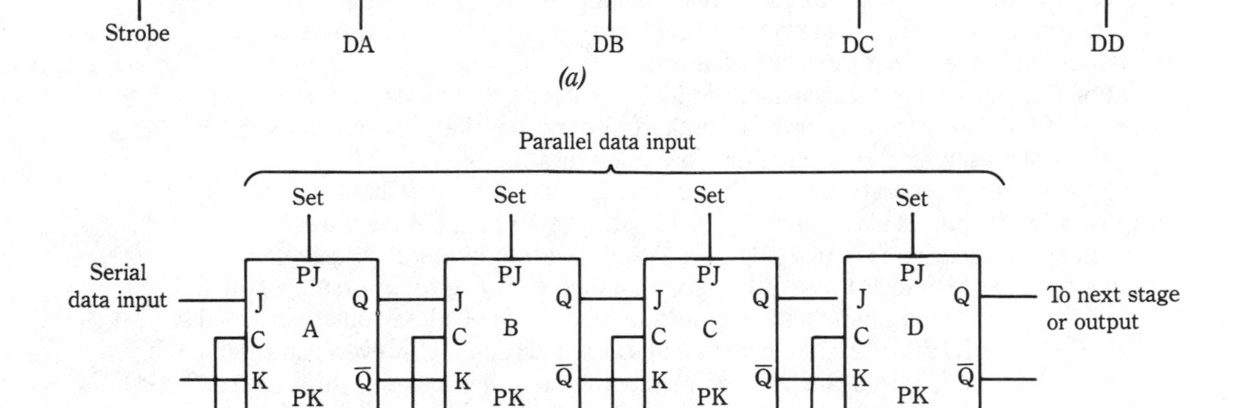

(b)

Fig. 2.20 Basic registers.

At the same time, the 1 input at the DA line is inverted by gate C and appears as a 0 at the input of gate D, together with a 1 from the STROBE line. The 1 and 0 inputs produce a 1 output from gate D, which is applied to the PK input. This assures that the Q output of FF A is 1 (with the $\overline{Q}$ output at 0). All of the remaining FFs receive opposite states: PJ at 1, PK at 0. The Q outputs of FFs B, C, and D go to a 0, and a binary 0001 results.

All information is cleared when the CLEAR line is enabled (all Q outputs move to 0), without regard to the existing state. Data can be taken from the Q or $\overline{Q}$ outputs, as required. Serial information is usually taken from the Q or $\overline{Q}$ outputs of FF D.

Basic shift registers A basic shift register is shown in Fig. 2.20*b*. (Note that the circuit is similar to the shift counters described in Sec. 2.8.4.) Circuits such as the one shown in Fig. 2.20*b* shift data contents *one position to the right* for each occurrence of the clock pulse (each stage switches once for each clock pulse).

Data can enter the registers in serial or parallel form from other registers or counters. The RESET (or clear) line is common to all stages and presets stages to 0. The individual SET lines allow parallel data transfer into the register. Both Q and $\overline{Q}$ outputs of all stages can also be brought out to permit parallel transfer of data from the register. Thus, data can be shifted serially in and out of the register.

When shifting in serial form, the most significant bit (FF D) is transferred out first. To clear the register automatically after all information is shifted out, the J and K inputs of the first stage (FF A) are tied to 0 and 1, respectively.

Multiple-function IC registers Shift registers available in IC form usually have more versatile features than those of the basic register shown in Fig. 2.20*b*. For example, a typical IC register is capable of shifting data stored in the register one position to the right (as with the basic register), or one position to the left, with each clock pulse. When shifting information out of the register, it might be necessary to replace the original information back into the register. In IC registers, this usually requires an *end-around shift* feature.

For high-speed digital systems, parallel transfer of data is essential. For this reason, IC registers usually provide both serial and parallel transfer, as well as combinations such as serial to parallel, and vice versa.

For arithmetic operations, such as subtraction and division (Chapter 3), a *complementation* feature is also desirable. IC registers often include a feature whereby a given set of commands cause the FFs to set up a toggle mode, and to complement with the next clock pulse (all FFs with a 0 go to 1, and vice versa).

IC registers accomplish these functions in many ways. Unless you are involved in the design of IC registers, the exact means are not crucial. However, you should be aware that the features (such as complementing, parallel-to-serial shift, etc.) do exist.

3

Typical digital IC and discrete circuits

This chapter describes the most common circuits found in a variety of digital electronic systems. All of these circuits are made up of the basic building blocks (gates, counters, and registers) discussed in Chapter 2. Most of these circuits are found in IC form. In a few cases, an IC contains one of the circuits. In most cases, several circuits are included in one IC.

A typical digital electronic device is actually a collection of ICs, all interconnected on a common PC (printed circuit) board, card or module (perhaps with a few miscellaneous resistors, capacitors, transistors, or gates on the same board). In more sophisticated digital systems involving microprocessors (Chapter 4), all or most of the circuits are combined on a single IC. Before getting into any circuit details, this chapter starts with a discussion of the various forms of digital ICs.

3.1 Digital IC forms

This section describes the various digital forms found in both present-day and older digital equipment. Some of these forms appear as discrete-component circuits in early digital systems. However, most of the forms are packaged as ICs in present-day equipment.

Note that the digital elements discussed here represent only a few of the many forms found in today's equipment. However, if you understand these forms, you should have no difficulty in understanding the many forms now in use and those that will be developed in the future.

3.1.1 Resistor-transistor logic (RTL)

Figure 3.1*a* shows a basic RTL circuit and symbol. RTL was derived from direct-coupled transistor logic (DCTL) and was the first digital IC form introduced. RTL is no longer found in present-day design, but might be of interest to students (or technicians who must service older equipment).

3.1.2 Diode-transistor logic (DTL)

Figure 3.1*b* shows a basic DTL circuit and symbol. DTL is another digital form that was translated from discrete design into IC elements, but is not generally used today.

3.1.3 High-threshold logic (HTL)

Figure 3.1*c* shows a basic HTL circuit and symbol. HTL is designed specifically for digital systems where *electrical noise* is a problem, but where operating speed is of little importance (because HTL is the slowest of all digital IC families).

Because of the slow speed, and because other digital forms (such as the MOS (metal oxide semiconductor) described in Sec. 3.1.6) offer similar *noise immunity*, HTL is not generally used in present-day design. However, HTL can be found in existing systems, particularly where digital circuits must be connected to industrial- and other heavy-duty equipment.

3.1.4 Transistor-transistor logic (TTL or T^2L)

Figure 3.1*d* shows a basic TTL circuit and symbol. TTL was one of the most popular digital IC families. Many manufacturers still offer a complete line of TTL devices, including *power gates* and *line drivers* to increase the fan-out (Sec. 2.6.1) or drive capability of the circuit.

Most TTL gates have an active pull-up resistor at the output. This does not permit using wired-OR (Sec. 2.6.4). Special gates, such as shown in Fig. 3.1*e*, are included in many TTL product lines to overcome this limitation. The output of the Fig. 3.1*e* circuit can be used for wired-OR or to drive discrete components.

One disadvantage of some TTL is the so-called *totem-pole* output. As shown in Fig. 3.1*d*, both output transistors are on during a portion of the switching time. Because the turn-off time of a transistor is normally greater than the turn-on time, a *current spike* can pass through both transistors and the load resistor. The *active bypass network* helps to limit this problem.

3.1.5 Emitter-coupled logic (ECL)

Figure 3.1*f* shows a basic ECL circuit and symbol. ECL operates at very high speeds, and produces both *true* and *complementary outputs*. The high operating speed is obtained because ECL uses transistors in the *nonsaturating* mode. That is, the transistors do not switch full-on or full-off, but swing above and below a given bias voltage. ECL generates a minimum of noise and has considerable noise immunity. However, as a trade-off for the nonsaturating mode (which produces high speed and low noise), ECL is the least efficient (ECL dissipates the most power for the least output voltage).

As an example (for positive logic), a 1 for the circuit of Fig. 3.1*f* is about −0.9 V. Logical 0 is about −1.7 V, which makes a nominal voltage switch or pulse of about 0.8 V. Unlike TTL (and other saturated logic), ECL requires a *bias voltage*

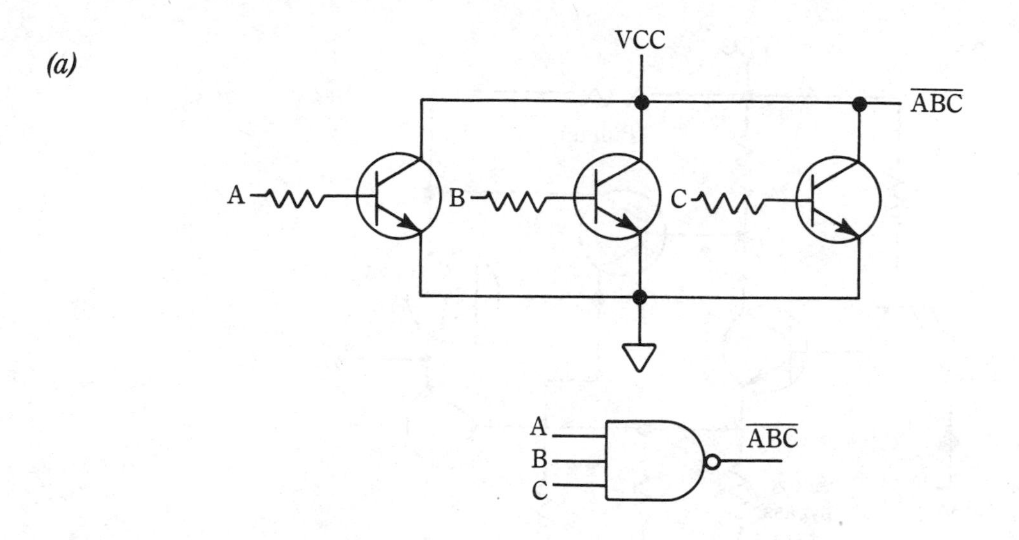

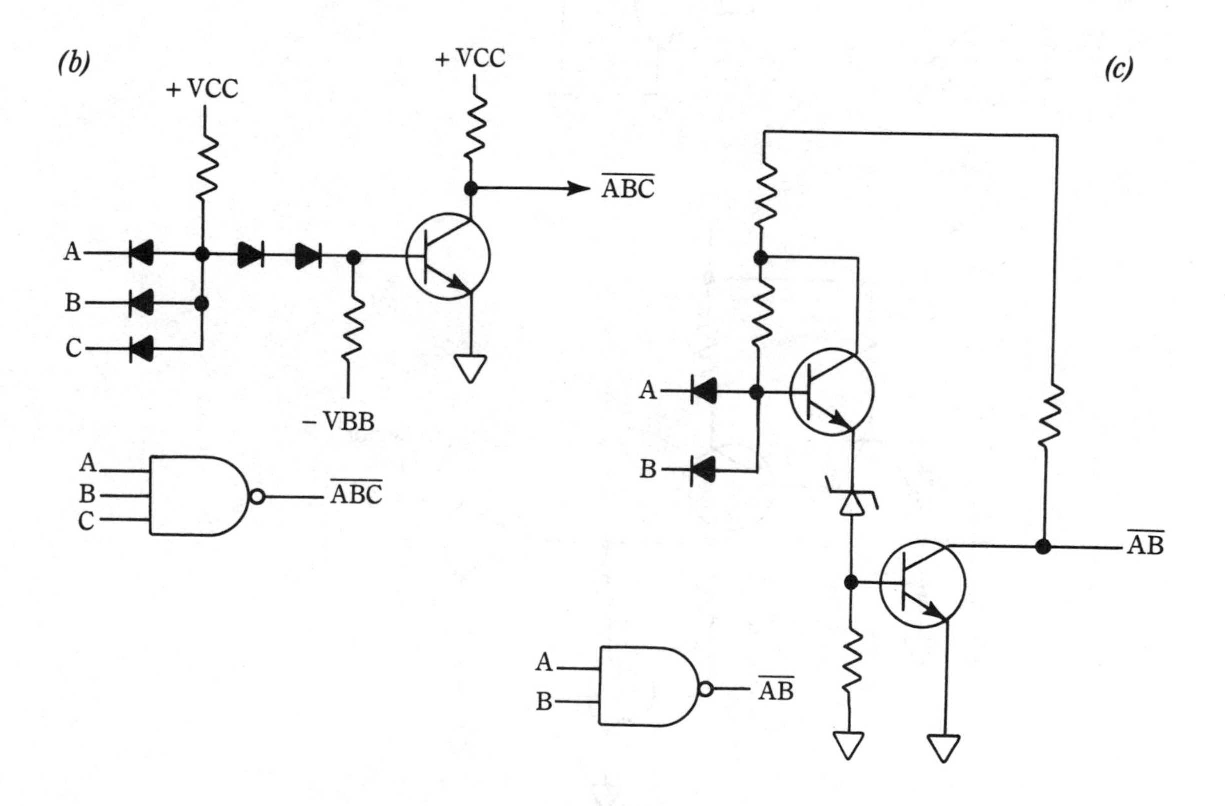

Fig. 3.1 Digital IC forms.

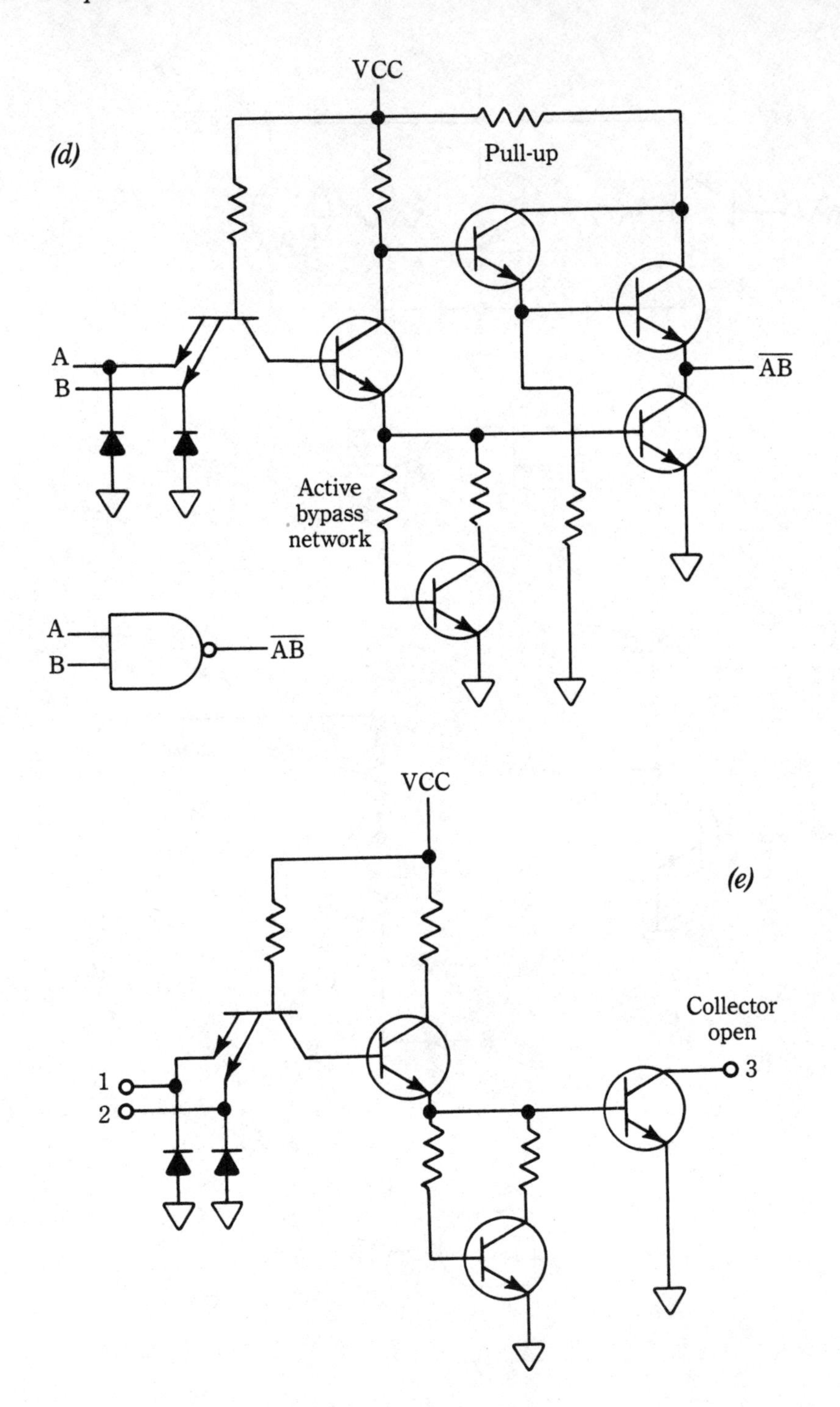

Fig. 3.1 Continued

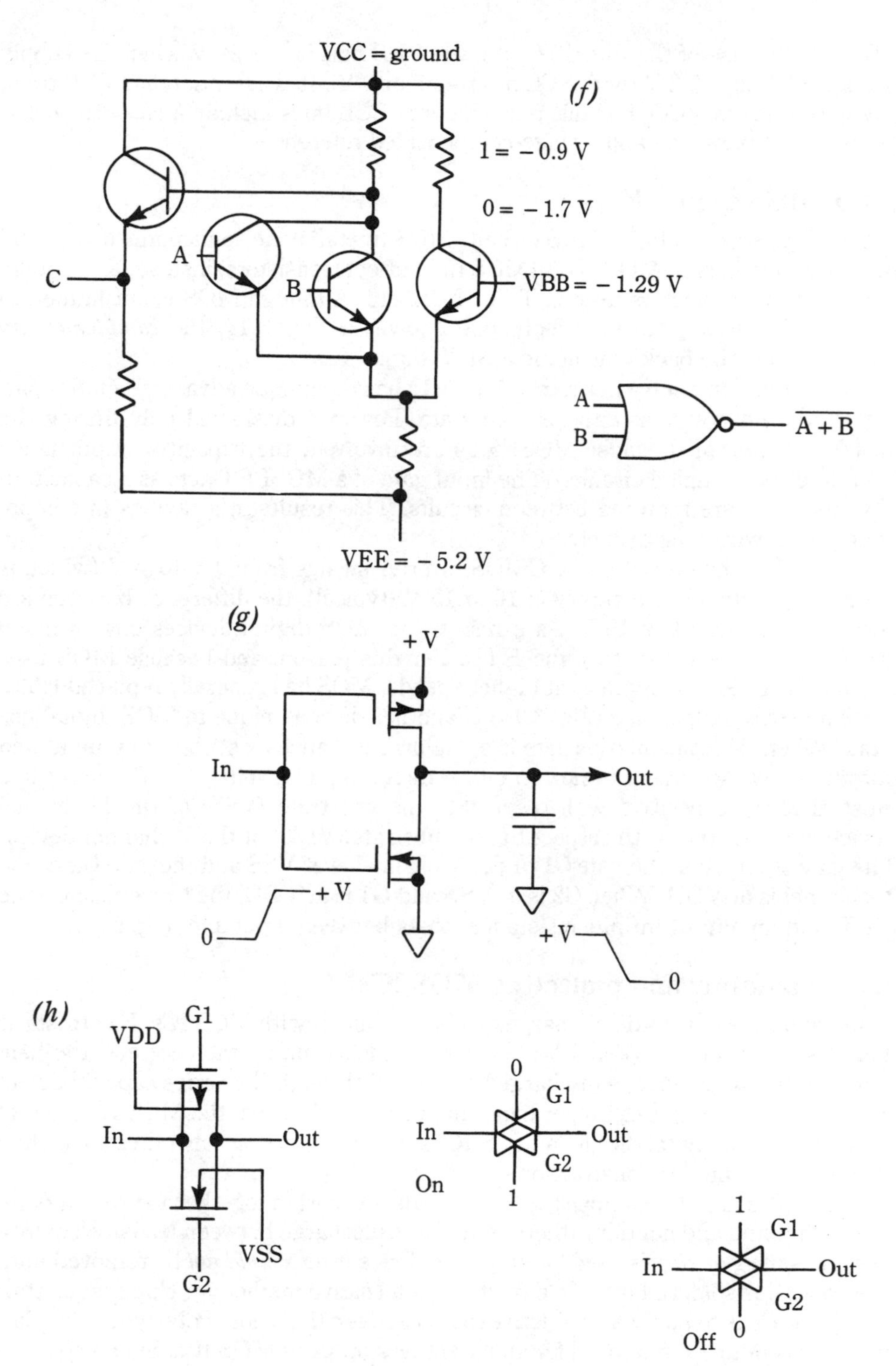

Fig. 3.1 Continued

VBB. In the case of the Fig. 3.1*f* gate, the VBB bias is −1.29 V when the supply voltage VEE is −5.2 V (with VCC at ground or 0 V). It is essential that VBB track any variations in VCC. For this reason, some ECL lines include a *bias driver* that provides temperature- and voltage-compensated reference.

3.1.6 MOS digital ICs

Figure 3.1*g* shows a basic MOS circuit. MOS (metal oxide semiconductor) digital ICs are formed using MOSFETs (MOS field-effect transistors) instead of two-junction (bipolar) transistors used in TTL, ECL, etc. Although MOS is not limited to the complementary (CMOS) technique shown in Fig. 3.1*g*, the *complementary inverter* forms the backbone of most MOS digital ICs.

The complementary inverter of Fig. 3.1*g* has the unique advantage of dissipating almost no power in either stable state. Power is dissipated only during the switching interval. Because MOSFETs are involved, the capacitive input lends itself to direct-coupled circuits. (The input gate of a MOSFET acts as a capacitor.) No capacitors are required between circuits. This results in a savings in component count, wiring, space, etc.

Note that the output of the CMOS inverter swings from 0 V to +V (which is the supply voltage). If a supply is 10 to 15 V (typical), the difference between a 0 and a 1 is about 10 or 15 V. As a result, the MOS digital devices have a much greater noise immunity than the HTL. For this reason, and because MOS uses much less power and operates at higher speeds, MOS has generally replaced HTL.

The *transmission gate* (Fig. 3.1*h*) is another device unique to MOS digital circuits. When the transmission gate is on, a low resistance exists between input and output, allowing current to flow in either direction. The voltage on the input line must always be positive with respect to the substrate (VSS) of the N-channel device and negative with respect to the substrate (VDD) of the P-channel device. The gate is on when the gate G1 of the P-channel is at VSS and the gate G2 of the N-channel is at VDD. When G2 is at VSS and G1 is at VDD, the transmission gate is off, and an almost infinite resistance exists between input and output.

3.1.7 Handling and protecting MOS ICs

Damage because of static discharge can be a problem with MOS ICs. Electrostatic discharges can occur when a MOS device is picked up by the case and the handler's body capacitance is discharged to ground through the series capacitance of the device. This requires proper handling, particularly when the MOS IC is out of the circuit. In a digital circuit, a MOS IC is just as rugged as any other solid-state component of similar construction.

MOS ICs are often shipped with the leads all shorted together to prevent damage in shipping and handling (there is no static discharge between leads). Usually a *shorting spring or ring* is used for shipping. The spring *should not* be removed until after the IC is soldered into the circuit. An alternative method for shipping or storing MOS ICs is to apply a conductive foam between the leads. Polystyrene insulating *snow* is not recommended for shipment or storage of MOS ICs. Such snow can acquire high static charges which could discharge through the device.

When removing or installing a MOS IC, first turn off all power. If the MOS IC is to be moved, your body should be at the same potential as the unit from when that IC is removed and installed. This can be done by placing one hand on the card or board before moving the MOS IC.

3.1.8 Comparisons of digital IC forms

This section summarizes the advantages and disadvantages for each of the digital forms discussed thus far. Again, these forms represent only a fraction of the available digital ICs. It is assumed that you will read this material and study all available datasheets for digital ICs that might suit your requirements.

Availability and compatibility of digital ICs TTL is often considered the universal IC family because there is an infinite number of gates, counters, registers, etc., available from more than one manufacturer. RTL and DTL were, at one time, the next most available. Today, both RTL and DTL have been replaced by TTL.

HTL is used only when a high logic swing or pulse (about 13 V) and high noise immunity are required. In most applications, HTL can be replaced by MOS, because MOS can provide the same logic swing with far less power and at higher speeds. For example, if MOS is operated at a supply power of 15 V, the logic swing can also be almost 15 V.

ECL is used primarily where high speed is essential. The disadvantages of ECL are high power consumption and a low logic swing (usually less than 1 V, but some ECL will provide nearly 2 V).

The advantages of MOS are low power consumption, a logic swing equal to any family, and small size. That is, you can get more MOS devices or functions on a given area than any other family. If VLSI (very large-scale integration) is required, MOS is a good choice.

TTL and DTL are directly compatible with each other. Because RTL is essentially an IC version of conventional solid-state digital circuits, RTL is most compatible with linear and analog (nondigital) systems or any discrete transistor application.

Because of their special nature, ECL and HTL are the least compatible with other ICs and with external devices. ECL requires a large supply voltage for a comparatively small logic swing, and HTL produces a very large logic swing, which is generally too high for other families. MOS can be made compatible with other families, even though MOS operating principles are quite different.

Remember that, barring some unusual circumstances, any digital IC can be adapted for use with other digital ICs, or with external equipment, by means of *interface circuits* (such as those discussed in Sec. 3.5).

Noise problems in digital ICs HTL has the highest noise immunity (or least noise sensitivity). MOS devices trigger at about 45 to 50% of the supply voltage. If MOS is operated at 15 V, the devices trigger at about 7 V, and are not affected by lower voltages. Thus, MOS can have a higher noise immunity than HTL if the supply voltage can be kept at 15 V. Typically, noise up to about 5 V does not affect HTL.

RTL has the lowest noise immunity. DTL, TTL, and ECL are about the same with regard to noise. In addition to signal-line noise, all ICs are affected by noise on the power-supply and ground lines. This problem can be minimized by adequate bypassing as described in Sec. 3.1.9.

In addition to noise immunity, noise generated by ICs must be considered. Whenever a transistor or diode switches from saturation to cutoff, and vice versa, large current spikes are generated. These spikes appear as noise on the signal, power-supply, and ground lines. Because ECL does not saturate, ECL produces the least amount of noise.

Digital system operating speed The speed of a digital IC system is inversely porportional to the delay of the IC elements. ICs with the shortest delay can operate at the highest speed. Because ECL does not saturate, delay is minimum and speed is maximum. TTL is the next-to-fastest, and can be used in any application except where extremely high speed is involved. MOS is slower than TTL and ECL.

Digital IC power requirements MOS requires the least power consumption of all digital ICs described here and can be operated over a wide range of power-supply voltages (typically 5, 10, or 15 V). CMOS is well suited for battery-operated systems, because little standby power is required. CMOS uses power when switching from one state to another, but not during standby (except for some power consumed by leakage). TTL and ECL generally operate with a 5-V supply and consume about 15 and 25 mW per gate. The power consumption of MOS is usually stated in microwatts.

Digital fan-out Some IC datasheets list *fan-out* (Sec 2.6.1) as a simple number (a fan-out of 3 means that the IC can drive three outputs or loads). Other datasheets describe fan-out (or load and drive) in terms of input and output current limits. Typical fan-outs for our digital families are: RTL 4 to 5, DTL 5 to 8, TTL 5 to 15, HTL 10, ECL 25, and MOS 10.

3.1.9 Digital IC layout

Figure 3.2 shows the general rules for digital IC layout. The following notes supplement this illustration. The information here is primarily for the experimenter and/or designer working with digital equipment for the first time. It is assumed that you are already familiar with practical considerations applicable to all electronic equipment, particularly IC equipment. For example, you should understand the basics of selecting IC packages, mounting and connecting ICs, and working with ICs (PC board repair, lead bending, solder techniques including SMD (surface mount design), and so on).

All digital circuits are subject to noise. Any digital circuit, discrete or IC, produces erroneous results if the noise level is high enough. Thus, it is recommended that noise and grounding problems be considered from the very beginning of layout design.

Whenever dc distribution lines run a long distance from the supply to a digital module or board, both lines (positive and negative) should be bypassed to ground with a capacitor, at the point where the wires enter the module or board.

The values for power-line bypass capacitors are typically 1 to 10 μF. If the digital

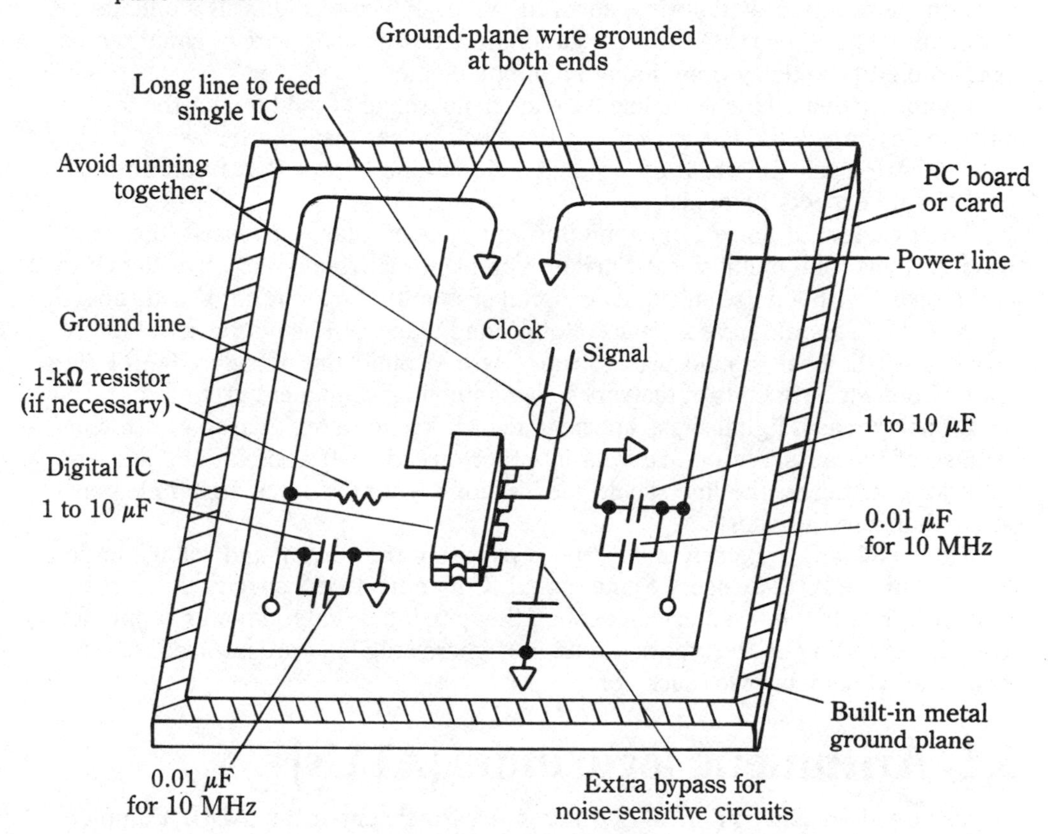

Fig. 3.2 Digital IC layout.

digital circuits operate at higher speeds (above about 10 MHz), add a 0.01-μF capacitor in parallel with each 1- to 10-μF capacitor. Remember that even though the system might operate at low speeds, harmonics are generated at higher speeds. The high-frequency signals might produce noise on the power lines and interconnecting wiring.

If the digital ICs are particularly sensitive to noise (TTL for example), extra bypass capacitors (in addition to those at the power and ground entry points) can be used effectively. The additional power and ground-line capacitors can be mounted at any convenient point on the board, provided that there is no more than a 7-inch space between any IC and a capacitor (as measured along the power or ground line). Use at least one additional capacitor for each 12 IC packages, and possibly as many as one capacitor for each six ICs.

The dc lines and ground-return lines should be large enough to minimize noise pickup and dc voltage drop. Unless otherwise recommended by the IC manufacturer, use AWG (American Wire Gauge) No. 20 or larger wire for all digital IC power and ground lines.

In general, keep all leads as short as possible, both to reduce noise pickup and to minimize the propagation time down the wire. Typically, digital ICs operate at speeds high enough so that the propagation time down a long wire or cable can be comparable to the delay time through a digital element.

Do not exceed 10 inches of line for each nanosecond of fall time for the fastest digital pulses involved. For example, if the clock pulses (usually the fastest in the system) have a fall time of 3 ns, no digital signal line (either PC or conventional wire) should exceed 30 inches.

The problem of noise can be minimized if ground planes are used (that is, if the circuit has solid metal sides). Such ground planes surround the active elements on the board with a noise shield. Any digital system that operates at speeds above about 30 MHz should have some form of ground plane. If it is not practical to use boards with built-in ground planes, run a wire around the outside edge of the board. Connect both ends of the wire to a common or equipment ground.

Do not run any digital signal line near a clock line for more than about 7 inches because of the possibility of *crosstalk* in either direction. If a digital line must be run a long distance, the line should feed a single gate (or other digital element) rather than several gates.

External loads to be driven must be kept within the current and voltage limits specified in the IC datasheets. Some digital IC manufacturers specify that a resistor (typically 1 kΩ) be connected between the gate input and the power supply (or ground, depending on the type of digital IC), where long lines are involved. Always check the IC datasheet for such notes.

3.2 Arithmetic logic units (ALUs)

As discussed in Chapter 4, most microprocessors contain an ALU. Complete ALUs are also available in IC form. In some cases, IC ALUs are interconnected to perform a special function, such as high-speed multiplication. Here, you will concentrate on explanations of basic principles and techniques used for addition, subtraction, multiplication, and division in digital equipment. A digital circuit that can perform these functions can solve almost any problem, because most mathematical operations are based on the four functions.

3.2.1 Adder circuits

Adder circuits are generally used in digital equipment to add binary numbers. There are two basic types of adders: the half-adder and the full-adder.

Half-adder Figure 3.3*a* shows the symbol and truth table for a half-adder or HA. The circuit has two inputs for the two bits to be added, and two outputs (one for the *sum bit* S and the other for the *carry-out bit C*. The half-adder sum is sometimes called the EXCLUSIVE-OR function (Chapter 2) and uses the corresponding symbols (Fig. 2.1). In a typical digital adder system, the HA receives inputs from an FF (or possibly a register).

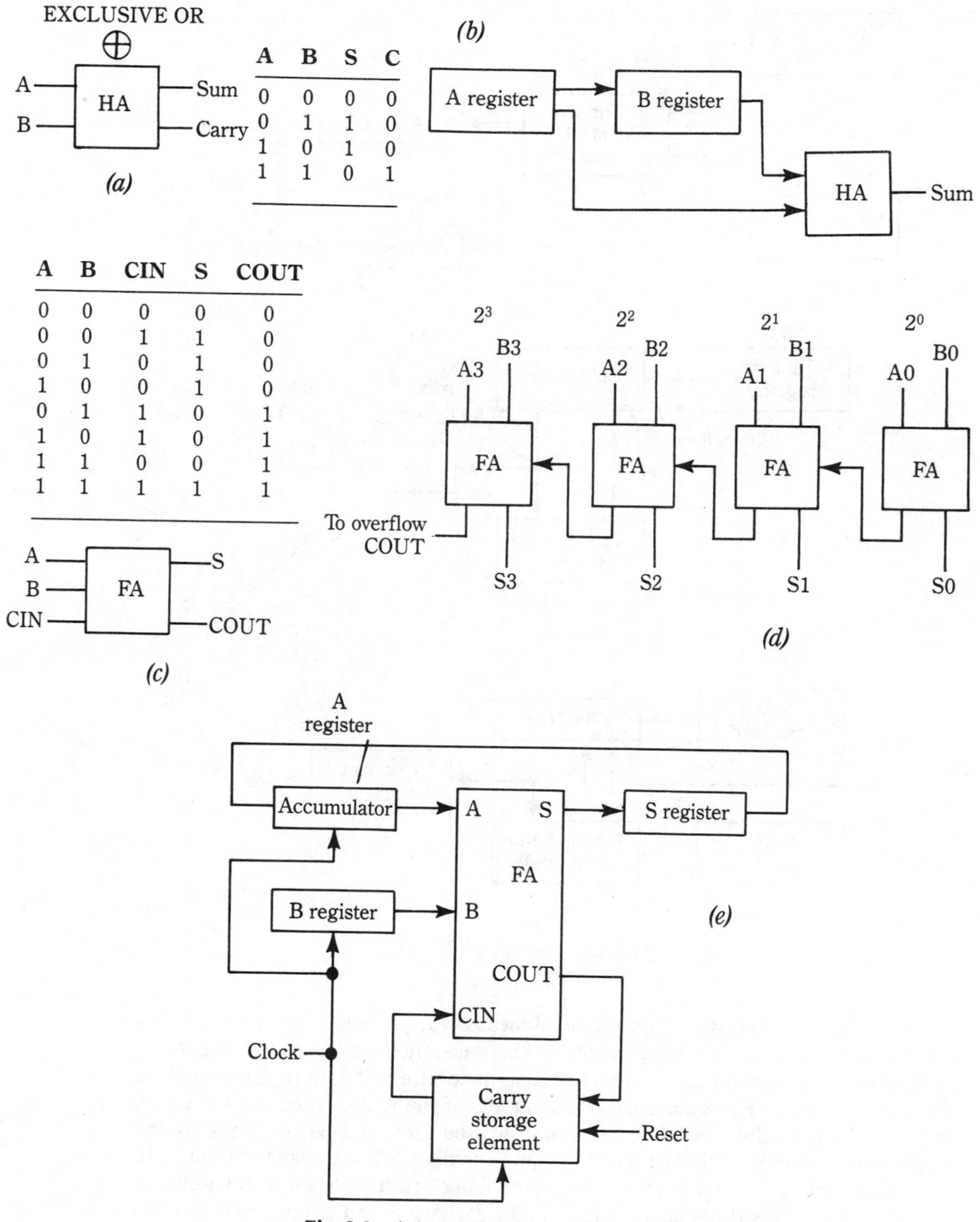

A	B	S	C
0	0	0	0
0	1	1	0
1	0	1	0
1	1	0	1

A	B	CIN	S	COUT
0	0	0	0	0
0	0	1	1	0
0	1	0	1	0
1	0	0	1	0
0	1	1	0	1
1	0	1	0	1
1	1	0	0	1
1	1	1	1	1

Fig. 3.3 Arithmetic logic units (ALUs).

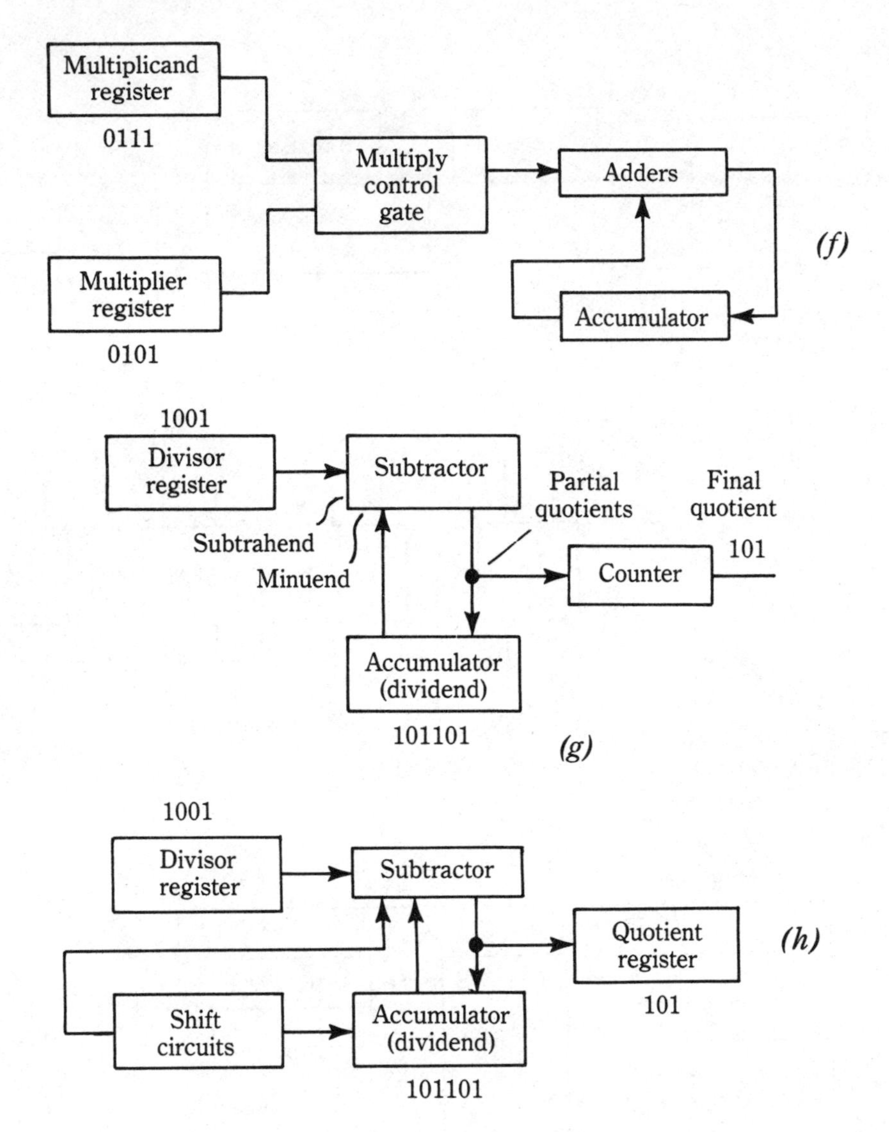

Fig. 3.3 Continued

Equality comparator Figure 3.3*b* shows how a half-adder can be used as a comparison circuit for detecting errors. In this case, the binary count in register A is transferred to register B, and the HA must make sure that both registers contain the same count. The HA compares the contents of the A and B registers, digit by digit. As long as the HA has no sum output, the digit in A is the same as the respective digit in B. If there is a sum output, the digit in S is unlike the digit in B.

Full-adder Figure 3.3*c* shows the symbol and truth table for a full-adder or FA. The circuit has three inputs (A, B, and C, where A and B represent the two bits to be added, and C represents the *carry-in* for that stage). The two outputs from the FA are the sum bit S (which is the sum of the three input bits, A, B, and CIN) and the *carry-out* bit COUT.

Parallel adder Figure 3.3*d* shows a typical parallel adder. This circuit adds binary numbers in parallel form (where each bit in a binary number appears on a separate line, but simultaneously with all other bits in the number). Parallel addition is an *asynchronous* operation, independent of a clock signal. In a practical parallel adder, the input binary numbers are taken from a register and are gated into the adder. At a predetermined time, the *sum number* S and any carry-out (COUT) are gated out to another register.

Serial adder Figure 3.3*e* shows a typical serial adder. This circuit adds binary numbers in a serial form (where all bits in a binary number appear on one line, in sequence, usually starting with the least significant bit). Binary serial addition is thus a time-sequential (*synchronous*) operation, performed one bit at a time.

In Fig. 3.3*e*, the A register contains the augend number and the B register contains the addend. The S register receives the sum number bits serially from the FA. The carry storage element provides temporary storage for COUT until COUT is used in the next addition, and the next COUT is placed in the storage element. The carry storage element is generally an FF and must be clocked by the same timing pulse applied to the registers. In some circuits, the S register is eliminated by feeding the sum bits back into the A register (which is then called an *accumulator*).

3.2.2 Subtractor circuits

Operation of a binary-number subtractor circuit is essentially the same as that for the adder circuit. However, the carry-in is replaced by a *borrow-in*, the carry-out is replaced by a *borrow-out*, and the sum is replaced by a *difference*. In some circuits, subtraction is done by complementing. In such circuits, the subtrahend is complemented (or inverted) before entering the inputs of a full-adder.

3.2.3 Multiplication circuits

Figure 3.3*f* shows a basic *repeated-addition multiplier* circuit where the value of the multiplicand is put into a multiplicand register, and the value of the multiplier is put into a multiplier counter. The counter counts down (from the multiplier value) to 0. For example, if the multiplier is 101, the counter registers 101 before the first count, 100 after the first count, and 011 after the second count and so on, down to 000. At count zero, the multiply-control gate is closed, halting the repeated-addition process. In some circuits, multiplication is done by adding and shifting, because multiplication is the process of adding the multiplicand to itself as many times as the multiplier dictates.

3.2.4 Division circuits

Figure 3.3*g* shows a basic *repeated-subtraction divider* circuit, where the value of the divider is put into a division register, and the value of the dividend is put into the accumulator. The divisor and accumulator output make up the subtrahend and minuend inputs to the subtractor circuit. The counter starts at zero and holds the partial quotients until the last subtraction takes place.

Figure 3.3*h* shows a basic *subtraction-shifting divider* circuit, where the dividend is put into the accumulator, and the divisor is put into the divisor register. The quotient register holds the quotients produced by repeated subtraction. The digits posted in the quotient register become the final quotient, as shown by the sample division problem in Fig. 3.3*h*.

3.3 D/A and A/D conversion

This section describes the various digital-to-analog (D/A) and analog-to-digital (A/D) techniques found in digital equipment. Here, we concentrate on explanations of the basic principles and techniques of D/A and A/D converters. By studying this information, you should be able to understand operation of all advanced D/A and A/D systems now in use.

3.3.1 Basic analog/digital conversion

Before discussing operation of the various conversion circuits, this section considers the signal formats for BCD data (Chapter 1) as well as the four-bit system.

Typical BCD signal formats Figure 3.4*a* shows the relationship of the three most common BCD signal formats: NRZL (nonreturn-to-zero level), NRZM (nonreturn-to-zero-mark), and RZ (return-to-zero).

In NRZL, a 1 is one signal level, and a 0 is another signal level. These levels can be 5 V, 10 V, 0 V, −5 V, or any other selected values, provided that the 1 and 0 levels are entirely different.

In RZ, a 1-bit is represented by a pulse of some definite width (usually a $1/2$ bit width) that returns to zero signal level, and the 0-bit is represented by a zero-level signal.

In NRZM, the level of the pulse has no meaning. A 1 is represented by a change in level, and a 0 is represented by no change.

Four-bit system in the conversion process Figure 3.4*b* shows the relation between two voltage levels to be converted, and the corresponding binary code (in NRZL form), in a basic A/D converter. In practice, a four-bit A/D converter (also called a *binary encoder*) samples the voltage level to be converted and compares the voltage to $1/2$ scale, $1/4$ scale, $1/8$ scale, and $1/16$ scale (in that order) of some given full-scale voltage. The A/D converter then produces four bits, in sequence, with the comparison made on the most significant ($1/2$ scale) first.

As shown in Fig. 3.4*b*, each of the two voltage levels is divided into four equal time increments. The first time increment is used to represent the $1/2$-scale bit, the second increment represents the $1/4$-scale bit, and so on.

In voltage level 1, the first two time increments are at binary 1, with the second two increments at 0. This produces a 1100, or decimal 12. Twelve is $3/4$ of 16. Thus, level 1 is 75% of full scale. For example, if full scale is 10 V, level 1 is 7.5 V.

In level 2, the first two increments are at 0, and the second two increments are at 1. This is represented as 0011, or 3. Thus, level 2 is $3/16$ of full scale (or 1.875 V). This can be expressed in another way. In the first or $1/2$-scale increment, the con-

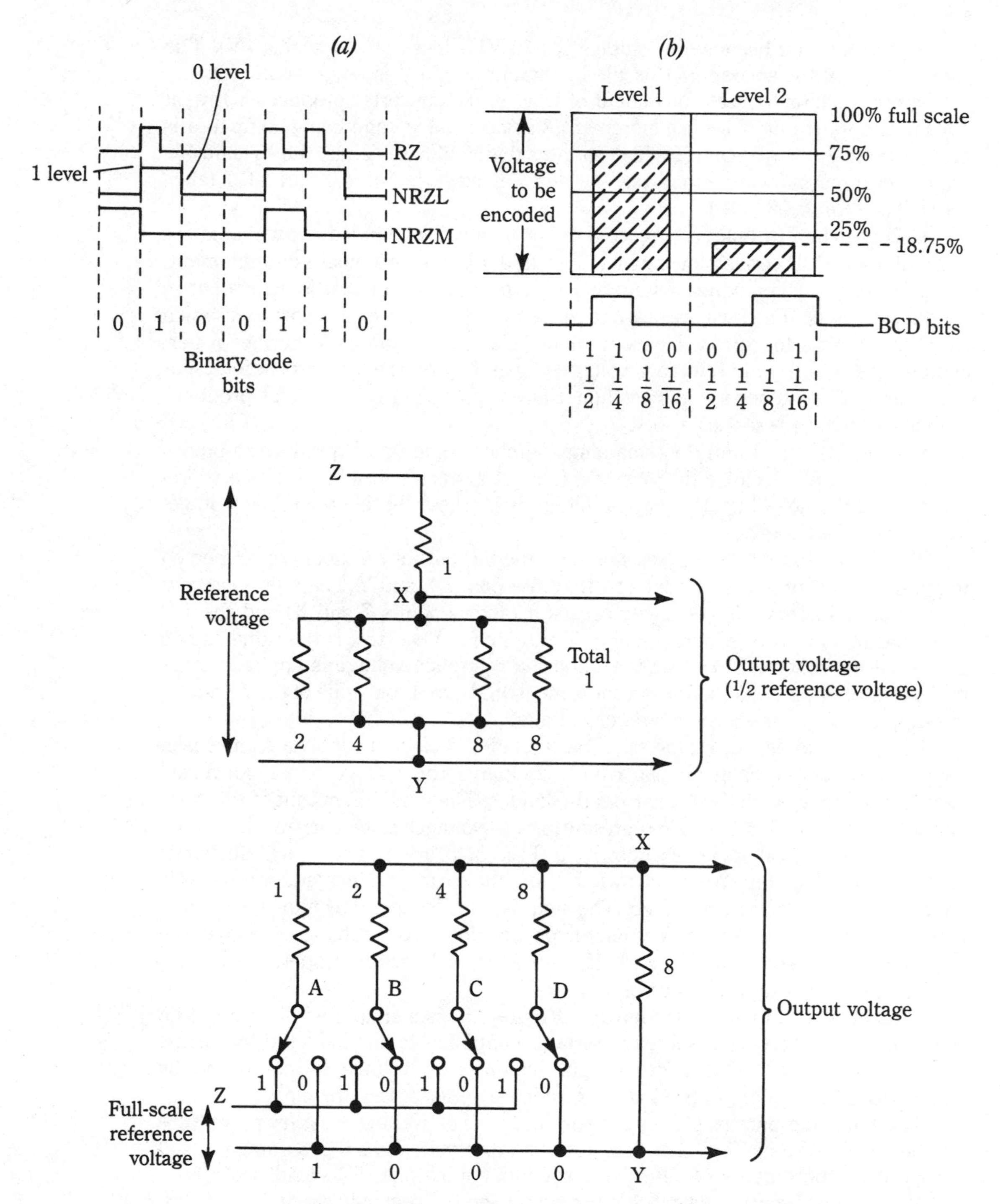

Fig. 3.4 D/A and A/D conversion basics.

verter produces a 0 because the voltage (1.875 V) is less than $^1/_2$ scale (5 V). The same is true of the second or $^1/_4$-scale increment (1.875 V is less than 2.5 V).

In the third or $^1/_8$-scale increment of level 2, the converter produces a 1, as it does in the fourth or $^1/_{16}$-scale increment, because the voltage being compared is greater than $^1/_8$ of full scale (1.875 or greater than 0.625 V). Thus, the $^1/_2$- and $^1/_4$-scale increments are at 0, and the $^1/_8$- and $^1/_{16}$-scale increments are at 1 (also, $^1/_8 + {}^1/_{16} = {}^3/_{16}$ or 18.75%).

A/D conversion ladder Figure 3.4*c* shows a conversion ladder, which is the heart of many A/D conversion circuits. The ladder provides a means of implementing a four-bit binary coding system and produces an output that is equivalent to switch positions. The switches can be moved to either a 1 or a 0 position, which corresponds to a four-place binary number. The output voltage describes a percentage of the full-scale reference voltage, depending on the switch positions. For example, if all switches are at 0 position, there is no output voltage. This produces a binary 0000, represented by 0 V.

If switch A is at 1 and the remaining switches are at 0, this produces a binary 1000 (decimal 8). Because the total in a four-bit system is 16 (0 to 15), $\times 8$ represents $^1/_2$ full scale. Thus, the output voltage is $^1/_2$ the full-scale reference voltage. This is done as follows.

The 2-, 4-, and 8-Ω switch resistors and the 8-Ω output resistors are connected in parallel. This produces a value of 1 Ω across points X and Y. The reference voltage is applied across the 1-Ω switch resistor (across points Z and X) and the 1-Ω combination of resistors (across points X and Y); in effect, this is the same as two 1-Ω resistors in series. Because the full-scale reference voltage is applied across both resistors in series, and the output is measured across only one of the resistors, the output voltage is $^1/_2$ the reference voltage.

In a practical converter, the same basic ladder is used to supply a comparison voltage to a comparison circuit, which compares the voltage to be converted against the binary-coded voltage from the ladder. The resultant output of the comparison circuit is a binary code representing the voltages to be converted.

The mechanical switches shown in Fig. 3.4*c* are replaced by electronic switches (usually FFs). When the switch is on, the corresponding ladder resistor is connected to the reference voltage. The switches are triggered by four pulses (representing each of the four binary bits) from the clock. An enable pulse is used to turn the comparison circuit on and off, so that as each switch is operated, a comparison can be made of the four bits.

Typical A/D operating sequence Figure 3.5*a* is a simplified diagram of an A/D converter. Here, the reference voltage is applied to the ladder through the electronic switches. The ladder output (comparison voltage) is controlled by switch positions which, in turn, are controlled by pulses from the clock.

The following paragraphs outline the sequence of events necessary to produce a series of four binary bits that describe the input voltage as a percentage of full scale (in $^1/_{16}$ increments). Assume that the input voltage is 75% of full scale.

When pulse 1 arrives, switch 1 is turned on and the remaining switches are off. The ladder output is a 50% voltage that is applied to the differential amplifier. The balance of this amplifier is set so that the output is sufficient to turn on one AND

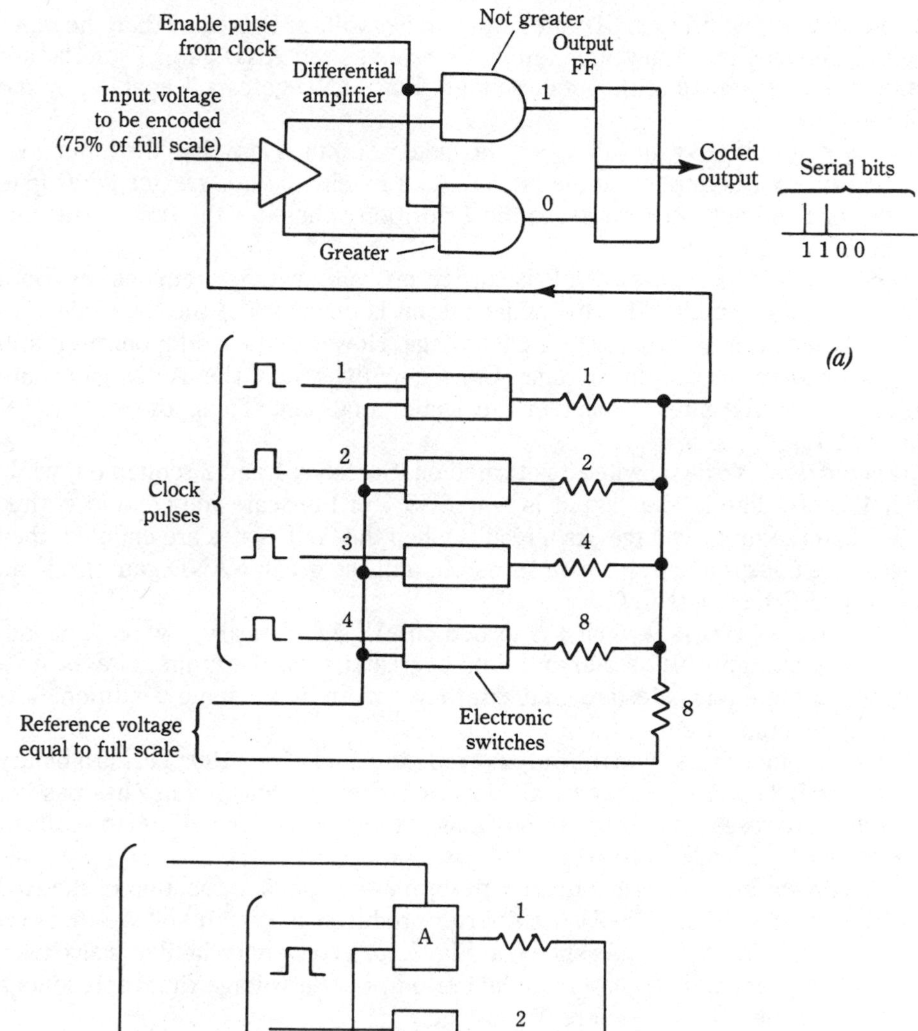

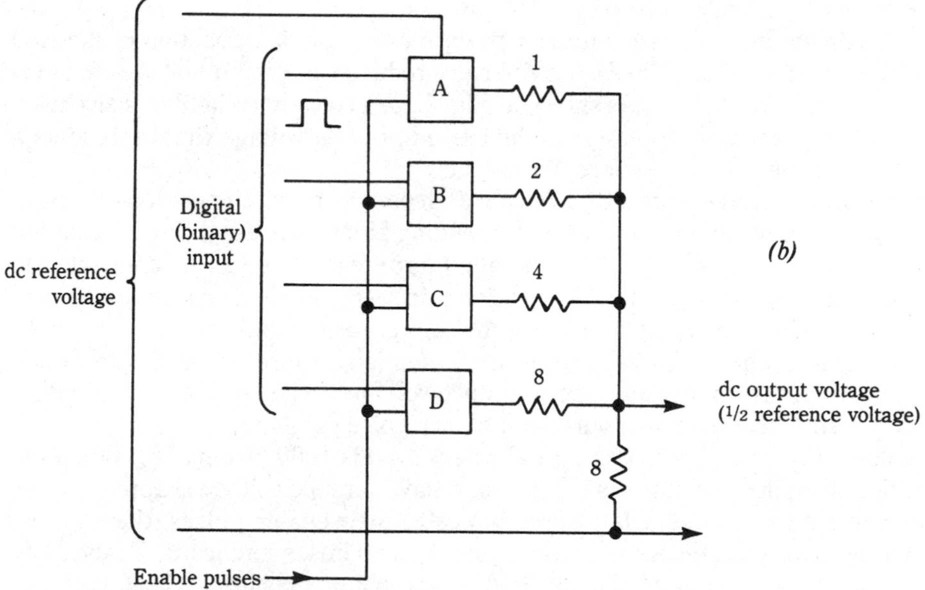

Fig. 3.5 D/A and A/D conversion circuits.

gate and turn off the other AND gate, if the ladder voltage is greater than the input voltage. Similarly, the differential amplifier reverses the AND gates if the ladder voltage is not greater than the input voltage. Both AND gates are enabled by the clock pulse.

In this example (75% of full scale), the ladder output is less than the input voltage when pulse 1 is applied to the ladder. As a result, the not-greater AND gate turns on, and the output FF is set to the 1 position. Thus, for the first of the four bits, the FF output is 1.

When pulse 2 arrives, switch 2 is turned on, and switch 1 remains on. Both switches 3 and 4 remain off. The ladder output is now 75% of the full-scale voltage. The ladder voltage equals the input voltage. However, the ladder output is still not greater than the input voltage. Consequently, when the AND gates are enabled, the AND gates remain in the same condition. Thus, the output FF remains at 1.

When pulse 3 arrives, switch 3 is turned on. Switches 1 and 2 remain on, while switch 4 is off. The ladder output is now 87.5% of full-scale voltage and is thus greater than the input voltage. As a result, when the AND gates are enabled, they reverse. The not-greater AND gate turns off, and the greater AND gate turns on. The output FF then sets to 0.

When pulse 4 arrives, switch 4 is turned on. All switches are now on. The ladder is now maximum (full scale) and thus is greater than the input voltage. As a result, when the AND gates are enabled, they remain in the same condition. The output FF remains at a 0.

The four binary bits from the output are 1, 1, 0, and 0, or 1100. This is a binary 12, which is 75% of 16. In a practical converter when the fourth pulse has passed, all switches are reset to the off position. This places them in a condition to spell out the next four-bit binary word.

D/A conversion A D/A converter performs the opposite function of the A/D converter just described. The D/A converter produces an output voltage that corresponds to the binary code. As shown in Fig. 3.5*b*, a conversion ladder is also used in the D/A converter. The conversion-ladder output is a voltage that represents a percentage of the full-scale reference voltage.

The output voltage from a D/A converter depends on switch positions. In turn, the switches are set to on or off by corresponding binary pulses. If the information is applied to the switch in four-line (parallel) form, each line can be connected to the corresponding switch. If the information is in serial form, the data must be converted to parallel by a register (shift and/or storage, Sec. 2.8.5).

The switches in the D/A converter are essentially a form of AND gate. Each gate completes the circuit from the reference voltage to the corresponding ladder resistor when both the enable pulse and binary pulse coincide.

Assume that the digital number to be converted is 1000 (decimal 8). When the first pulse is applied, switch A is enabled and the reference voltage is applied to the 1-Ω resistor. When switches B, C, and D receive their enable pulses, there are no binary pulses (or the pulses are in the 0 condition). Thus, switches B, C, and D do not complete the circuits to the 2-, 4-, and 3-Ω ladder resistors. These resistors combine with the 8-Ω output resistor to produce a 1-Ω resistance in series with the

1-Ω ladder resistance. This divides the reference voltage in half to produce 50% of full-scale output. Because 8 is 1/2 of 16, the 50% output voltage represents 8.

3.3.2 High-speed A/D converters

Although there are a number of A/D converter schemes, there are only three basic types: parallel, serial, and combination. In parallel, all bits are converted simultaneously by many circuits. In serial, each bit is converted in sequence, one at a time. The combination A/D conversion includes features of both types. Generally, parallel is faster but more complex than serial. The combination types are a compromise between speed and complexity.

Parallel (flash) A/D Figure 3.6*a* shows the basic parallel (or flash) A/D conversion circuit, where all bits of the digital representation are determined simultaneously by a bank of voltage comparators. For N bits of binary information, the system requires 2^{N-1} comparators, and each comparator determines one LSB level. This requires a great number of circuits. Another disadvantage of parallel A/D is that the comparator output is not directly usable information. The output must be converted to binary information using a decoder.

Tracking A/D Figure 3.6*b* shows the basic tracking A/D conversion circuit. Tracking A/D continuously tracks the analog input voltage and is often used in communications systems or similar applications where the input is a continuously varying signal. The accuracy of the system is no better than the D/A converter used in the feedback path (typically a 6- to 16-bit converter).

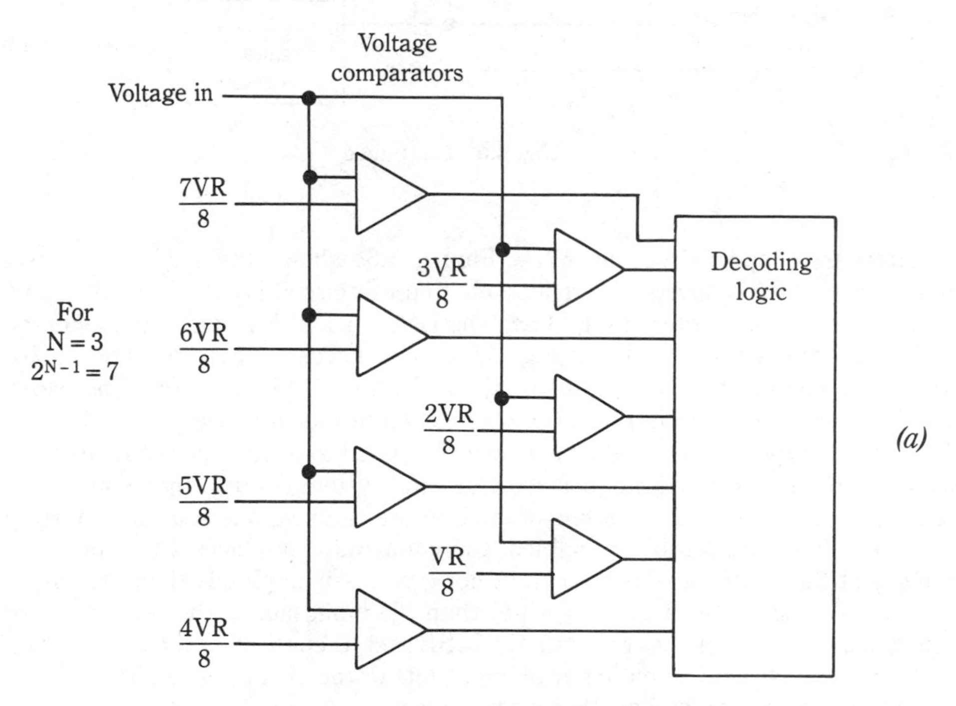

Fig. 3.6 High-speed A/D converters.

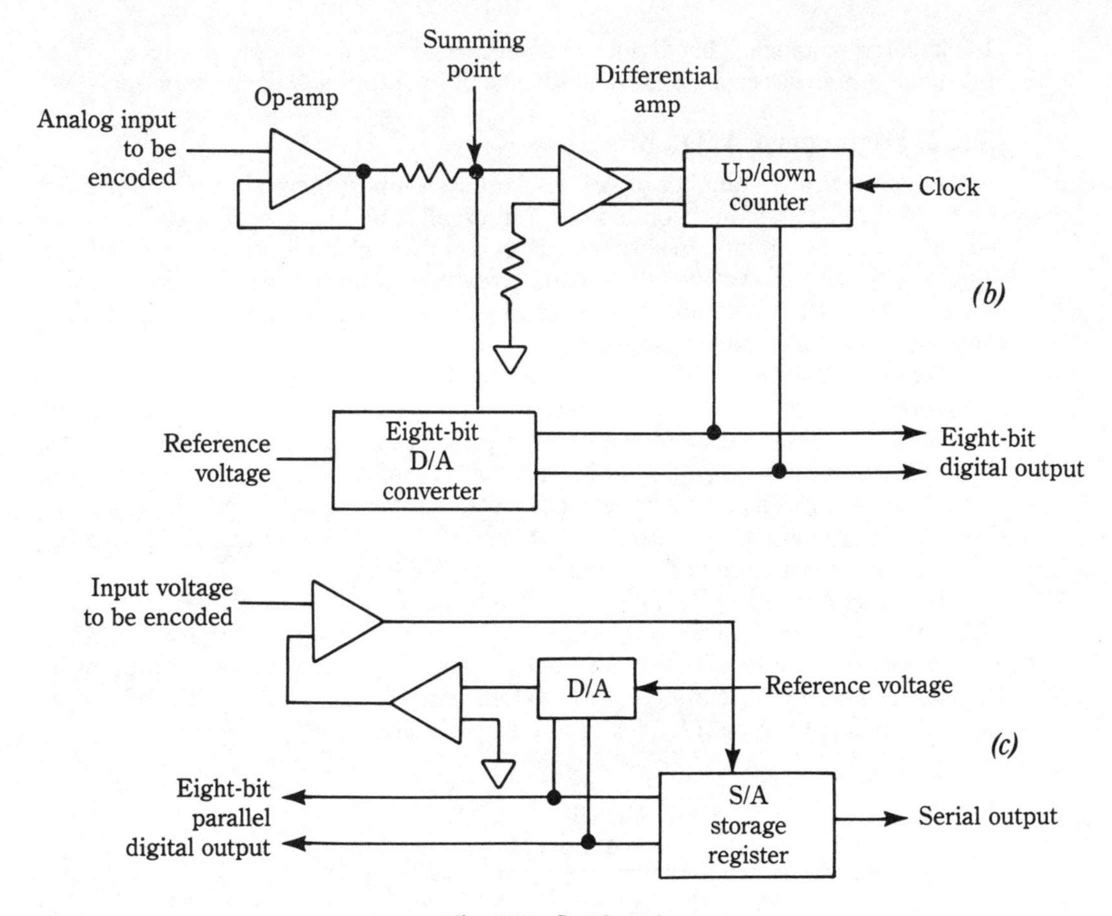

Fig. 3.6 Continued

Successive approximation A/D Figure 3.6*c* shows the basic successive approximation A/D conversion circuit. Note that this circuit is essentially the same as the basic A/D converter of Fig. 3.5*a*. The D/A block of Fig. 3.6*c* represents the electronic switches and ladder of Fig. 3.5*a*. The successive approximation (S/A) storage registers of Fig. 3.6*c* represent the AND gates and FF of Fig. 3.5*a*. However, four bits are shown in Fig. 3.5*a*, whereas eight bits are used in Fig. 3.6*c*.

The S/A type of A/D is relatively slow compared to other types of high-speed A/Ds, but the low cost, ease of construction, and system features make up for the lack of speed. With S/As, eight bits of the D/A are enabled, one at a time, starting with the MSB. As each bit is enabled, the comparator produces an output indicating that the input signal is greater, or not greater, in amplitude than the output of the D/A. If the D/A output is greater than the input signal, the bit is reset or turned off. The system does this with the MSB first, then the next most significant bit, then the next, and so on. After all eight bits of the D/A are tried, the conversion cycle is complete, and another cycle is started.

The serial output of the system is taken from the output of the comparator. While the system is in the conversion cycle, the comparator output is either 0 or 1, corresponding to the digital state of the respective bit. In this way, the S/A type of A/D gives serial output during conversion and a parallel output between conversion cycles.

3.4 Digital readouts (numeric displays)

There are two basic types of digital readout systems: *direct drive* and *multiplex*. These systems (also called numeric displays) generally use the *seven-segment format* in most present-day equipment. Although there are many types of readouts now available, we cover only those in most common use here.

3.4.1 Direct-drive displays

The simplest type of display system, shown in Fig. 3.7*a*, consists of four lines of BCD information, feeding a decoder/driver. In turn, the decoder/driver drives the seven-segment display. This direct-drive system does not have information-storage capability and thus reads out in real time.

Another display system, shown in Fig. 3.7*b*, contains a decade counter, a quad or four-line latch (FFs), a decoder/driver, and the display, one such channel for every digit. This alternate system has storage capability (the FF latches), which allows the counter to recount during the storage time.

Both systems have decoder/drivers, which convert the BCD count into voltages suitable for operating the seven-segment displays. Figure 3.7*c* shows the relationship between the decoder and seven-segment display. As shown by the truth table, the segments are illuminated in accordance with the decimal number applied at the BCD input. For example, for a decimal 3, the BCD input to the decoder is 0011, and segments a, b, c, d, and g are turned on by the decoder. Segments e and f are not illuminated, and the display forms a numeral 3.

3.4.2 Multiplex displays

The most commonly used system for multidigit displays is the multiplexed (or time-shared or strobed) system shown in Fig. 3.8. By time sharing the one decoder/driver, the parts count, interconnections, and power can be saved. The N-stage data register (one stage for each digit) feeds a scanned multiplexer or MUX. In turn, the sequenced BCD output of the MUX drives like segments of the display. The digit-select elements are sequentially driven by the scan circuit, which also drives the MUX. Thus, each display is scanned or strobed in sync. With this system, the BCD information is presented to the decoder at a sufficiently high rate (usually greater than 50 scans per second) to appear as a continuously energized multidigit display.

3.4.3 Light-emitting diode (LED)

There are two possible connections when LEDs are used as seven-segment displays. *Common-cathode* LEDs require the drive circuit to supply current (source) to

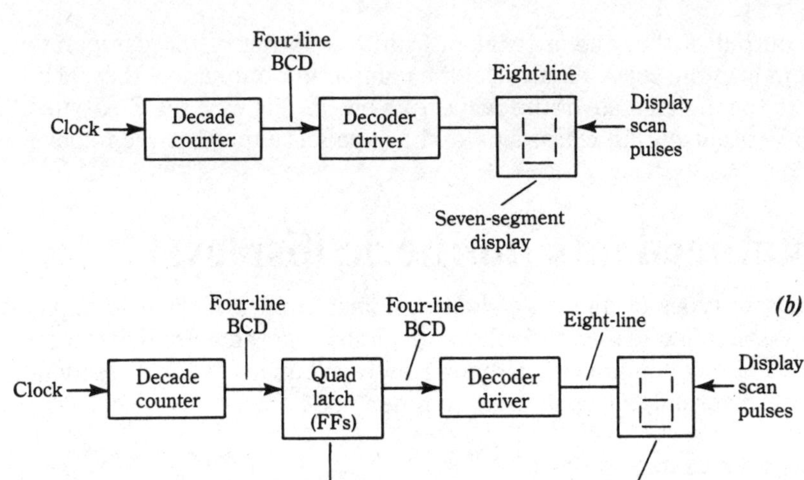

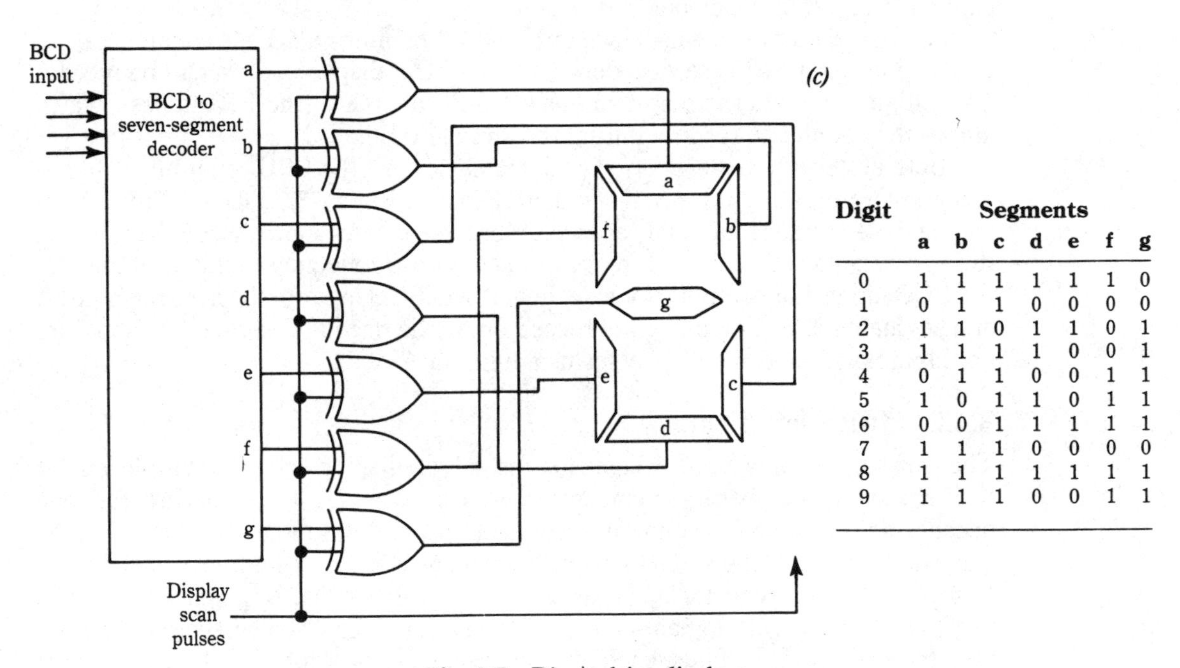

Digit	Segments						
	a	b	c	d	e	f	g
0	1	1	1	1	1	1	0
1	0	1	1	0	0	0	0
2	1	1	0	1	1	0	1
3	1	1	1	1	0	0	1
4	0	1	1	0	0	1	1
5	1	0	1	1	0	1	1
6	0	0	1	1	1	1	1
7	1	1	1	0	0	0	0
8	1	1	1	1	1	1	1
9	1	1	1	0	0	1	1

Fig. 3.7 Direct-drive displays.

the segments. *Common-anode* LEDs require segment-drive, circuit-sink capability (the drive circuit must dissipate the current). Because of the low voltage (typically 1.6 V) and relatively small current requirements, LED displays can be readily

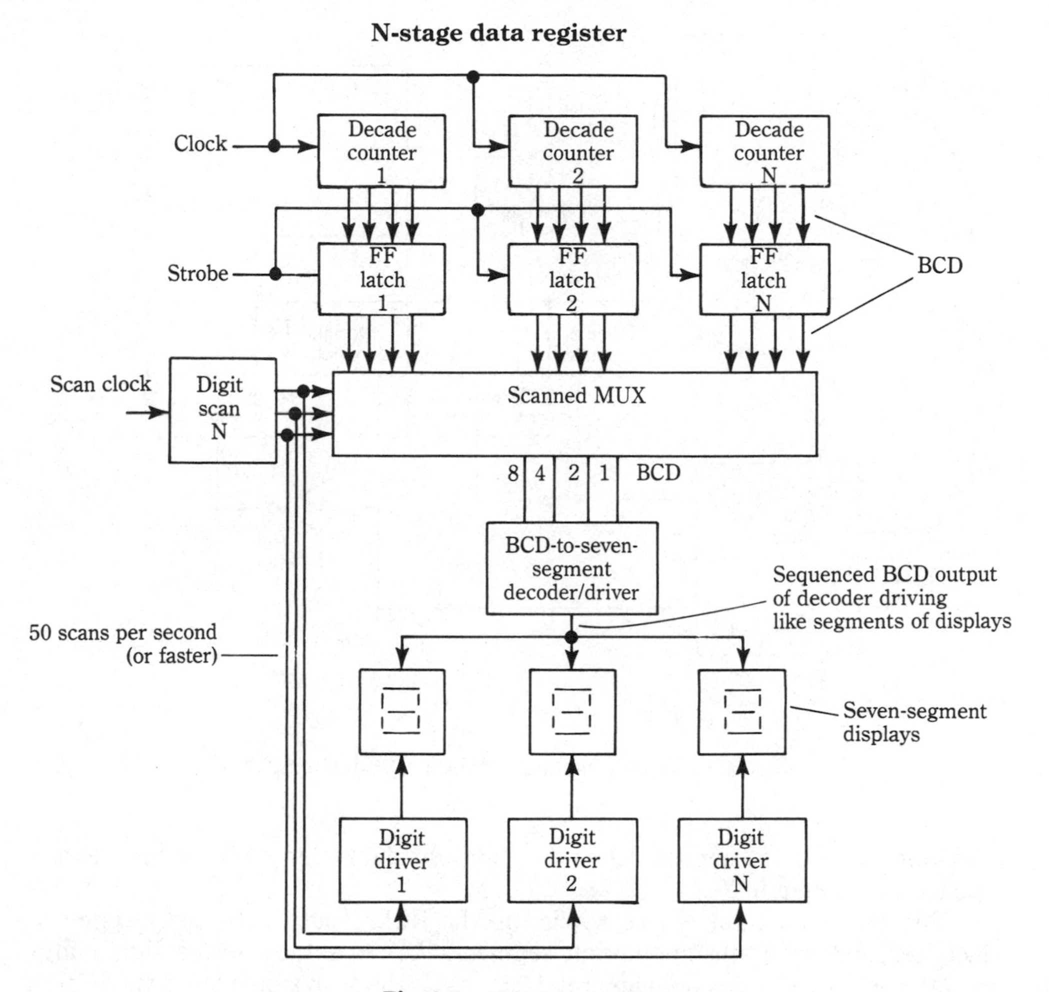

Fig. 3.8 Multiplex displays.

interfaced with most IC families. When such ICs lack the drive capability, transistors can be easily interfaced between the ICs and the display.

Example of LED display Figure 3.9 shows a typical five-digit real-time LED display system. This display uses a BCD-to-seven segment decoder, and a five-digit counter, as the basic control elements. The system also uses quad drivers and hex drivers to drive the LEDs. All like-anode segments of the common-cathode displays are driven by the emitter outputs of the quad drivers. The cathode elements are driven by the hex drivers.

The digits are selected, in turn, when a drive signal is applied to the corresponding common cathodes by the five-digit counter (through the hex drivers).

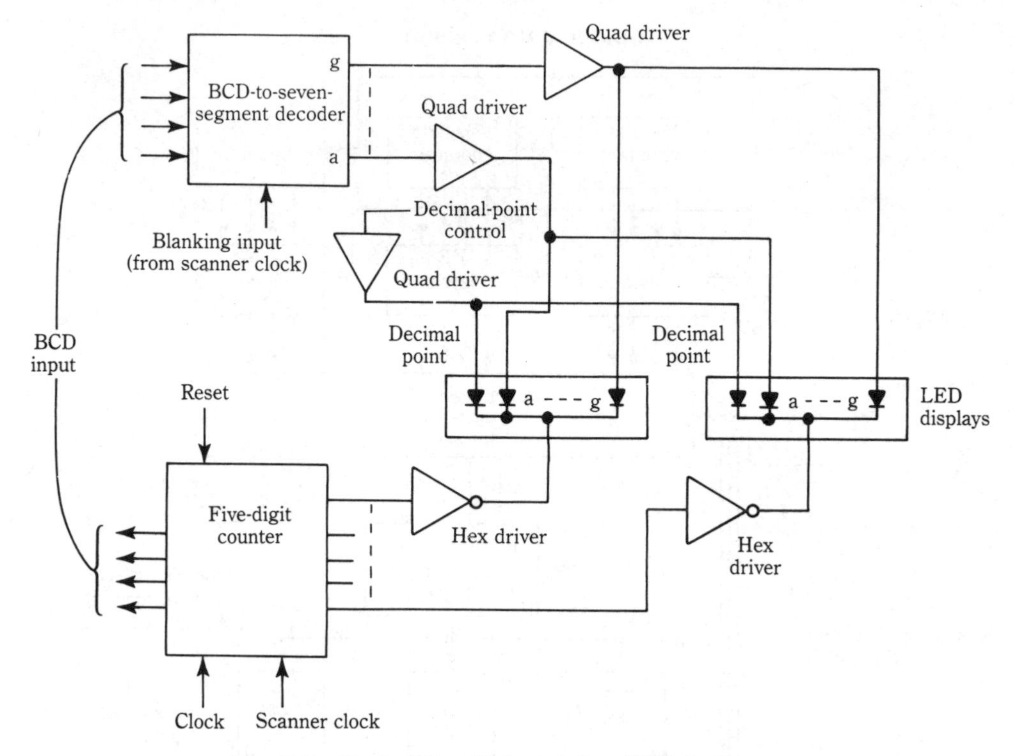

Fig. 3.9 Typical five-digit real-time LEd display.

Each digit is selected in order when the appropriate number of scanner pulses are applied to the counter.

The counter output is also applied to the BCD input of the seven-segment decoders, and causes the appropriate segments (LEDs) of the selected digit to illuminate and form the desired numeral. Note that if the scan signal fails, the display is blanked, thus preventing damage to the displays.

3.4.4 Fluorescent displays

Fluorescent displays, both diode and triode types, are electrically similar to diode and triode vacuum tubes, except that the anodes are coated with a phosphor. When a positive voltage is placed across the anode and the directly heated cathode (or filament), the electrons hitting the anode cause the phosphor to fluoresce and emit light. The light output peaks in the blue-green range, which when appropriately filtered, can display other colors.

Example of fluorescent displays Figure 3.10*a* shows a partial circuit for a six-digit triode fluorescent display. The digits are selected, in turn, when an appropriate drive signal is applied to the corresponding control grid by the counter through the drive transistor. Each digit is selected in order when the appropriate number of scanning pulses are applied to the counter. The counter output is also applied to the BCD input of the decoder, and causes the appropriate segments (anodes) of the selected digit to illuminate. This operation is similar to that of the five-digit LED display described in Sec. 3.4.3.

3.4.5 Liquid crystal display (LCD)

LCDs consist of organic compounds that change characteristics when placed in an electric field. When the fields are applied to the compounds in a certain pattern (a seven-segment numerical display, for example), images are formed by the compounds according to the pattern. In *dynamic scattering* LCDs, the field rearranges the compound molecules to scatter the available light. This causes the compound to change from a transparent state to an opaque state, and thus form the desired image.

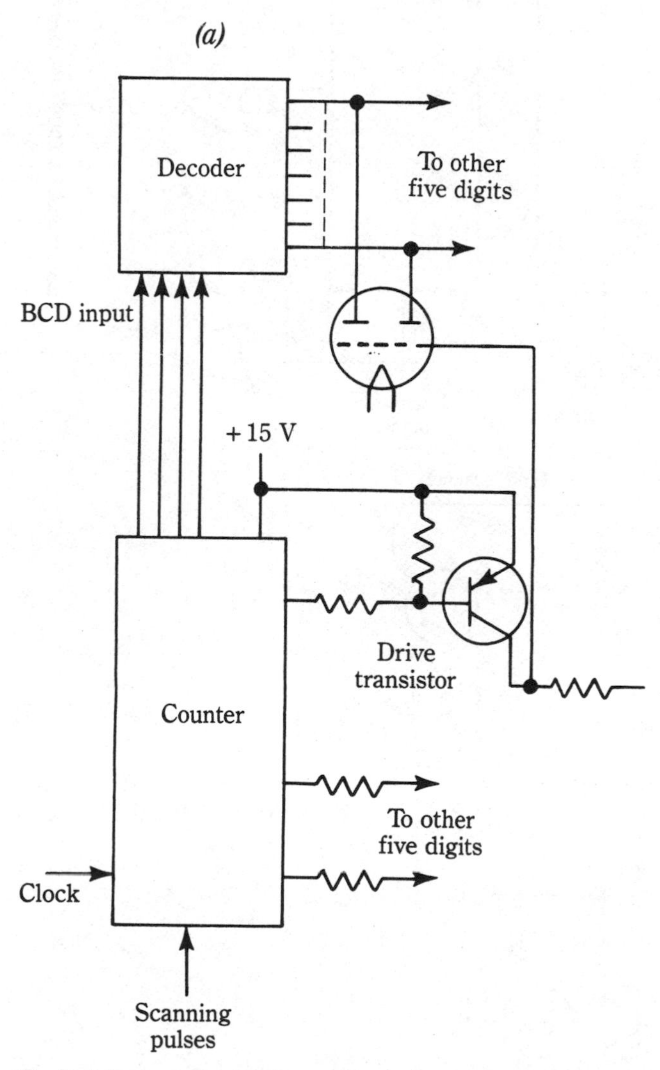

Fig. 3.10 Typical fluorescent, LCD, gas discharge, and incandescent displays.

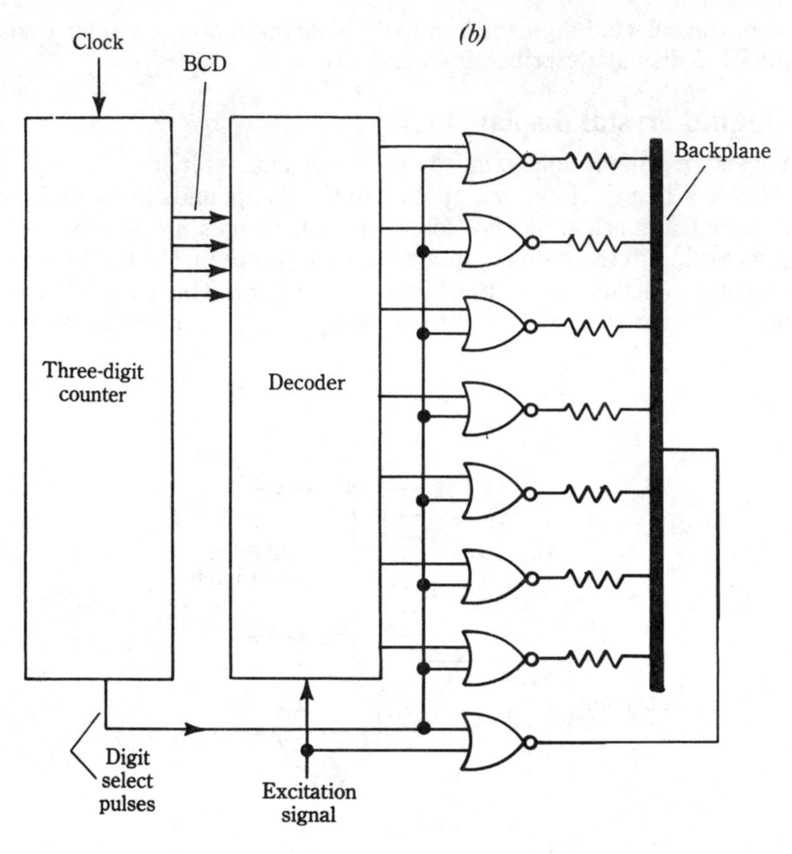

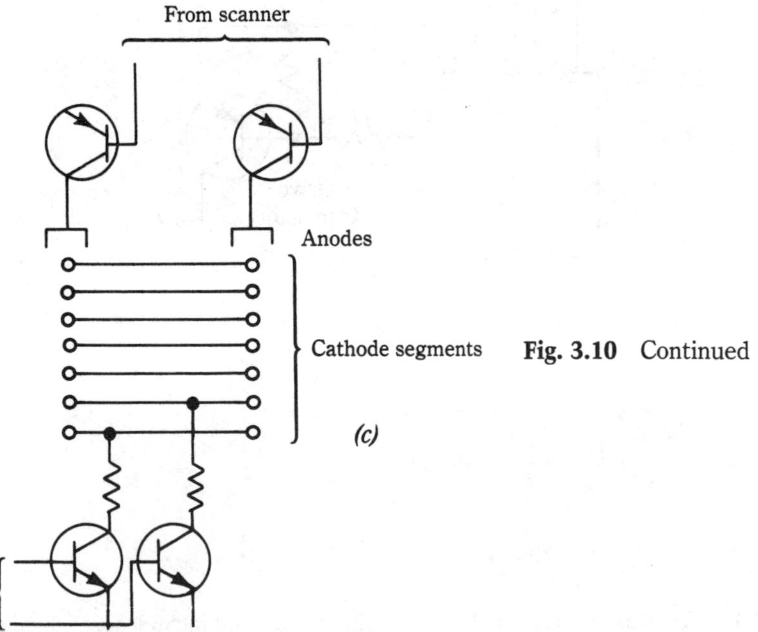

Fig. 3.10 Continued

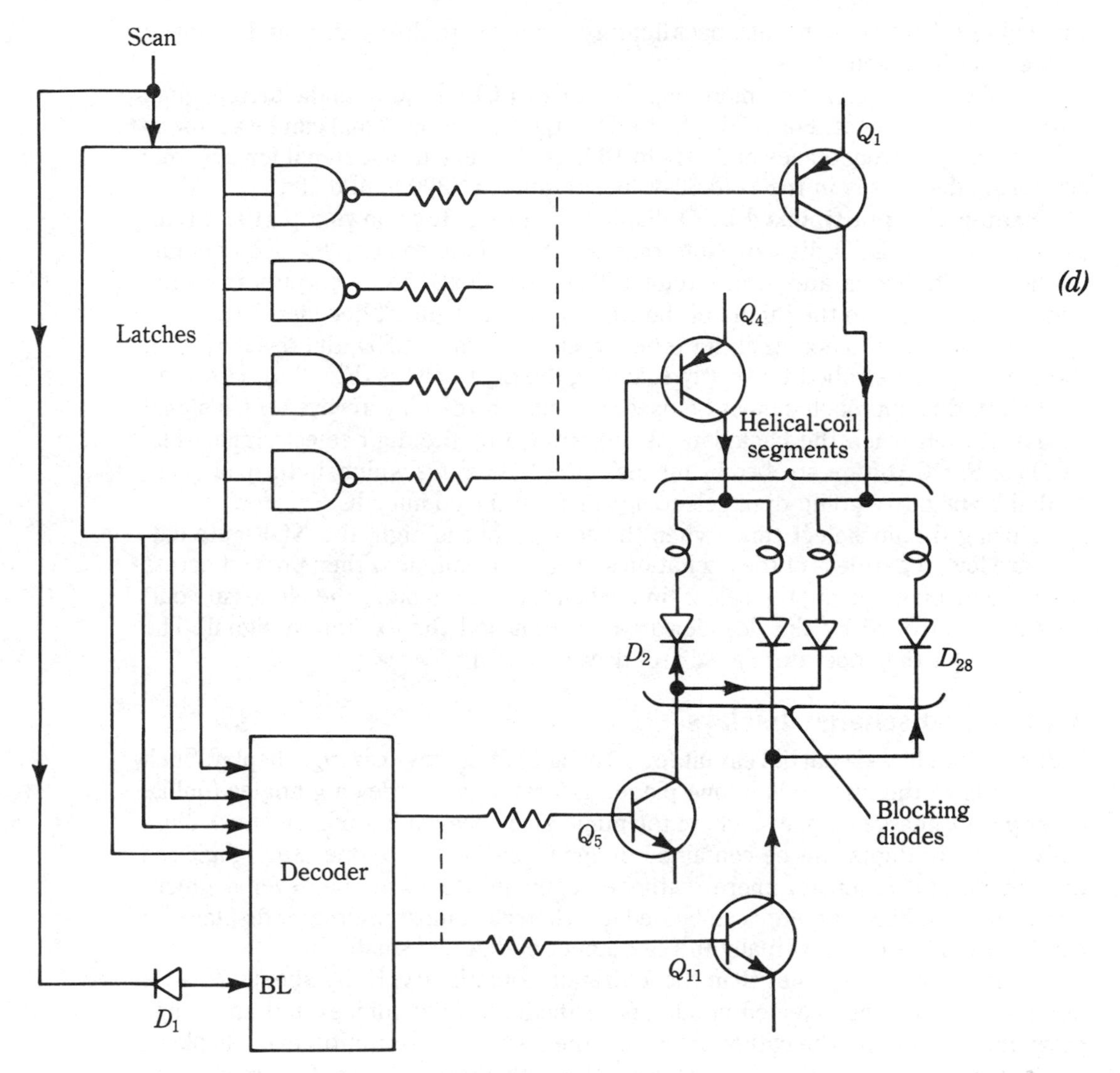

Fig. 3.10 Continued

With *field-effect* LCDs, a liquid-crystal material is injected between two plates of glass. A metal pattern (with the desired image) is etched onto the glass. In the case of a seven-segment display, the liquid crystal under the selected segment is activated when a field is placed on that particular segment. Polarizers are attached to the front and back of the display. Light striking the polarizer is either reflected or absorbed, depending on the relative direction of the polarizers.

If a reflective display is wanted, reflective material is adhered to the back polarizer. Reflective displays are generally used where ambient light is available. A *backlighted display* is often used where only a small amount of light is available.

Instead of reflective materials, backlighting is done with diffused incandescent or fluorescent light sources.

Unlike the simpler (and more popular) LED, LCDs require an ac drive signal, with no dc component. For field-effect LCDs, the excitation signal can be as low as 2 V, typically at frequencies of 60 Hz to 10 kHz. The excitation signal for dynamic scattering displays is in the 7- to 30-V (peak) range, at 200 to 400 Hz.

Example of multiplexed LCD display Figure 3.10*b* shows a partial circuit for a three-digit LCD display. Note that the three-digit counter IC also contains latches, multiplexers and scan circuits. The multiplexed BCD outputs from the counter IC are fed to the inputs of the BCD-to-seven segment decoder/driver.

It is important to note that, for other types of displays (LED, fluorescent, etc.), the digits can be strobed by simply scanning the digit drivers (Fig. 3.9). However, for LCDs, this approach cannot be used, because the display always sees a signal across it, even when the backplane is not strobed by the digit-select signal. The LCDs of Fig. 3.10*b* are strobed by means of the NOR gates which, in turn, are controlled by negative-going digit-select signals from the counter IC.

During the nonselect time, when the scan output is high, the NOR-gate outputs are low, regardless of the excitation signal. Zero voltage is then present across the LCD display. During the select time when the scan is high, the NOR-gate outputs are controlled by the decoder/driver output and the excitation signal. The selected digit then operates as a direct-driven system.

3.4.6 Gas-discharge displays

Figure 3.10*c* shows a partial circuit for a 12-digit *planar gas discharge* display. Such displays have the numerals in one plane, facilitating wide viewing angles (unlike older gas discharge displays, where the numerals appear to jump in and out). Typically, up to 16 digits can be contained in one neon-filled envelope. Each digit has one anode and seven (or more) cathode segments (to form the seven-segment numeral). The like segments can be tied together, as in most multidigit displays, or can be brought out individually (if the number of digits is small).

When a voltage greater than the ionization potential (typically about 170 V) is applied between the selected anode and cathode, the gas ionizes and an orange glow appears around the cathode (forming the numeral). For multiplexed displays, a blanking period is required between the cathode-select and the anode-scan pulses. This ensures that the previous digit is completely deionized before the following digit is strobed and thus prevents erroneous readouts.

In the circuit of Fig. 3.10*c*, the digits are selected, in turn, when an appropriate voltage is applied to the corresponding digit anode from a scanner (counter/decoder) circuit through level-shifting transistors. Each digit is selected in order. A BCD-to-seven segment decoder output is applied (through other level-shifting transistors) to the appropriate segments (cathodes) of the selected digit to be illuminated.

3.4.7 Incandescent displays

Figure 3.10*d* shows a partial circuit for a four-digit incandescent display. The smaller direct-view incandescent displays have seven helical-coil segments (to form the seven-segment numeral) fashioned from tungsten alloy. The power requirements for incandescent displays (typically 1.5 to 5 V at about 8 to 24 mA) make incandescent compatible with LED drivers or decoder/drivers. However, when multiplexing incandescent displays, greater filament peak power is required to maintain a brightness equivalent to LEDs. Also, *blocking diodes* are required, one for each segment, to provide erroneous display indications through sneak electrical paths.

In the circuit of Fig. 3.10*d*, the digits are selected (in turn) when appropriate scan pulses are applied through transistors $Q_1 - Q_4$ to all coil segments of the selected digit. The BCD-to-seven segment decoder output is applied through transistors $Q_5 - Q_{11}$ and diodes $D_2 - D_{28}$ to the appropriate segments of the selected digit to be illuminated.

Note that if the scan signal fails, the display is blanked, thus preventing damage to the displays. Diode D_1 converts scan pulses into a dc voltage. This voltage is applied through an inverter to the blanking (BL) input of the decoder. If the scan oscillator stops, the dc voltage developed by D_1 is removed and the display is blanked.

3.5 Interfacing digital equipment

This section concentrates on explanations of the basic principles and techniques for interfacing between digital circuits, particularly between the digital IC families. In Chapter 4, you will read about interfacing complete digital equipments such as computers, video terminals, and so on.

No matter what drive and load characteristics are involved for a digital IC, it might be necessary to provide the necessary drive current, to change the logic levels, and so on with an *interfacing-circuit* or device. It is not practical to have a universal interfacing circuit for all families of all digital IC manufacturers. It is not even possible to have a universal circuit for interface between a given family of one manufacturer and all other families of the same manufacturer. For one thing, the digital voltage levels (for 0 and 1), the supply voltages, and the temperature ranges vary with manufacturers, and with digital families. For this reason, most digital IC manufacturers publish interfacing data for their particular lines.

There is no attempt to duplicate this information here. Instead, the interface requirements for a few popular lines and families are covered. A careful study of this information should provide you with sufficient background to understand the basic problems involved with interfacing digital systems, and to interpret interfacing data that appears on digital IC datasheets.

3.5.1 Basic interfacing circuit

Figure 3.11*a* shows the basic circuit for interfacing between digital ICs. The equations shown in Fig. 3.11*a* are used to find the approximate or trial value of pull-up resistor R_1. The following is an example of how to use the equations.

Assume that the common-collector circuit is used, the supply voltage is 5 V, the VOH (high or logic-1 state voltage) is 3 V, there are seven gate inputs, each with

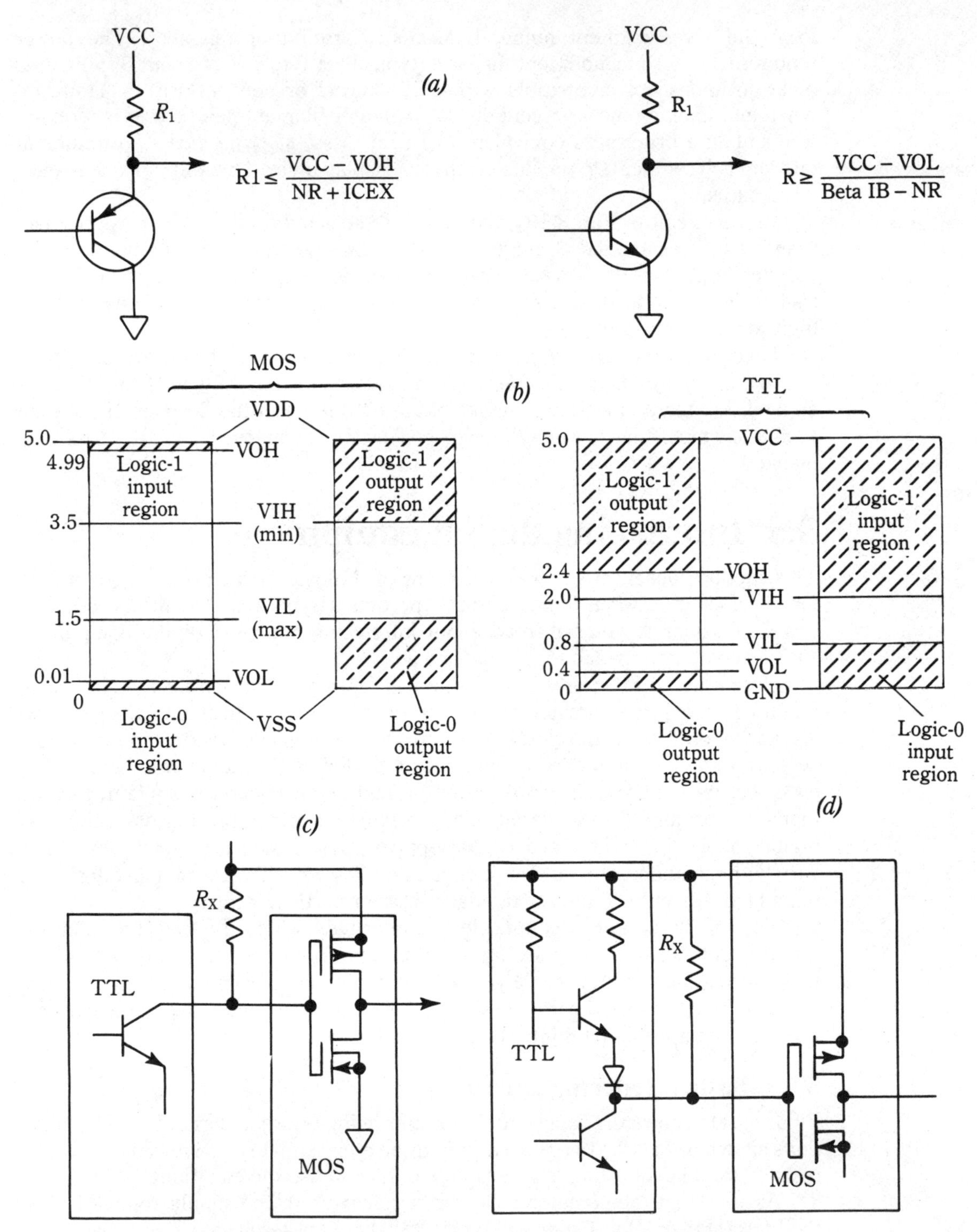

Fig. 3.11 Basic interlacing circuits and characteristics.

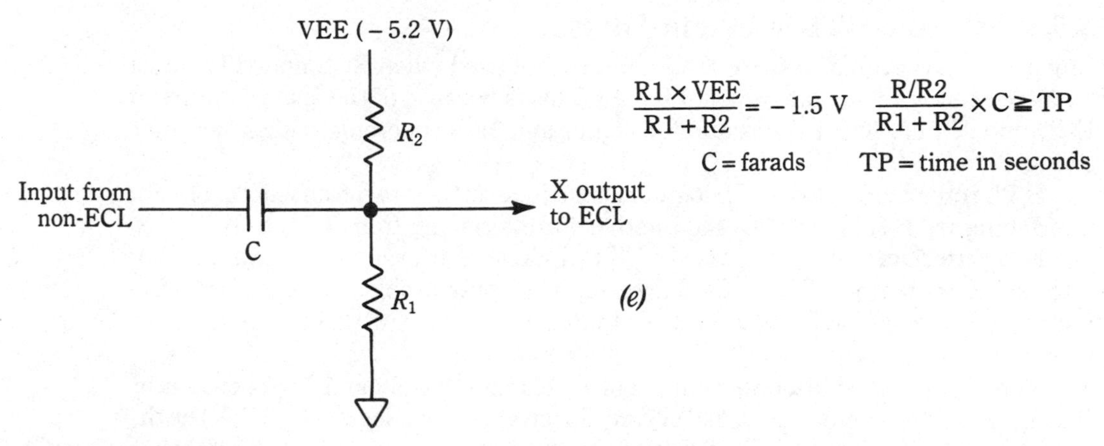

Fig. 3.11 Continued

a current of 1 and that ICEX (*input leakage current*) is 3 mA. Using the equation of Fig. 3.11*a*, the value of R_1 is:

$$R_1 = \frac{5 - 3\ V}{(7 \times 1) + 3\ mA}$$
$$= \frac{2}{10}$$
$$= 200\ \Omega$$

The next lowest standard value is 180 Ω.

3.5.2 MOS/TTL interface

When interfacing one digital IC family to another, attention must be given to a logic swing, output drive, dc input current, noise immunity, and speed of each family. Figure 3.11*b* shows a comparison for MOS and TTL. Typical MOS characteristics are: supply voltage 3.0 to 15 V, logic swing VSS to VDD, dc input current 10 pA, noise immunity 1.5 V, delay 33 ns. Typical TTL characteristics are: supply voltage 5 V, logic swing 0.4 to 2.4 V, dc input current 1.6 mA max in 0 state, noise immunity 0.4 V, delay 20 ns.

When TTL (or any two-junction device) is used to drive a MOS device, the output drive capability of the driving device, as well as the switching levels and input currents of the driven device, are important considerations. There are three two-junction output configurations to consider: *resistor pull-up*, *open collector*, and *active pull-up*.

Devices with resistor pull-ups can use the basic interface circuit of Fig. 3.11*a* when used to drive MOS. Devices with open collectors require an external pull-up resistor as shown in Fig. 3.11*c*. When an active pull-up is used, a circuit similar to that of Fig. 3.11*d* is often required. In many cases, it is necessary to use a *level-shifter* or *level-translator* IC (as described in the manufacturers' IC datasheets).

3.5.3 ECL and HTL level translators

In general, level translators (or level shifters) are used when ECL and HTL must be interfaced with any other digital family. This is because of the special nature of ECL and HTL (ECL is nonsaturated logic; and HTL operates with large logic swings).

HTL interface Most HTL logic lines include at least two translators: one for interfacing from HTL to TTL and another for interfacing from TTL to HTL.

ECL interface As in the case of HTL, ECL level translators are available for interfacing with other digital ICs. However, if the only problem is one of interfacing from an input to ECL, it is possible that the circuit of Fig. 3.11*e* can solve the problem.

The input (either discrete component or IC) must be about 1 V. As shown by the equations, the values of R_1 and R_2 are selected to give a logic 0 (– 1.5 V) at the output (to the ECL input). The value of C should be selected so that the RC time constant should be several times greater than the input pulse duration.

3.5.4 Summary of digital IC interface problems

As discussed, each digital IC line or family has its own set of design or usage problems. Most of these problems can be overcome using the datasheets and brochures supplied with the device. The following is a summary of this information.

If MOS is involved, follow the recommendations in Sec. 3.5.2. If either ECL or HTL is involved, use the various level translators supplied with most ECL and HTL lines, or use the specified translator circuit shown in the datasheets and brochures. In the absence of such information, use the circuits described in Sec. 3.5.3.

TTL is generally compatible with DTL, thus eliminating the need for special translocators or interface circuits. However, the load and drive characteristics might be altered somewhat. These changes are generally noted on the datasheets. In the absence of any TTL interface data, use the information in Sec. 3.5.1 through 3.5.3.

3.5.5 Digital line drivers and receivers

Line drivers and receivers are used to transmit digital data from one system to another. The distances involved might vary from a few feet to several thousand feet. This section summarizes the basic characteristics associated with IC line drivers and receivers. Note that these IC devices are part of the IEEE-488 digital data-communications systems described in Chapter 4, and are not to be confused with modems used to transmit digital data over telephone lines (also discussed in Chapter 4).

Figure 3.12*a* shows a basic driver/receiver digital data transmission system. Here, an input data stream VIN feeds a driver which, in turn, drives a line. Information at the other end of cable or line is detected by a receiver that provides an output data stream VOUT. Usually, VOUT is of the same digital level as VIN. The line can be a single line, a coaxial line, a twisted pair, or a multiline cable (ribbon cable type, multitwisted pair type, etc.). The line can be operated in a single-ended mode, or a differential mode requiring a pair of lines.

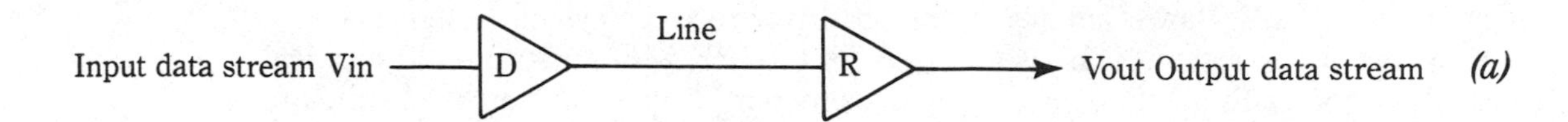

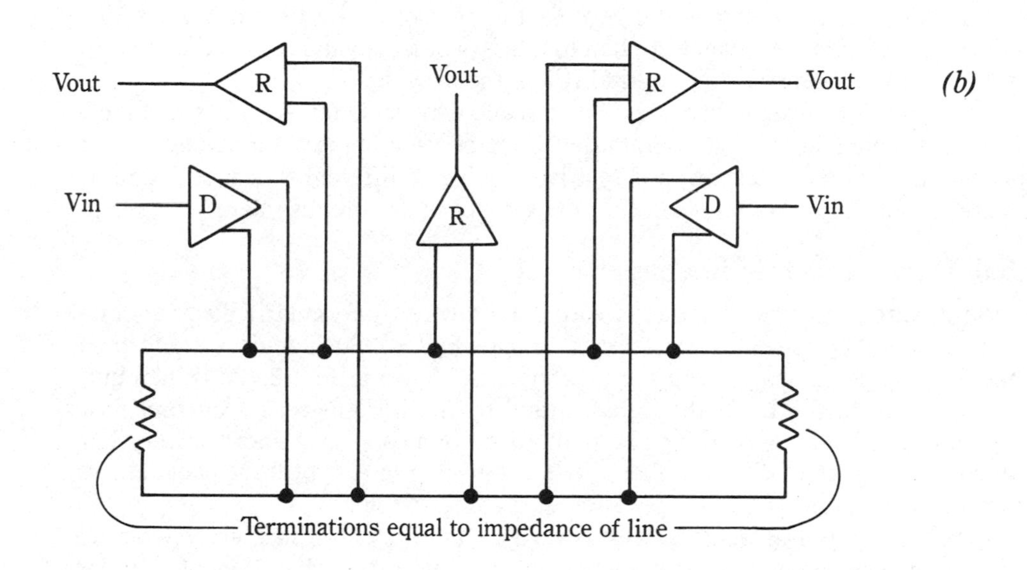

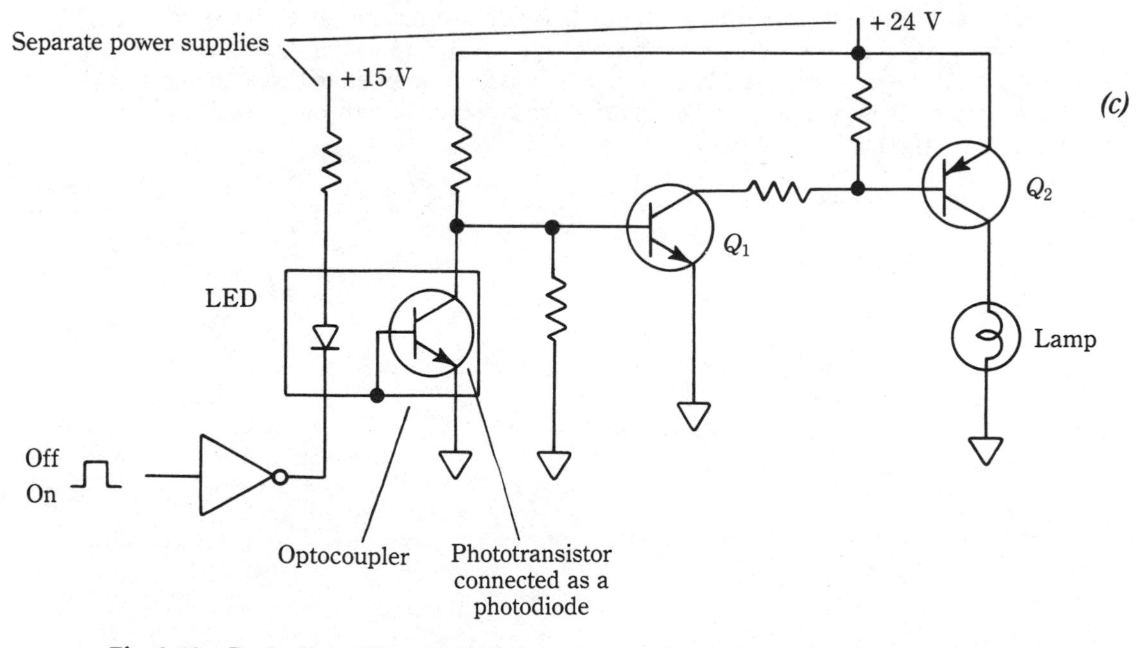

Fig. 3.12 Basic digital line-driver, line-receiver, and optoelectronic interface circuits.

Figure 3.12*b* shows another common driver/receiver system, often called a *party-line* or *bus* system. (This is the basis for the IEEE-488 (Institute of Electrical and Electronics Engineers) bus system discussed in Chapter 4.) Note that although any driver can be used to drive the line, only one driver is used at one time.

The line driver translates the input digital levels (TTL, MOS, and so on) to a signal more suitable for driving the line. An important exception to this is found in the ECL family, where ECL gates are often used to drive the line directly.

The *line receiver* provides the reverse function of a line driver. With a line receiver, the voltage previously applied to a line is detected and restored to an output level compatible with other digital ICs in the system.

There are ICs designed as line receivers and drivers. Often such ICs include a strobe or enable feature that permits the ICs to be used in party-line systems. For example, if several drivers are used to drive the line at different times, each can be enabled when desired (while the other drivers not in use are disabled).

3.5.6 Optoelectronic interface

Digital ICs are sometimes interfaced with other devices (particularly discrete-component circuits) by *optoelectronic couplers* (or *optocouplers*). Such devices are formed when an LED is packaged with a photodiode, as shown in Fig. 3.12*c*. When current is applied to the LED, the light emitted by the LED increases current flow through the photodiode, causing the photodiode to appear as a short or very low resistance. Both the LED and photodiode are sealed in a lightproof package, so external light has no effect on operation of the optocoupler.

In the circuit of Fig. 3.12*b*, a MOS inverter is interfaced with a discrete-component lamp power circuit. The lamp is normally de-energized when the optocoupler is energized. A high (1) level on the input of the MOS inverter energizes the optocoupler, clamping the base-emitter junction of Q_1 to off. Transistor Q_1 removes drive to Q_2, thus de-energizing the load. When a digital 0 is applied to the optocoupler, the clamp to Q_1 is removed, and Q_2 is driven into conduction through the load, thus energizing the load.

4

Microprocessors, computers, and controllers

This chapter introduces microprocessors (or μPs) and microprocessor-based digital equipment (such as computers, video terminals, and so on). A microprocessor is an IC that performs many of the functions performed by a digital computer. A single microprocessor IC is capable of performing all the arithmetic and control functions of a computer. By itself, a typical microprocessor does not contain the memories and input/output (I/O) functions of a computer. However, when these functions are provided by additional ICs or peripherals, a *microcomputer* or *computer* is formed.

A microprocessor is not always used in digital-computer applications. Instead, the microprocessor is used as a *controller*. Many such controller applications are presented in Chapter 9. As a matter of interest, the microprocessor was originally developed as the control element for those applications where digital-computer functions (the ability to store and execute a complete program automatically) are required, but where a computer (of even the most basic type) is too large or expensive.

When used in control-type applications, the microprocessor is often called a controller or microcontroller, although the term processor can also be used. In some literature, microprocessors are referred to as microprocessor units (MPU) or control processor units (CPU), although CPU can also mean central processor unit. Do not be surprised at any terms you find in digital-equipment literature!

When the microprocessor is used as the main element in a computer, the system also requires a read-only memory (ROM) to store the computer program or instructions, a random access to store temporary data (the information to be acted upon by the computer program), and an I/O IC to make the system compatible with outside (or peripheral equipment), such as an interactive video terminal, CD-ROM, and so on. There are some ICs that contain some or all of these functions. In effect, when an IC contains all of the basic functions, the IC is a computer on a chip. However, this is not the typical case.

4.1 Typical microprocessor-based device

The familiar pocket calculator is typical of the many microprocessor-based devices now in use. As shown in Fig. 4.1, the calculator has a keyboard and a numeric display. When a key is pressed, the corresponding number should appear on the display. This system is a natural application for a microprocessor and can be considered as a basic computer.

The microprocessor is the brains of the system and contains all of the logic to recognize and execute the list of instructions (or *program*). The memory stores the program and may also store data.

The microprocessor needs to exchange information with the keyboard and display. The *input port*, from which the microprocessor reads data, connects the microprocessor to the keyboard. The *output port*, to which the microprocessor sends data, connects the microprocessor to the display.

The blocks within this "computer" are interconnected by three buses. A *bus* is a group of wires that connect the system devices in parallel. The microprocessor uses the *address bus* to select memory locations or input/output ports. Addresses identify the locations in memory that are used to hold data.

Once the microprocessor selects a particular location via the address bus, the microprocessor transfers the data on the *data bus*. Information can travel from the microprocessor to memory. Note that the microprocessor is involved in all data

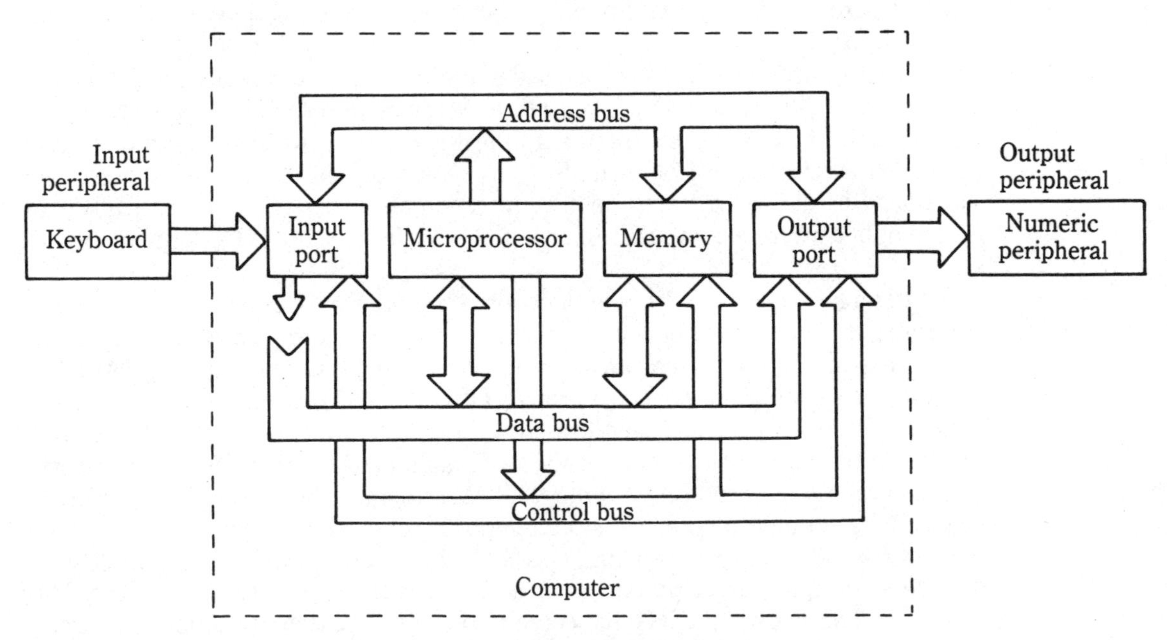

Fig. 4.1 Pocket calculator as a basic computer.

transfers. Information usually does not go directly from one port to another, or from the memory to a port.

The third bus (*control bus*) carries a group of signals used by the microprocessor to notify memory and I/O devices that the microprocessor is ready to perform a data transfer. Some signals on the control bus allow I/O or memory devices to make special requests from the microprocessor.

A single digit of binary information (1 or 0) is called a *bit* (a construction of binary digit). One digital signal (high or low) carries one bit of information. Typically, a binary 1, or high, is represented by the presence of a voltage (usually +5 V), and a binary 0, or low, is represented by the absence of voltage (0 V or ground).

Note that the (high +5 V, low 0 V) system is not true for every device, so always look for any notes on troubleshooting schematics regarding electrical representation of binary or digital information. (As a practical matter, it is always a good idea to read all notes on the service literature before troubleshooting any type of equipment.)

Microprocessors used in computer applications handle data in groups of bits called *words*. Control-type microprocessors often operate with single bits of data, as described in Chapter 9. The oldest (and still most common) microprocessor-based computers use eight-bit words called *bytes*. The microprocessors involved are 8-bit microprocessors. For an eight-bit microprocessor, byte and word are often interchanged. *Word* (or even *byte*) can also mean a group of 16 (or even 32) or more bits. When a microprocessor uses more than one bit, but less than the full 8- or 16-bits, the term *nibble* is sometimes used.

4.2 Microprocessor structure

This section discusses the basic structure of microprocessor-based digital devices, primarily computers and video terminals.

4.2.1 System addresses and memories

Computer-type microprocessors are often used with RAM or ROM (random-access memory) memories, typically both, that hold data bytes to be manipulated by the microprocessor. These memories also hold instructions to be followed by the microprocessor during the execution of a program. Microprocessors communicate with these memories and other system elements by means of electrical signals, often arranged in binary form (Fig. 1.6).

A ROM is a memory that can only be read, not altered. The data bits are programmed into the ROM at the time of manufacture, or by a special programming procedure prior to installation in the circuit. A program recorded into a ROM is sometimes called *firmware*.

A RAM is a memory in which data can be stored and then retrieved. The term RAM is somewhat confusing. In a strict sense, *random access* means that the time to access any memory location is the same. Read/write (R/W) memory is a more accurate term for what are called RAMs, but RAM is widely used to mean an IC read/write memory.

An important characteristic of semiconductor RAMs is that they are volatile (they lose data when power is turned off). When power is turned back on, RAMs contain unknown data. ROMs do not have this problem, so ROMs are used for permanent program and data storage. Because the contents of a ROM cannot be modified, RAMs must be used for temporary storage.

Memories are divided into locations called *addresses*. Each address is identified by a number (usually decimal). In a typical computer program, the microprocessor selects each address in a certain order (determined by the program) and reads the contents of the address. Such contents can be an instruction, data, or a combination of both. Likewise, it is possible to write information into memory using electrical signals arranged in binary form.

4.2.2 Memory arrangement

Figure 4.2 shows a basic arrangement where a microprocessor is connected to a RAM and ROM by data and address buses, forming an extremely simple computer.

The data bus has eight lines (which is typical) and the address bus has eight lines (which is not typical). Generally, the address bus has 16 lines in a microprocessor-based computer, but the eight-line system is used here for simplicity.

The term *highway* is sometimes used when a bus interconnects many system components. The term *handshake bus* is sometimes used to indicate a bus that interconnects a microprocessor system with the outside world.

The term *port* is often applied to the point or terminals at which the bus enters the microprocessor or other IC element. So, in Fig. 4.2, there is an address port and a data port for the microprocessor.

Buses are generally bidirectional. That is, the electrical pulses (representing data bytes, addresses, and so on) can pass in either direction along the bus. For example, data bytes can be written into memory from the microprocessor, or read from memory into the microprocessor, on the same data bus. Ports may or may not be directional, depending on design.

4.2.3 Transferring pulses on the bus

In Fig. 4.2, the microprocessor is selecting address number 77. This is done by arranging the electrical pulses on the address bus to produce a binary 0100 1101 (decimal 77). Both the ROM and RAM receive the same set of electrical pulses (or binary word) because these pulses appear on the address bus.

Because address 77 is located in the ROM, the contents of the ROM at that address are read back to the microprocessor via the data bus. The pulses on the address bus have no effect on the RAM, so no data bits are obtained from the RAM. Likewise, the address pulses have no effect on other addresses in the ROM. Only the data bits at the selected address are read back on the data bus.

In Fig. 4.2, the electrical pulses on the data bus are arranged to form the binary word 0010 0001, which can be converted to hex 21. In this particular microprocessor, hex 21 is an instruction to add the contents of a register within the

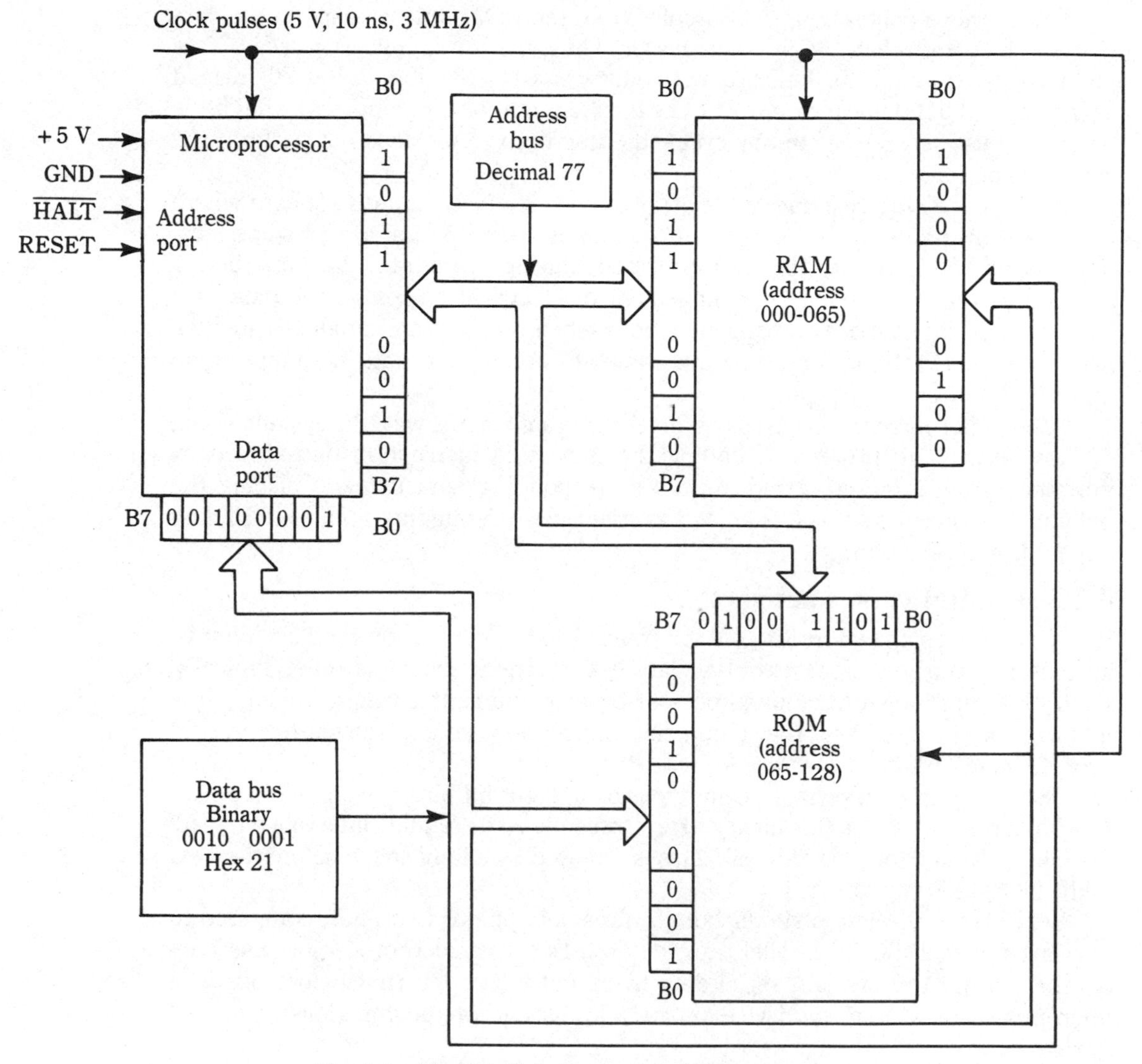

Fig. 4.2 Typical digital memory arrangement.

microprocessor to the contents at some other address in the RAM (read out during a previous step in the program).

4.2.4 Troubleshooting versus debugging

In this book, *troubleshooting* is a term used to find electrical or mechanical equipment faults in microprocessor-based equipment. Troubleshooting implies that the equipment once performed the program properly. For example, referring to Fig. 4.2, assume that the electrical line labeled B0 on the data bus breaks loose (after months of operation) at the microprocessor terminal.

Under these conditions, the B0 pulse from the ROM still appears on the data bus but does not reach the microprocessor. Thus, the microprocessor sees 0 V or binary 0, on terminal B0. This produces binary word 0010 0000, or hex 20, instead of the desired 0100 0001, or hex 21. The hex 20 might be meaningless or might be an erroneous instruction. In any event, the microprocessor does not perform the correct function.

Microprocessors respond to the arrangement of pulses and do not care where the pulses originate. If an *undesired instruction* is stored in memory at some location that is addressed by the microprocessor during a program, that instruction appears as the corresponding arrangement of electrical pulses on the data bus. The microprocessor follows the instructions when received and produces an erroneous result. (Probably, the program comes to a complete halt, or jumps to an undesired address.)

When microprocessor-based equipment operates in this way, the system is said to have *bugs*. The process of finding the undesired instruction (introduced by improper programming), removing the corresponding data byte, or placing the byte at the correct address, is known as *debugging*. Debugging applies to finding any fault in a program.

4.2.5 Parallel versus serial

All of the pulses shown in Fig. 4.2 are transmitted in parallel on the bus. That is, all pulses in a given binary word or data byte arrive at the same time. Typically, binary information within microprocessor-based equipment is transmitted in parallel form. On the outside of the equipment, information is often transmitted in *serial* form (Sec. 4.7.5).

As an example, in serial transmissions of an eight-bit binary word, eight pulses (for binary 1) or spaces (for binary 0) are transmitted at regular intervals on a single electrical line (or pair of lines). This is followed by a long space before the next eight-bit byte is transmitted.

Serial transmission is slower but requires only one or two lines, compared to one line for each bit in parallel transmission. Because microprocessors use only parallel within the system, serial data bytes must be converted before use in a microprocessor system (and vice versa). This is one of the functions of an I/O device.

4.2.6 System clock

Microprocessor-based equipment requires some form of system clock. The clock can be internal (part of the microprocessor) or external. In Fig. 4.2, the clock is an external 5-V pulse of 10-ns duration at a frequency of 3 MHz. The clock circuit is an oscillator that produces pulses of fixed amplitude and duration at regular intervals.

ROM and RAM do not actually generate the binary pulses but produce their binary word pulses on the address and data lines when the clock pulses are received. It might take many clock pulses to form a binary word.

In most cases, the microprocessor, ROM and RAM must also receive other signals before the binary word (data byte) is produced. For example, a RAM usually requires a read signal, plus the clock pulses, before the contents of an address are read onto the data bus from the RAM. Such control signals are discussed in later paragraphs.

4.2.7 System power supply and control signals

All signals applied to a microprocessor are not necessarily in pulse form. For example, as shown in Fig. 4.2, there are two lines into the microprocessor labeled +5 V and GND. These are the power-supply lines connected to an external 5 V power supply.

There are also the $\overline{\text{HALT}}$ and RESET lines that receive a +5 V signal from various circuits in the system. These signals might be a momentary pulse identical to those on the address and data buses or might be a fixed +5 V that remains on the line for some time. When a fixed signal (sometimes called a *level*) is applied, the line is said to be *high* and the function is turned on.

For example, if +5 V is applied to the RESET line, the RESET line is high, and all circuits within the microprocessor are reset to zero, regardless of their condition before the line goes high. When the fixed voltage is removed, the RESET line goes low (at 0 V), and the RESET function is no longer in effect.

In most microprocessors, when a reset signal is received, all circuits return to zero and then resume their normal function (counting, etc.). Note that in some literature, the terms *true* and *false* are used instead of high and low, respectively. Likewise binary 1 and 0 are used for high and low in other literature.

An overbar is used on the word $\overline{\text{HALT}}$ to indicate that the $\overline{\text{HALT}}$ function operates on the reverse of all other lines. That is, the $\overline{\text{HALT}}$ function is in effect when the line is at 0 V (the normal low condition). When the line is at +5 V (normal high), the $\overline{\text{HALT}}$ function is removed.

In Fig. 4.2, when the $\overline{\text{HALT}}$ line is at 0 V, all functions within the microprocessor (counting through the program, etc.) are stopped as long as the $\overline{\text{HALT}}$ line remains low. All functions resume normal operation when the $\overline{\text{HALT}}$ line is made high by a +5-V level. If the microprocessor is in the middle of some operation when $\overline{\text{HALT}}$ is applied (by 0 V on the line), the operation stops but continues from the same point when the +5 V is reapplied.

This shows the need for a thorough knowledge of microprocessor functions and control signals when troubleshooting microprocessor-based equipment. For example, should the microprocessor $\overline{\text{HALT}}$ line be broken, or shorted to ground accidentally, the microprocessor would stop in the middle of the program and possibly in the middle of an instruction. This brings the entire system to a halt (the system crashes).

4.3 Microprocessor functions

Because microprocessors are in IC form, and you cannot change the way they work (nor do you have access to the internal circuit elements), it is not essential that you understand every detail of the internal circuits (such as those discussed in Chapters 1 through 3). But it is essential that you know what registers and counters are, and how they operate, to effectively troubleshoot microprocessor-based equipment. So run through some typical circuits found in microprocessors, particularly computer-type microprocessors.

4.3.1 Manipulating information within the microprocessor

For microprocessors used in computers, binary numbers are held and manipulated (in electrical form) by counters, registers, accumulators, and pointers. Typically, such circuits have one stage for each binary bit to be held or manipulated. That is, an eight-bit counter/register has eight stages.

Flip-flops are used for counter/register stages. This is because FFs can be in only one of two electrical states, 1 or 0. If you measured the instantaneous state of a typical microprocessor FF, you would find the state at +5 V if the stage or FF represents a 1. You would find the state at 0 V if binary 0 is represented. (As a practical matter, you will not measure the state of any circuit within a microprocessor. However, this concept might help you to understand the operation of counters and registers.)

In computer-type microprocessors, a *counter* is used to count events (program steps, address-selection sequence, etc.). This is usually done by counting pulses. Counters are sometimes used as *pointers*, because they point to another event or location. For example, a typical program counter counts each step of the program and then advances to the next address to be used in the program. Thus, the counter points to the next step of the program.

Registers are used to hold binary numbers (or words) so that the numbers can be manipulated. For example, a register might hold some binary number taken from a particular address in memory so that the number can be added to another number in memory. Also, a register can hold a binary number to be added to another number in another register. When a register is used primarily for arithmetic operations, the register is often called an *accumulator*.

Serial manipulation Figure 4.3*a* shows the operation of a typical counter used to count serial pulses. Assume that the circuit is used as a *program counter* and that one pulse is received for each step in the program. That is, the counter is *incremented* (or advanced) for each pulse representing a step and *decremented* (or reduced by 1) when each step or pulse removes one count.

All eight stages are reset or cleared to low (binary 0 or 0 V) by a reset signal (this is sometimes known as *initializing* the counter). Then the first, or LSB, stage is set to high (binary 1 or 5 V) by the first pulse to be counted. All other stages receive a pulse from the stage ahead.

As each pulse to be counted is applied, the stage representing the LSB is changed from 0 to 1 (0 V to +5 V). When the next pulse arrives, the first stage returns back to 0 V (binary 0) and sends a signal to the next stage, which moves to binary 1. This process continues until all pulses are counted. For example, if there are four pulses as shown in Fig. 4.3*a*, the third stage moves to 1, and the first two stages are at 0. All remaining stages are at 0.

The counter now indicates a binary 00000100 (decimal 4) that corresponds to the number of pulses applied (and the number of steps in the program). The instantaneous count corresponds to the program step just accomplished or to be

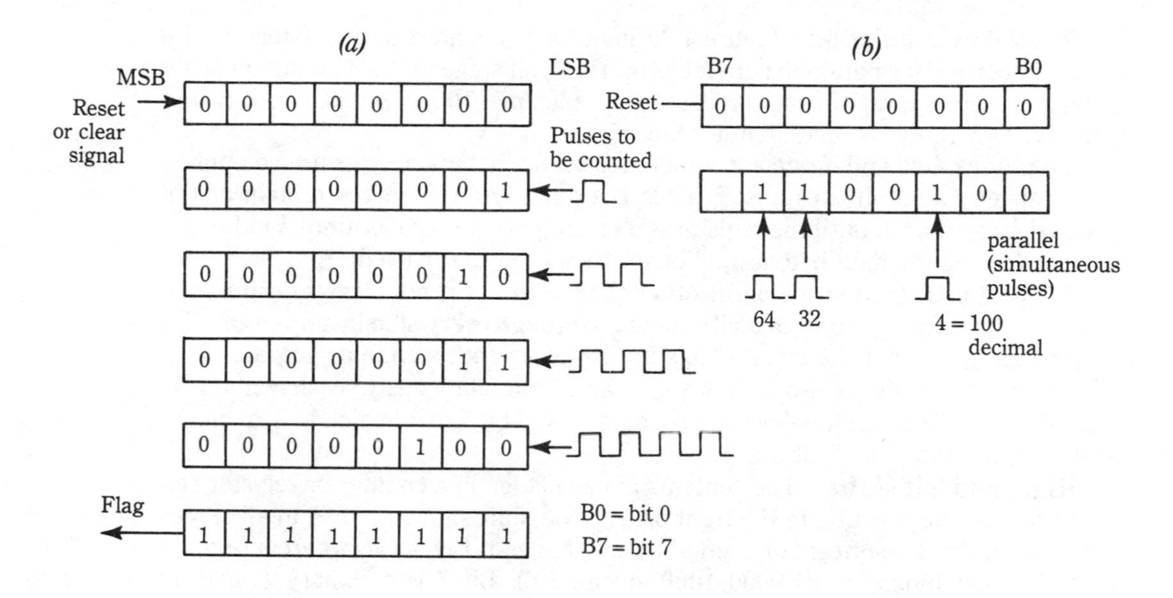

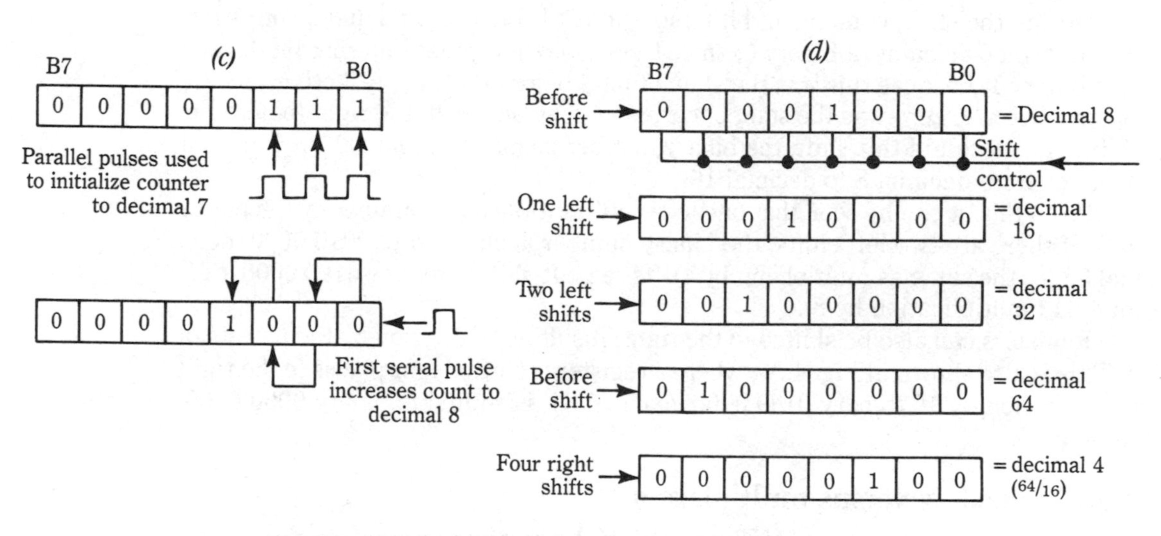

Fig. 4.3 Manipulation of data in microprocessor circuits.

done next, depending on design. When all stages are moved by sufficient pulses to a 1 (binary count of 0111 1111, or decimal 127) the counter is full.

In computer-type microprocessors, a full counter causes a *flag* signal to be sent to other circuits. The flag can be used for any number of functions. For example, the flag can be used to reset operation of the microprocessor if the program has only 127 steps. Note that the term *flag* can also be used to mean many other conditions such as a full count, error request for further information, temporary interruption, etc.).

Parallel manipulation Figure 4.3*b* shows how counters and registers can be used to receive information in parallel form. Here, all stages are set simultaneously by pulses used to form binary word 0110 0100 (decimal 100). That is, bits 2, 5, and 6 receive +5 V pulses, and all other bits remain at 0 V.

The terms *load* and *dump* are sometimes used when data bytes are so applied to a register. Generally, a register holds the data byte (all stages remain at the selected state, 1 or 0) until the register is cleared (reset to zero) or until another set of pulses forming a data byte is applied (or the power is removed).

Starting counts at some point other than zero It is not always desirable to start all counts from zero or that all counts go through every step in a program. For example, assume that the count is to start at the seventh step in a program. This can be done by applying +5 V to bits 0, 1, and 2 simultaneously, as shown in Fig. 4.3*c*. Then the first serial pulse to be counted moves bit 0 to 0, which in turn moves bit 1 to 0, bit 2 to 0, and bit 3 to 1.

Right and left shifts The contents of each stage in a counter or register can be shifted by one position to the right or left by a shift signal applied to all stages simultaneously. The effects of a left shift are shown in Fig. 4.3*d*, where a register is holding the binary word 0000 1000 (decimal 8). Bit 3 is at binary 1, and all remaining bits are at 0.

During the shift, contents of bit 0 move to bit 1, bit 1 to bit 2 and so on. After the shift, bit 0 remains at binary 0, since there is no new pulse entering bit 0. Bit 4 is at binary 1, because this was the state of bit 3 before the shift. All other bits are at zero, because zero was the state of corresponding stages to the right (before the shift). As a result of this shift, the binary number changes from 0000 1000 to 0001 0000, or from decimal 8 to decimal 16.

From this, it can be seen that one left-shift multiplies the number by the power of 2. If there are two left shifts, the binary number is changed to 0010 0000, decimal 32 (or the same as multiplying by 4). Three left-shifts produce 0100 0000, decimal 64 (multiplication by 8).

Registers can also be shifted to the right, resulting in division by the powers of 2. This is also shown in Fig. 4.3*d*, where a register is shifted four places to the right for a division by 16 (binary 0100 0000, decimal 64, is shifted to binary 0000 0100, decimal 4).

4.3.2 Decoders versus multiplexers

The terms *decoder*, *encoder*, *MUX* and *DEMUX* are often interchanged in digital literature. In a strict sense, a decoder or encoder converts from one code to another, such as from hex to decimal, and so on. In an equally strict sense, a MUX or DEMUX is a *data selector* and/or *distributor*. However, in digital literature, the terms can be applied to any circuit that converts data from one form to another.

As an example, a certain microprocessor system contains a decoder in each ROM that makes it possible to select one of 128 memory addresses with an eight-bit word supplied on an eight-bit address bus. Another microprocessor-based system contains a circuit (designated by the manufacturer as a MUX) that converts 16 lines of information from a register into one line (in serial form) for transmission to another circuit.

4.3.3 Buffers, drivers, and latches

A *buffer* can be considered any circuit between two other circuits that serves to isolate the circuits under certain conditions and to connect the circuits under other conditions. A buffer circuit between an internal register and the data bus (at the data port) is a classic example, as shown in Fig. 4.4*a*. Here, the buffer can be switched on or off by a bus-enable signal or pulse. When the enable signal is present, the buffer passes data, instructions, or whatever combination is on the data bus, to the register within the microprocessors.

The use of a buffer in this application is necessary when the register is holding old data not yet processed, but there are new data bits on the bus. The buffer closes the data port until the register is ready to accept new data. This raises the obvious problem of what happens when the new data bits are momentary. In such cases, the buffer has a *latch* function, permitting each bit in the buffer to be latched to a 1 or 0 by the data pulses. The latched word is held in the buffer until the register is ready to accept new data.

Some buffers also include a *driver* function, particularly where the buffer must feed many devices on a bus. The register or counter output might be sufficient to drive one other register but not many registers or devices. The driver function amplifies the drive capability of the buffer.

In a typical computer-type microprocessor, the buffers are *three-state devices*. Each stage in the buffer can be in one of three states, *in, out*, or at a *high-impedance level*. This high-impedance level, or state, makes it appear that the circuit is closed to the passage of data (no data in or out).

4.3.4 Matrix, intersect, and vector

The term *matrix* can be applied to many circuits (such as a group of registers within a microprocessor) but is most often used to identify memory arrangements. A typical *memory matrix* is divided into sections or locations, and each section is identified by an address, as shown in Fig. 4.4*b*. In turn, the addresses are divided into a number of stages with one stage for each bit to be held in memory. Thus, each address in an eight-bit memory has 8 bits.

The number of addresses in a memory matrix depends on use. Generally, the number is based on the powers of 2 because the addresses are selected by a number system based on binary (such as hex). A typical ROM matrix has 128 addresses, each with 8 bits, and is described as a 128×8 matrix. A typical RAM matrix has 512, 1024, or 4096 addresses (or possibly more). Sometimes a 1024×8 memory is described as a 1K memory, because there are approximately 1000 addresses available. A 4096×8 memory matrix is likewise called a 4K memory.

In a computer-type microprocessor system, the addresses of the matrix are selected by some form of *vector* or *intersect*, as shown in Fig. 4.4*c*. The memory address to be read in or read out is selected when a combination of two appropriate bit lines have a +5 V pulse present (are at binary 1). Only one of the X lines and one of the Y lines can be binary 1 at any given time. All other lines are at binary 0. This action is controlled by some form of decoder or MUX.

Assume that address 0003 (decimal 3) is to be selected in the vector system of

Fig. 4.4*c*. The desired decimal three is converted to hex 81 (1000 0001) by a decoder/multiplexer. This causes bits B0 and B7 to receive pulses (+5 V or binary 1). All remaining bit lines are at binary 0 (0 V). So the selected address (number 3) at the intersection or vector of the two lines is selected. Remember that there are

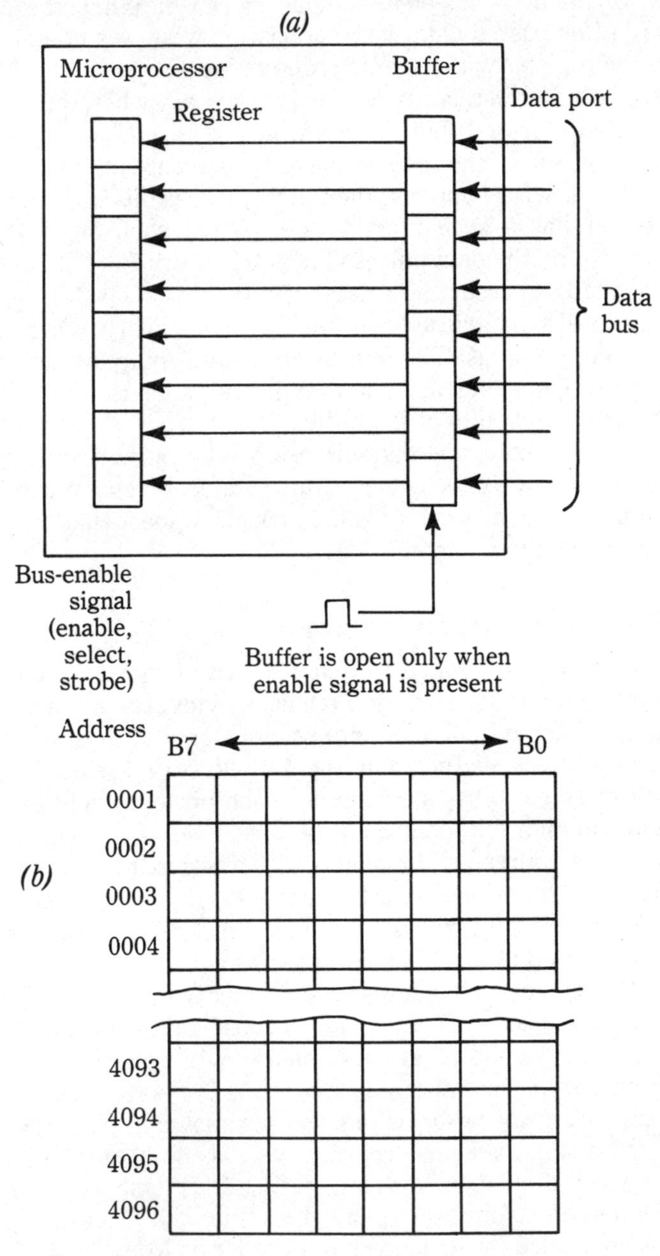

Fig. 4.4 Typical microprocessor functions.

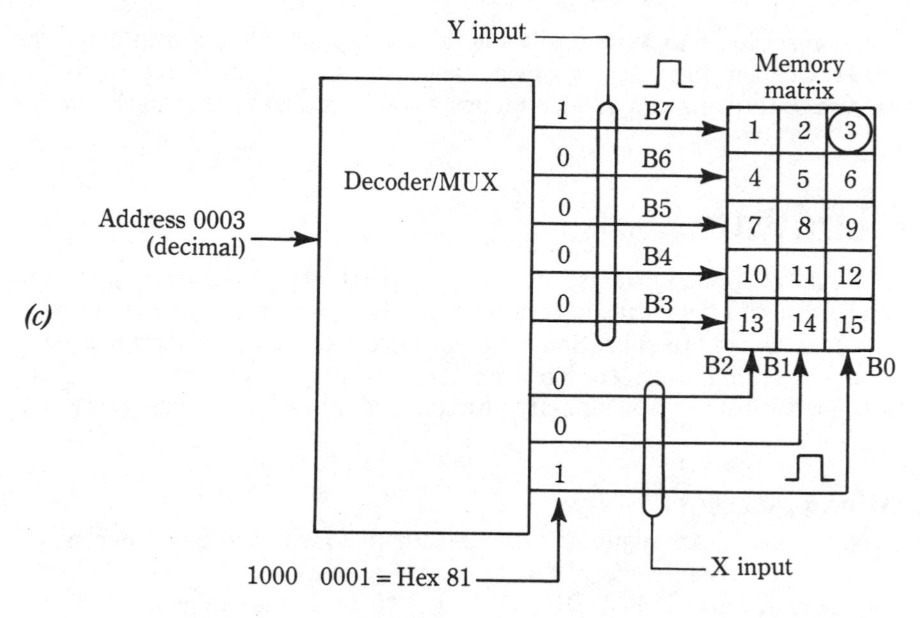

Fig. 4.4 Continued

eight bits in each address and that each of the 8 bits is fed back to the microprocessor or other destination on a bus (usually the data bus).

4.3.5 ALU and adders

Those microprocessors used in computer applications have an ALU (Sec. 3.2), although it may not be called an ALU. Many microprocessors are not used in computer applications and, therefore, do not need the ALU function. When required, the ALU performs the arithmetic and logic operation on data bytes.

The basic symbol for an ALU is a simple box with lines or arrows leading in or out. On rare occasions, the box contains some hint as to the functions capable of being performed by the ALU. However, these functions are usually described only in the microprocessor instruction set. As a minimum, a typical ALU has an adder circuit that is capable of combining the contents of two registers.

4.3.6 Microprocessor control

All microprocessors have some form of control circuit. As an example, in a computer-type microprocessor, after an instruction is taken from memory (or *fetched*) and decoded, the control circuit issues the appropriate signals for initiating the proper processing action (such as a write signal to write data in memory, a read signal to read data from memory, and so on).

Because microprocessors are controllers, it is difficult to generalize about control functions. However, one control function in all computer-type microprocessors is the capability of responding to an *interrupt* signal or *service request* (from a keyboard, video terminal, disk drive, CD-ROM, or modem). An interrupt request causes the control logic to temporarily interrupt the program, jump to a special routine to service the interrupting device, and then automatically return to the main program. Interrupts and service requests are discussed further in Sec. 4.7.

4.4 Hardware

The term *hardware* applied to digital equipment refers to the physical components, wiring, and so on. This contrasts with the term *software*, which applies to programs, instructions, and the like. Some manufacturers also use the term *firmware* to describe something between hardware and software. For example, when instructions (software) are permanently programmed into a ROM (hardware), the result is firmware.

4.4.1 Microprocessors

The microprocessor of one manufacturer has little to do with those of other manufacturers, with the possible exception of outward physical appearance. Typically, microprocessors are dual-in-line (DIP) ICs in the 40- to 64-pin range.

Compare the microprocessors shown in Figs. 4.5 and 4.6. Figure 4.5 shows the simplified block diagram of a typical microprocessor used primarily in computer applications (where the equipment operates on a program). The microprocessor of Fig. 4.6 shows a control-type microprocessor IC_{402} used to control operation of the circuits in a CD player.

Note that neither of these diagrams shows how the microprocessor performs the functions (in the same sense that a TV set block diagram shows the function of the circuits). The diagram shown in Figs. 4.5 and 4.6 are typical for microprocessor literature.

In the case of microprocessors used in computer applications, you must consult the instruction set to find what the microprocessor does in response to commands. In control-type microprocessors, you must consult the service literature to discover what outputs are produced by the microprocessors for a given input or inputs.

Do not be surprised if the service literature does not give this information for all microprocessors. In fact, do not be surprised at what you find, or do not find, in the service literature of microprocessor-based equipment.

This section has no detail concerning operation of the microprocessors shown in Figs. 4.5 and 4.6. Operation of a typical computer, using a microprocessor similar to that of Fig. 4.5, is described in Sec. 4.5. Operation of many control-type microprocessors is described in Chapter 9.

4.4.2 Memories

As discussed, the most common types of memories used in microprocessor-based equipment are the ROM and RAM. Typically, each of these memories is contained in a separate IC package. However, there are microprocessor-based systems where one or both the memories are contained in the same IC as the microprocessor.

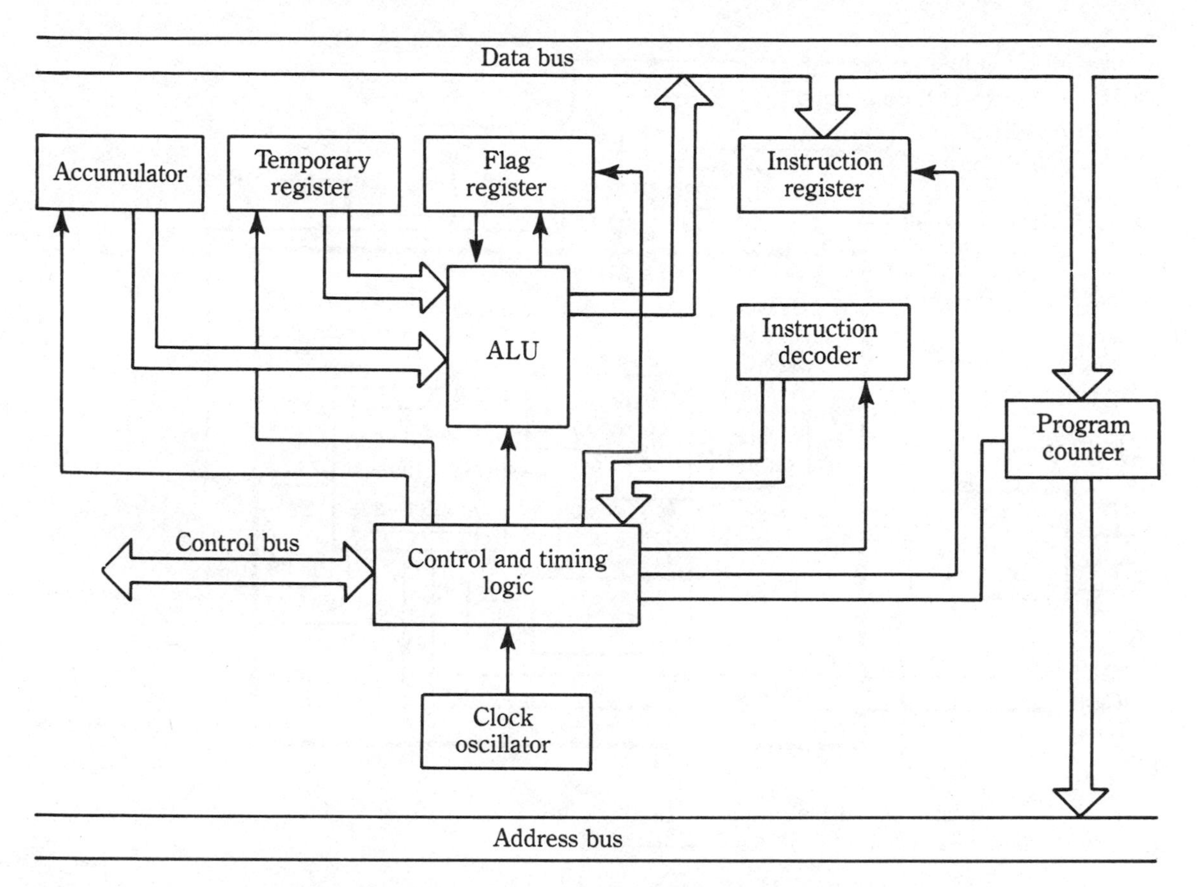

Fig. 4.5 Microprocessor used in computer applications.

4.4.3 RAMs

Figure 4.7*a* shows the functional block diagram of a typical RAM, described as a 128×8 static RAM. (This means that the matrix has 128 addresses, each with eight bits, so eight-bit bytes can be read into, or out of, 128 locations.)

The data bytes appear on the data lines D0 through D7 and are usually connected to a data bus. The data byte is read into the matrix if the buffer is in the in state and is read out when the buffer is in the out state. The data lines are disconnected from the matrix if the buffer is in the high-impedance state.

Note that the state of the buffer is controlled by two factors: the read/write signal at pin 16 and the control signals at pins 10 through 15. If any one of the control signals (called *chip-select inputs*) is absent, the buffer remains in the high-impedance state, and no data bits pass between the matrix and data lines. Note that some of the chip-selected (CS) inputs are *active high* (turned on when the line is at binary 1), whereas other inputs are *active low* (turned on at binary 0). Active lows are indicated by the overbar (CS5 and so on). This chip-select arrangement permits one memory IC to be turned on, with other memories turned off, by the same signal on the same address line.

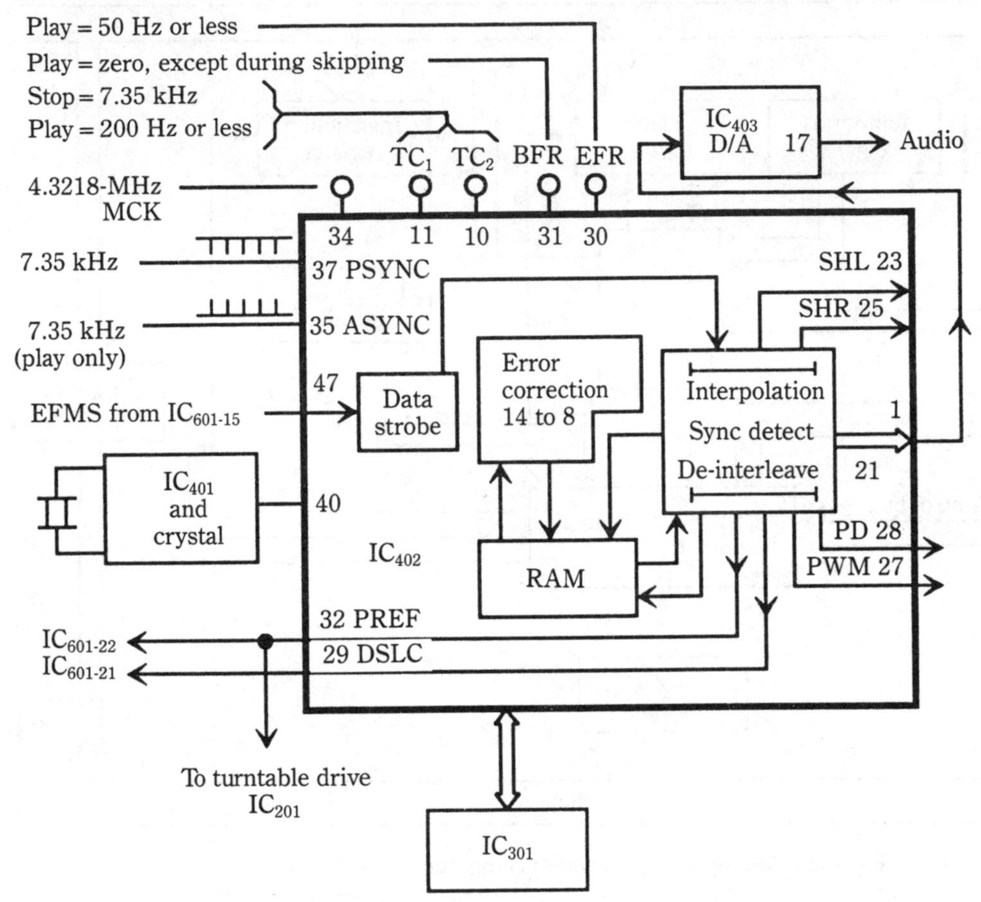

Fig. 4.6 Microprocessor used in control-type applications.

When all of the chip-select inputs are present (all active highs are at 1, all active lows at 0), the buffer can be placed in the read (data into the matrix) or write (data from the matrix to the data line) condition by a signal on the read/write line (pin 16).

In some cases, the address lines are used as chip-select inputs. Figure 4.7*b* shows an example of this where a 128 × 8 RAM is used on the same data and address buses with a 1024 × 8 ROM. The RAM is assigned to lower-number addresses (from 0000 to 0127, in decimal), and the ROM contains the higher-number addresses (49152 to 50175, in decimal). When the highest address line A15 is at 1, the ROM is made active. When A15 is at 0, the RAM is active.

Of course, this system is not used in all microprocessor-based equipment. Figure 4.7*c* shows the block diagram of a 32-word by 8-bit RAM together with the operational modes of the buffer. From this discussion, note that there is no common method to control RAM hardware during the read/write function of a microprocessor-based computer.

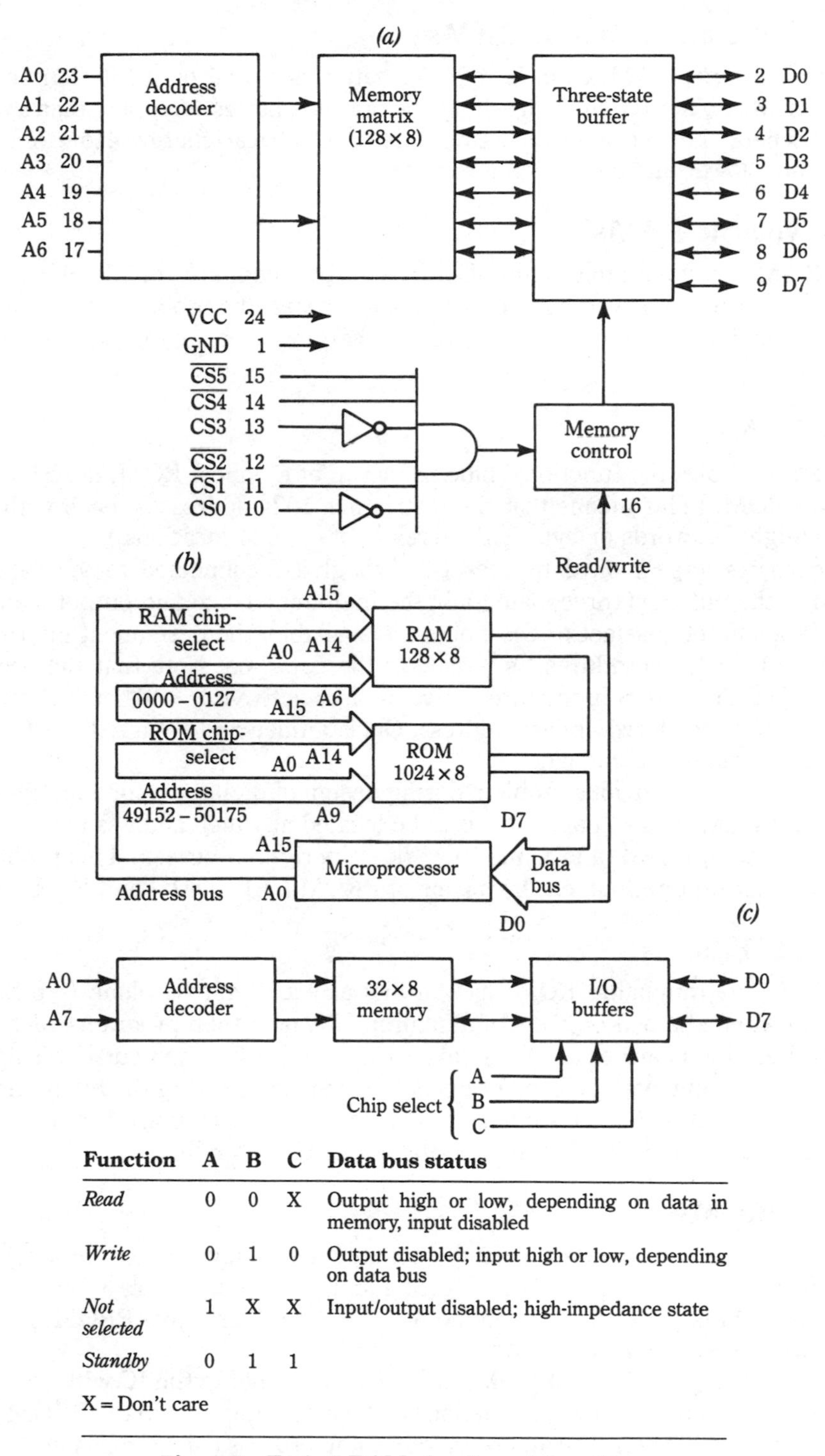

Function	A	B	C	Data bus status
Read	0	0	X	Output high or low, depending on data in memory, input disabled
Write	0	1	0	Output disabled; input high or low, depending on data bus
Not selected	1	X	X	Input/output disabled; high-impedance state
Standby	0	1	1	

X = Don't care

Fig. 4.7 Typical RAM circuits and functions.

4.4.4 Static and dynamic RAMs

In a typical *static* RAM, each bit of information is stored on a flip-flop or latch. Static RAMs do not require refreshing and are thus far less complex than *dynamic* RAMs where the information is stored as an electrical charge. However, static RAMs are slower and consume more power.

4.4.5 Volatile RAMs

Most RAMs used with microprocessor-based equipment are volatile (where information is lost when power is removed). In a few cases, the problem is avoided with a battery-maintained power supply (or by charging a capacitor, as described in Chapter 9).

4.4.6 ROMs

Figure 4.8*a* shows the functional block diagram of a typical ROM, described as a 1024×8 ROM. (This means that the matrix has 1024 addresses, each with eight bits, so eight-bit words or bytes can be read out of 1024 locations.)

The bytes appear on data lines D0 through D7 connected to the data bus. Although the buffer is three state, only the high-impedance and output states are used. When all chip-select or CS inputs are available, the permanent information stored at the selected address is passed to the data bus. Note that the user can define whether the CS inputs are active-high or active-low and must define the binary word to be stored at each address. Once defined and set into the ROM, the information cannot be changed.

ROMs have an obvious problem during design of digital equipment. Typically, you do not know what binary word is to be located at which address until the program is written, tested, and found to work properly in all cases. This problem is overcome during development by means of PROMs, EPROMs, and EAROMs.

4.4.7 PROMs

A *PROM* (programmable ROM) has all bits at each address blank (typically at binary 0) when shipped by the manufacturer. The user then programs each bit at each address by means of an electrical current. Typically, when current is applied to a bit, a nichrome wire is fused or opened by current, making the bit assume the electrical characteristics of a binary 1. Bits to be at binary 0 are left untouched. The program in the PROM is then permanent and irreversible.

4.4.8 EPROMs

Although there are automatic and semiautomatic devices for programming PROMs, there is still the problem of testing and debugging before a final program is obtained. This program is overcome by an *EPROM* (erasable PROM) in which information is stored as a charge in a MOSFET.

As shown in Fig. 4.8*b*, EPROMs are erased by flooding the IC with ultraviolet radiation. Once erased, new information can be programmed into the PROM in the normal manner. The erasure and reprogramming process can be repeated as many

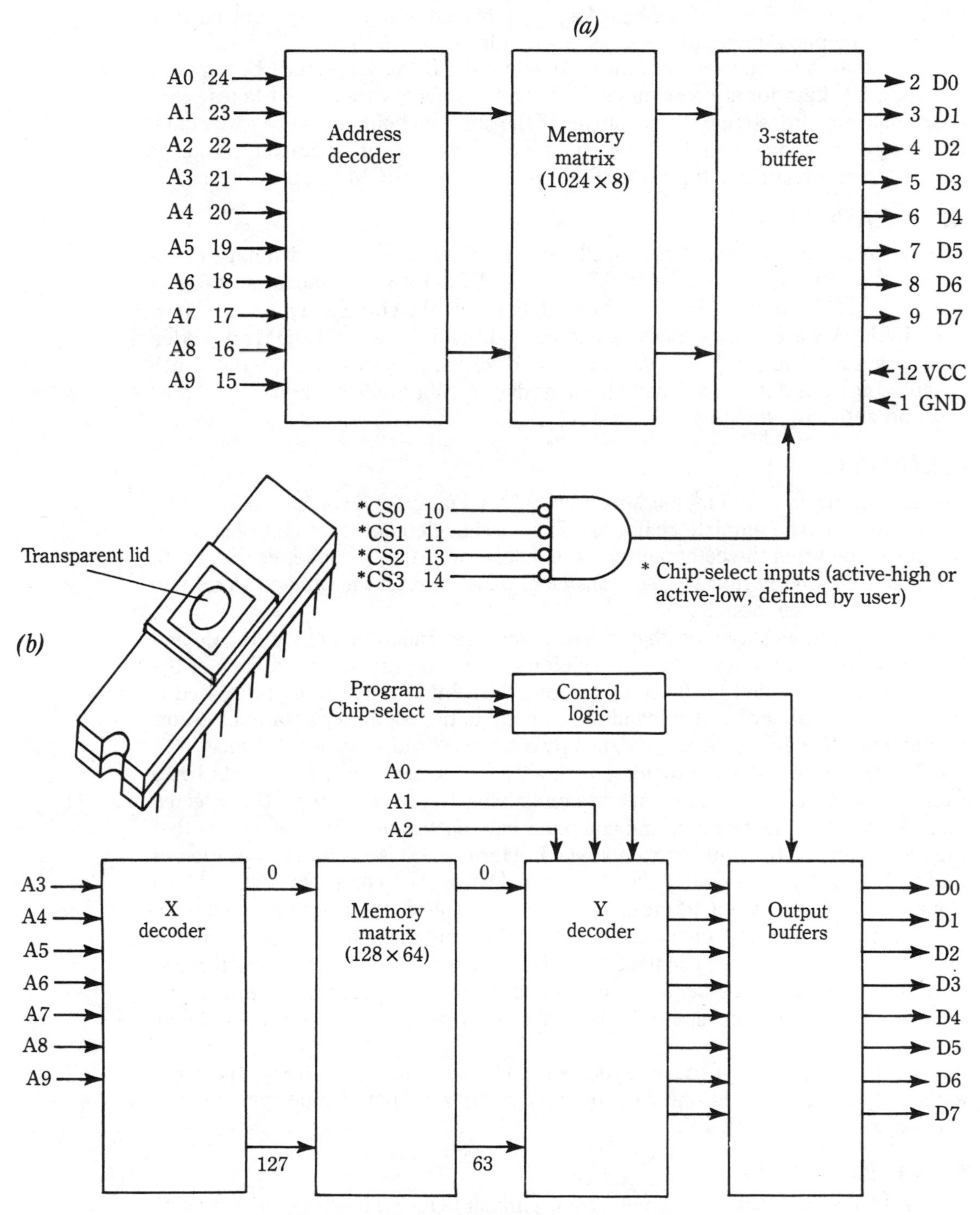

Fig. 4.8 Typical ROM circuits and functions.

times as required. The EPROM package is provided with a transparent lid that allows the memory content to be erased with ultraviolet (*UV*) light.

Note that for proper erasure the chip within the IC package must be exposed to strong UV light for a few minutes. Exposure to ordinary room light takes years to produce erasure, so there is no danger of the program being erased accidentally. Even exposure to direct sunlight does not produce erasure for several days. Except for the erasure feature, a typical EPROM is similar to a ROM in function.

4.4.9 EAROMs

The *EAROM* can be erased electrically while in the circuit. One advantage of the EAROM (electrically alterable ROM) over the EPROM is that small sections of the EAROM can be erased, whereas EPROMs must be completely erased. However, EAROMs are more expensive and more difficult to use. EAROMs are often used in systems that require data to be stored for long periods of time, but which must change this data occasionally (such as digital TV tuners, calibrated transducers, and automatic telephone dialers).

4.4.10 I/Os

The microprocessor and memories described thus far can be connected to form a simple and almost complete computer. The missing element is an input/output or I/O device between the computer and the outside world, or peripherals (Sec. 4.7). Without some I/O hardware, the transfer of data between the outside world and the computer is impossible.

As an example, assume that a basic computer (microprocessor, ROM, and RAM) is used with a keyboard and video terminal, and that the data lines of the terminal are connected to the computer data bus. Information to be processed is typed on the keyboard and transmitted directly to the computer data lines. After processing, the data bytes are returned directly to the video-terminal display.

There are three basic problems with such an arrangement. First, data bytes from the keyboard can easily appear simultaneously with data from the selected memory. Second, the terminal and computer have no way of telling each other that they are ready to transmit or receive data. Third, there is no synchronization of timing between the computer and terminal. That is, the computer and terminal clocks might be operating at different frequencies, phase relationships and so on.

If the video terminal operates with serial data and the computer required parallel data, the problem of compatibility is increased further. Likewise, there is always the problem of *interfacing* electronic devices (to accommodate different voltage levels, impedances, etc.). These and other problems are overcome by an I/O device.

Most microprocessor manufacturers supply one or more I/O ICs in their systems. Typically, there is one I/O for interfacing with parallel peripherals and another for serial peripherals.

4.4.11 Parallel I/O

Figure 4.9*a* is the block diagram of a basic parallel I/O IC. The circuit is described as an eight-bit I/O port and consists essentially of an eight-bit register and eight-bit

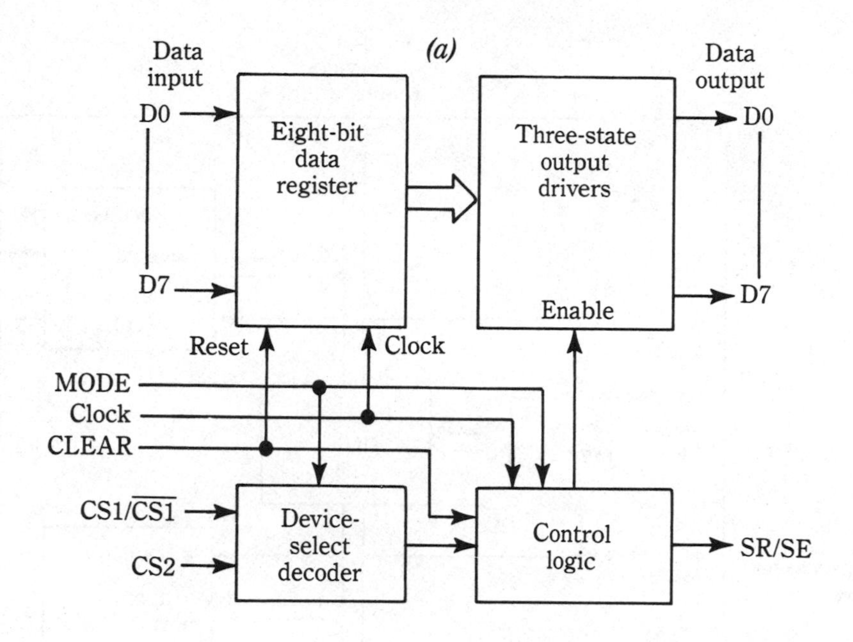

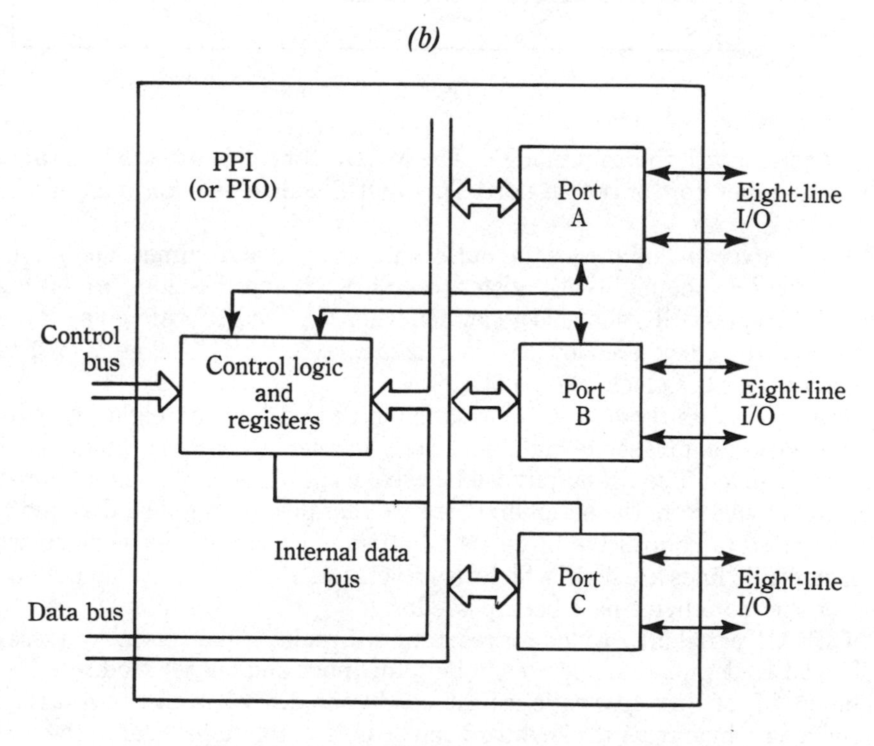

Fig. 4.9 Typical I/O circuits and functions.

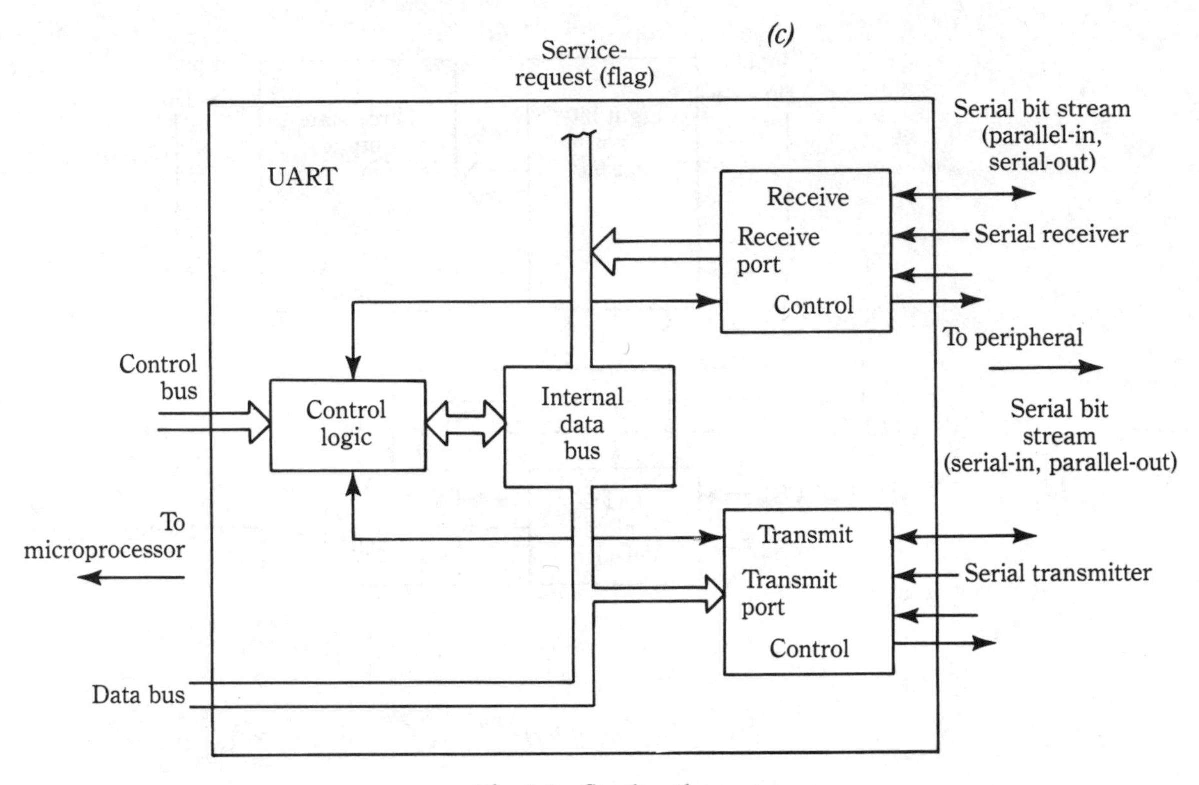

Fig. 4.9 Continued

buffers together with the control logic. The MODE control is used to program the device as an input port or output port. The MODE control is 0 for input and 1 for output.

When used as an output port, the buffers are enabled at all times, and information is passed into the eight-bit register (if CS1, CS2, and the clock are all high). The service request (SR) signal is generated when CS1 and CS2 are going low and remains until the clock goes low (SR is generated when CS1/CS2 go from 1 to 0, and remains until the clock goes to 0).

When used as an input port, information is passed into the eight-bit register when the clock-line pulses are high. The clock also sets the SR circuit and latches data in the register. The SR output can be used to signal, or flag, the microprocessor that data bytes from the peripheral are ready for processing. The CS1 and CS2 inputs are used to control the three-state buffers. The buffers are enabled when the CS1 and CS2 lines are high, which also resets the SR circuits (to flag the microprocessor that data bytes have been passed).

A CLEAR signal is provided for resetting the register and service-request circuit. The CLEAR function operates in both the input and output modes.

The circuit of Fig. 4.9*a* satisfies the conditions described in Sec 4.4.10. For example, a data byte from the keyboard can be held in the register until the microprocessor is ready to accept new data. This condition is signaled, or flagged, to the microprocessor by the SR line. Then the buffers are enabled by the microprocessor, and the data byte is passed to the computer system. The buffers are set to the high-impedance state, and the registers are reset.

After processing in computer, the data byte is placed in the registers, and this condition is flagged to the video terminal by the SR line. When the terminal is ready to accept data, the buffers are enabled by the terminal, the byte is passed to the display, the buffers are again returned to the high-impedance state, and the registers and SR line are reset.

4.4.12 Serial I/O

Serial I/O circuits are generally more complex than those of the basic parallel I/O circuit described in Sec. 4.4.11. In addition to all the control, timing, and data-transfer functions, a serial I/O must also convert from parallel to serial and vice versa for each exchange of information between the peripheral and computer system.

4.4.13 PPIs

The *parallel peripheral interface*, or PPI, also called a *parallel input output*, or PIO, is one of many IC I/Os used in digital equipment. These ICs are commonly used in microprocessor-based equipment to simplify many interfacing tasks, reduce component count, and increase system cost-effectiveness.

Figure 4.9*b* shows a typical PPI that contains three I/O ports. Each port is used either as an input or output port. The direction of each port is controlled by a control register on the IC. An initialization program, contained in the system ROM, sets the control register to select the desired combination of input and output ports.

Parallel peripheral interface devices have the advantage of providing several ports in one IC. Because the nature of each port may be changed by programming, PPIs are also very flexible. PPIs usually include some control logic for synchronizing communication and interrupt control.

4.4.14 UARTs

Figure 4.9*c* shows a typical UART, or *universal asynchronous receiver and transmitter*. The UART is a common type of interface IC that provides serial inputs and outputs, and is also known as a serial input output (SIO) or asynchronous communications interface adapter (ACIA).

UARTs accept a byte of data from the microprocessor and then output the byte one bit at a time. In this application, UARTs are essentially a parallel-in/serial-out shift register. Also, start, stop, and other synchronization bits can be automatically inserted by the UART. The format is controlled by registers similar to these described for the PPI (Sec. 4.4.13).

UARTs can also handle data in the other direction and convert a serial bit stream into parallel form suitable for direct use by the microprocessor. In this mode, UARTs are essentially a serial-in/parallel-out shift register.

Note that serial I/O is most commonly used for communication between a microprocessor-based system and a peripheral (Sec. 4.7), such as a video terminal or modem. Because the information is in a serial format, only two wires are needed to interconnect the devices.

4.5 A basic computer system

As discussed, not all microprocessors are used in computer applications. For example, all of the control-type microprocessors covered in Chapter 9 have no programs as such. However, most present-day computers (virtually all personal computers or PCs) use some form of microprocessor. So, before going into the details of troubleshooting, let's consider a microprocessor and related ICs operating as a typical computer system. This is difficult because there are many different types of microprocessors and even more computer configurations.

The computer described here contains the basic principles on which all programmed microprocessor-based systems operate. Once you understand how this simplified equipment operates, you will be better able to understand the need for specialized troubleshooting equipment and techniques described throughout this book.

The next section starts with a quick review of basic programming. Although this is not a programming book, it is essential that you have some knowledge of programming to troubleshoot any programmed device.

4.5.1 Programming fundamentals

Programs are first written in a way that is convenient for the person writing the program (the *programmer*). The program is then rewritten and stored in a code that the microprocessor understands. The microprocessor then reads the codes from memory, one at a time, and performs the indicated operations.

Microprocessor-based systems are often used to replace circuits composed of standard logic devices. To show the differences between a programmed logic device and a conventional logic device, consider using a microprocessor as a basic AND gate (Chapter 2).

A microprocessor-based AND gate requires an input port for the gate inputs and an output port for the gate output, as shown in Fig. 4.10*a*. Using instructions stored in memory, the microprocessor performs the AND function. Because an AND gate has only one output, only one bit of the port is needed.

The microprocessor must be programmed to perform the basic AND-gate function. A list of instructions for such a program is as follows:

1. Read the input port.
2. Go to step 5 if all inputs are high; otherwise continue.
3. Set output to low.
4. Go to step 1.
5. Set output to high.
6. Go to step 1.

First, the input port is read. Then the inputs are examined to see if they are all high, since that is the function of an AND gate. If the inputs are all high, the out-

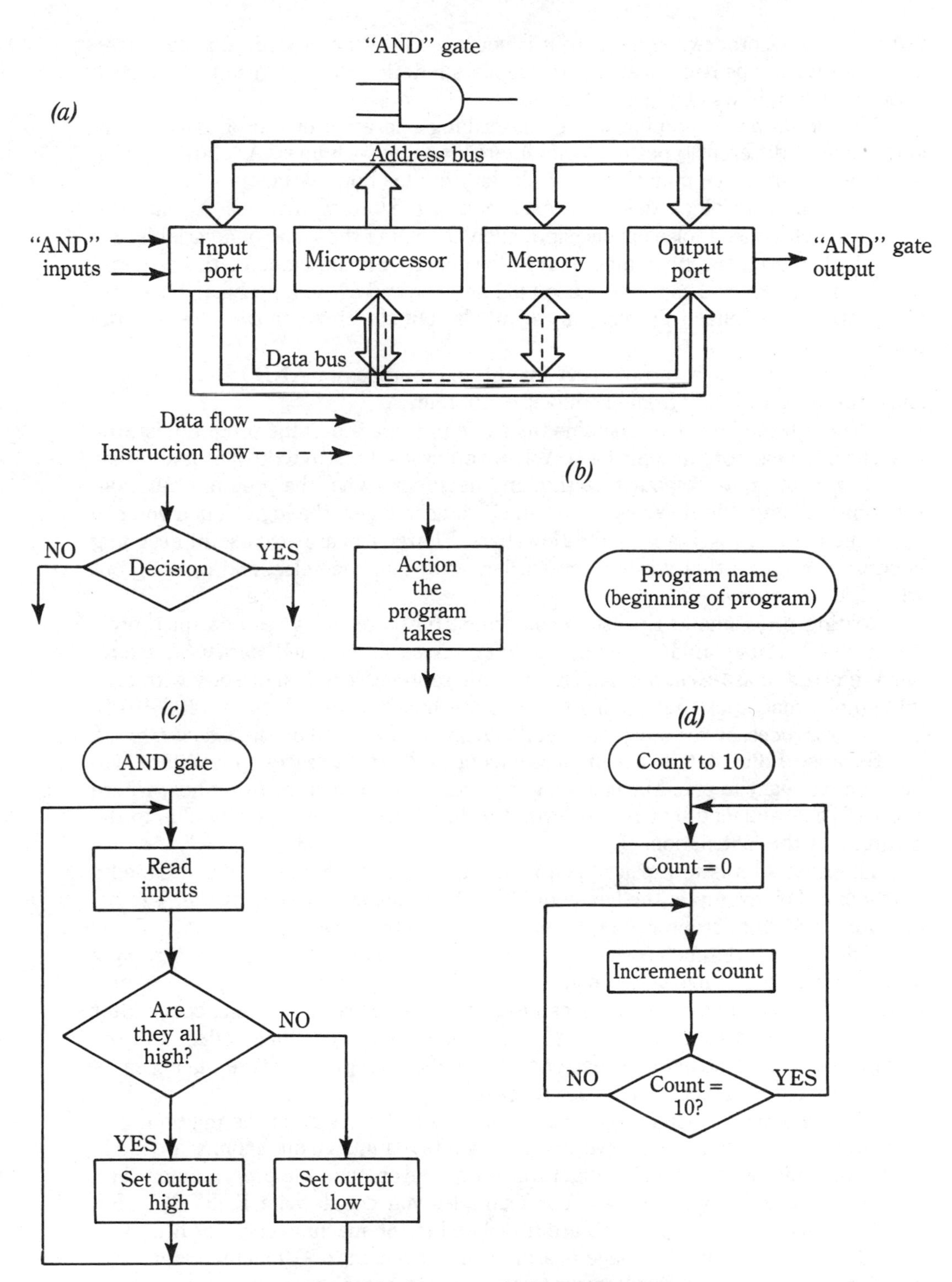

Fig. 4.10 Programming fundamentals.

put is set high; otherwise, the output is set low. Once the procedure is complete, the program jumps back to step 1 and repeats indefinitely. As a result, the output continuously follows changes in the inputs.

Flowcharts are a graphic way of describing operation of a program and are composed of different types of blocks interconnected with lines. As shown in Fig. 4.10*b*, there are three major types of blocks used for flowcharts.

A rectangular block describes each action the program takes. A diamond-shaped-block is used for each decision, such as testing the value of a variable. An oval block marks the beginning of the flowchart with the name of the program placed inside. An oval can also be used to mark the end of the flowchart. There are many other specialized flowcharting symbols, but the three shown here are the most common.

Figure 4.10*c* shows a flowchart for the microprocessor AND gate. The lines show the program flow from one block to another.

Although the flowchart contains the same information as the program list, the flowchart is in a more graphic form. When you first set out to write a program, use a flowchart to organize your thoughts and document what the program must do. By going through the flowchart "by hand," you can check the logic. Then you can write the actual program from the flowchart. Flowcharts are also useful for going back to a program that was written earlier, or by someone else, and figuring out what the program does.

Writing programs in English is convenient but is meaningless to a microprocessor. The language understood by the microprocessor is called *machine language* (also referred to as *machine code*). Because microprocessors deal directly with digital signals, machine language instructions are binary codes (such as 1100 1010). The microprocessor recognizes a specific group of codes called the *instruction set*.

Because 1100 1010 is meaningless except to the microprocessor, machine language is not easy to use. Machine language can be simplified by using hex, where 1100 1010 is replaced by CA. However, this still does not provide any clue to the meaning of the instruction.

The next step is to replace each instruction code with a short name called a *mnemonic*. For example, the hex code 3C, which means "increment the A register" for an 8085 microprocessor, is represented by the mnemonic INR A.

Mnemonics are much easier to remember than machine codes. By assigning a mnemonic to each instruction code, you can write programs using mnemonics instead of codes. The mnemonics can easily be converted to machine codes after the program is written. So you need not remember the machine codes, and the meaning of each instruction is easier to remember. Programs written using mnemonics are called *assembly-language programs*.

The machine language is generally determined by design of the microprocessor and cannot be changed. However, the assembly-language mnemonics are made up by the microprocessor manufacturer as a convenience for programmers and have nothing to do with design. For example, you could write BESTSELLER instead of INR A as long as both are translated to the machine code for hex 3C.

Although assembly language is a great improvement over machine language, assembly language is difficult to use when writing complex programs. Computer languages, such as BASIC, are used to make programming simpler. These computer languages are similar to English and are generally independent of any particular microprocessor. A typical computer instruction might be "PRINT COUNT" or "LET COUNT = 10". Computer languages are sometimes called *high-level languages*.

Computers using languages such as BASIC contain long, complex programs (permanently stored in memory) that translate the computer-language program into a machine-language program. A single computer instruction might translate into dozens of machine-language instructions. Such translator programs are called *compilers*.

The following is an example of microprocessor programming. Figure 4.10*d* shows the flowchart for a program that counts to 10. There is no input or output in this program. Instead, the contents of a designated memory location counts from 0 to 10 and repeats.

Using BASIC The translation from a flowchart to a computer language is relatively easy. The following example uses a variation of BASIC.

Figure 4.11*a* shows the program listing in BASIC. The first two lines of the program correspond exactly to the first two action blocks of the flowchart (Fig. 4.10*d*). In the first line, the memory location called COUNT is set to zero. The second line, LET COUNT = COUNT + 1, is another way of saying "increment the count".

Lines three and four perform the function of the decision block (Fig. 4.10*d*), then the next instruction executed should be line one. If COUNT does not equal 10, then the instruction in line three has no effect, and the program continues with line four; the instruction says "go to line two." So these two instructions perform the actions required by the decision block in the flowchart. Try following the program, step-by-step, to see the flow as the count reaches 10.

Using assembly language In reality, assembly language is not one specific language but a class of languages. Each microprocessor has its own machine language and its own assembly language (as defined by the manufacturer).

Figure 4.11*b* shows the assembly language listing (of an 8085 microprocessor) for the count-to-10 program. Note that this program performs the same function as the BASIC program of Fig. 4.11*a*, but is somewhat more difficult to understand.

Also remember that assembly-language characteristics are related to the microprocessor characteristics. So the program of Fig. 4.11*b* is different from the BASIC program (which is related to English rather than to the microprocessor machine language).

The *label* column of Fig. 4.11*b* provides the same function as the line number in the BASIC listing. Instead of numbering every line, the programmer simply makes up a name (called a *label*) for each line to which a reference is required. A colon (:) is used to identify the label. A line needs a label only if there is another instruction in the program that refers to that line. The label makes it easy to identify a line that is to be referenced (or "jumped" to) during program execution.

The *comments* in Fig. 4.11*b* are an aid to understanding the program. A semicolon (;) identifies the beginning of a comment. Computer programs do not need many comments because the instructions themselves are more descriptive. How-

(a)

LINE NUMBER	INSTRUCTION	DESCRIPTION
1	Let count = 0	Set count to 0
2	Let count = count + 1	Increment count
3	If count = 10 then 1	Go to 1 if count = 10
4	Go to 2	Otherwise go to 2

(b)

LABEL	INSTRUCTION	COMMENTS
Start:	MVI A.0	;Set A register to 0
Loop:	INR A	;Increment A register
	CPI 10	;Compare A register to 10
	JZ start	;Go to beginning if A = 10
JMP loop ;Repeat		

(c)

MEMORY ADDRESS		MEMORY CONTENTS
Hex	Opcode	Binary
07F0	3E	00111110
07F1	00	00000000
07F2	3C	00111100
07F3	FE	11111110
07F4	0A	00001010
07F5	CA	11001010
07F6	F0	11110000
07F7	07	00000111
07F8	C3	11000011
07F9	F2	11110010
07FA	07	00000111

(d)

BASIC Line No. Instruction	ASSEMBLY Label Instruction	MACHINE Address contents
1 Let count = 0	Start:MVI A.0	07F0 3E Opcode
		07F1 00 Data
2 Let count = count + 1	Loop:INR A	07F2 3C Opcode
3 If count = 10 then 1	CPI 10	07F3 FE Opcode
		07F4 0A Data
	JZ start	07F5 CA Opcode
		07F6 F0 Address
		07F7 07 Address
4 Go to 2	JMP loop	07F8 C3 Opcode
		07F9 F2 Address
		07FA 07 Address

Fig. 4.11 Using BASIC in microprocessor programming.

ever, comments are an invaluable aid in assembly-language programs. Comments are useful to people other than the programmer (such as the troubleshooter) who need to understand the program. Comments also help the programmer who must go back to a program at a later data.

The first instruction in Fig. 4.11*b* is MVI A.0 (move immediate to accumulator the data 0). The *accumulator* (also called the A register) is a storage location inside the microprocessor. The instruction MVI A.0 is the equivalent of LET COUNT = 0, except that a preassigned name (A) is used for the register instead of making up a name for the variable (COUNT). There is more discussion of the accumulator (or register A) in Sec. 4.5.3.

The next instruction in Fig. 4.11*b* is INR A (increment the value in the accumulator). The accumulator contains this count, so this is the equivalent of LET COUNT = COUNT + 1.

The next three instructions together implement the decision function of the program (shown as the diamond-shaped block in Fig. 4.10*d*).

The instruction CPI 10 (compare immediate) in Fig. MS 4.11*b* means "compare the value in the accumulator with the value 10." CPI 10 does not directly cause any jumps, regardless of the comparison. Instead, CPI 10 sets a special flip-flop (called a *flag*) in the microprocessor if the value in the A register (accumulator) is equal to 10.

The next instruction JZ START in Fig. 4.11*b* tests the flag. If the values are equal, JZ START detects that the flag (a flip-flop) is set and causes a jump to the line with the label START (the first line in Fig. 4.11*b*). These two instructions together (CPI 10 and JZ START) perform the function of the BASIC statement IF COUNT = 0 THEN 1.

The last instruction in Fig. 4.11*b*, JMP LOOP, causes a jump to the line with the label LOOP (second line), JMP LOOP is the equivalent of the BASIC statement GOTO 2.

Using machine language As the last step in this example, Fig. 4.11*c* shows a listing of the machine language that corresponds to the assembly-language program of Fig. 4.11*b* (as well as the flowchart of Fig. 4.10*d* and BASIC listing of Fig. 4.11*a*). All of the programs are based on using the 8085 microprocessor (which is discussed further in Sec. 4.5.3).

Figure 4.11*d* compares the three programs. Each memory location holds eight bits of data (which can be represented by two hex characters). Each instruction begins with an *opcode* (short for operation code). The opcode specifies the operation to be performed. All 8085 opcodes are eight bits (one byte) each, and thus occupy one memory location. In this program, an opcode may be followed by none, one, or two bytes of data, depending on the instruction.

The first byte (3E) at address 07F0 is the opcode, for instruction MVI A.0. The first byte is only part of the complete instruction, and specifies that you want to move some data into the accumulator. You now need another memory location to specify this data, so the next memory location (address 07F1) contains 00 (the data to be moved to the accumulator).

The third location (07F2) contains the opcode for the second instruction, INR A. This opcode (3C) tells the microprocessor to increment the accumulator. Because there is no additional information associated with this instruction, the instruction occupies only one memory location.

The code FF (at location 07F3) is the opcode for the compare instruction, CPI. Just as with the MVI A.0 instruction, the memory location (07F4) that follows the opcode contains data required by the instruction. Because the machine language is shown in hex, the data (10 decimal) appears at 0A (hex). This instruction compares the value in the accumulator with the value 10 and sets a flag (indicating that the two are equal).

The JZ instruction (07F5) has the opcode CA that tells the microprocessor to jump if the flag is set. The next two memory locations (07F6, 07F7) tell the microprocessor to what address it must jump. Because addresses in an 8085 are 16 bits long, it takes two memory locations (eight bits each) to store an address.

The two parts of the address are stored in the reverse order. That is, the least significant half is stored first and then the most significant half. So the address 07F0 is stored as F0 07.

The assembly-language instruction JZ START means that the microprocessor should jump to the instruction labeled START on line 1 (07F0).

The last instruction, JMP loop, is coded the same way (with the address in reverse order). The only difference is that this jump is independent of any conditions (an *unconditional jump*). The code for such an unconditional jump is C3. (Compare this to CA at 07F5, which is a *conditional jump* and occurs only if the flag is set.)

Note that the machine-language program consists of a series of bytes, each of which might have one of three meanings. Some bytes are opcodes, some are data, and some are jump addresses. You must know the context of the information to know which type of byte is involved. Circuits within the microprocessor determine if a particular opcode should be followed by data or an address, so the microprocessor can keep track of the three types.

Comparison of the programming languages As can be seen from the discussion, each of the languages (computer or high-level, assembly and machine) has certain advantages and disadvantages. The following summarizes the programming language characteristics.

Machine language is the only language directly understood by the microprocessor, but people have a hard time with machine language. It is clumsy, at best, to program directly in machine language. Programs are usually written in assembly language and then translated to machine code. The translation may be performed by a special program called an *assembler*.

Assembly language is widely used for programming computers. It is more difficult to write programs in assembly language than in a computer language. However, it is much easier to translate from assembly to machine language than from computer to machine language.

Assembly language is usually the best choice when programs must run very fast or must fit into a small memory. Assembly language also gives you a much better idea of how the microprocessor system works.

Computer or high-level languages are the easiest for programmers, and they can be independent of any particular microprocessor. However, lengthy translation programs (compilers) must be stored in computer memory to translate programs to machine code. Computer languages are also less efficient in terms of operating speed and memory use. An equivalent program written in assembly language normally runs faster and uses less memory space than in computer language.

4.5.2 Interrupts

Sometimes a microprocessor-based system must react to events that are very infrequent and unpredictable. For example, consider a microprocessor-based voltmeter with a CALIBRATE button on the front panel. When CALIBRATE is pressed, the microprocessor stops whatever program is in process and jumps to a calibration *subroutine*. After the calibration is performed, control returns to the program that was interrupted.

This function is known as an *interrupt*. Many microprocessors have inputs that cause interrupts. (The microprocessor of Fig. 4.5 has five interrupt inputs.) A high on any of these inputs causes the microprocessor to stop and jump to a selected subroutine. The subroutine performs the task required by the interrupting device. A return instruction at the end of the interrupt routine causes the program flow to return to the interrupted program.

4.5.3 Microprocessor operation during a program

Throughout most of this book, the microprocessor is treated as a *black box*, a device with known characteristics whose internal structure is of no concern. However, you must have some knowledge of what goes on inside the microprocessor to understand operation of a basic computer system.

Internal registers The microprocessor of Fig. 4.5 contains a number of internal registers used for various purposes. Some registers, such as the accumulator, are used for storing and manipulating data. For example, in Sec. 4.5.1, the accumulator is used to store the number of counts in the count-to-10 program.

The contents of the accumulator can also be modified (in this case, incremented). Keep in mind that internal registers are within the microprocessor and are not selected by the address bus as are the external ROM and RAM.

Internal registers are selected directly by a particular instruction (as an example, MVI A selects the accumulator). The control logic within the microprocessor controls the registers directly (without using buses). As a result, internal registers are more convenient for temporary or intermediate data storage.

Program counter Some internal registers are not used for general-purpose storage, but have a very specific function. The program counter is the most important of such registers. In the microprocessor of Fig. 4.5, the program counter is a 16-bit register that keeps track of the instruction being executed.

When an instruction is read from memory, the contents of the program counter are placed on the address bus. The addressed instruction then appears on the data bus. After the microprocessor reads the instruction, the program counter is incremented, the contents of the program counter are again placed on the address bus, and the next instruction is read. This process continues unless a jump instruction is executed (which causes the program counter to be loaded with the jump address).

Reading an opcode Figure 4.12*a* shows the relationship of the internal registers to the external when reading the opcode from memory for an MVI A instruction (in the microprocessor of Fig. 4.5). Before you get into this instruction (which is a classic instruction for most computers), look at the register functions.

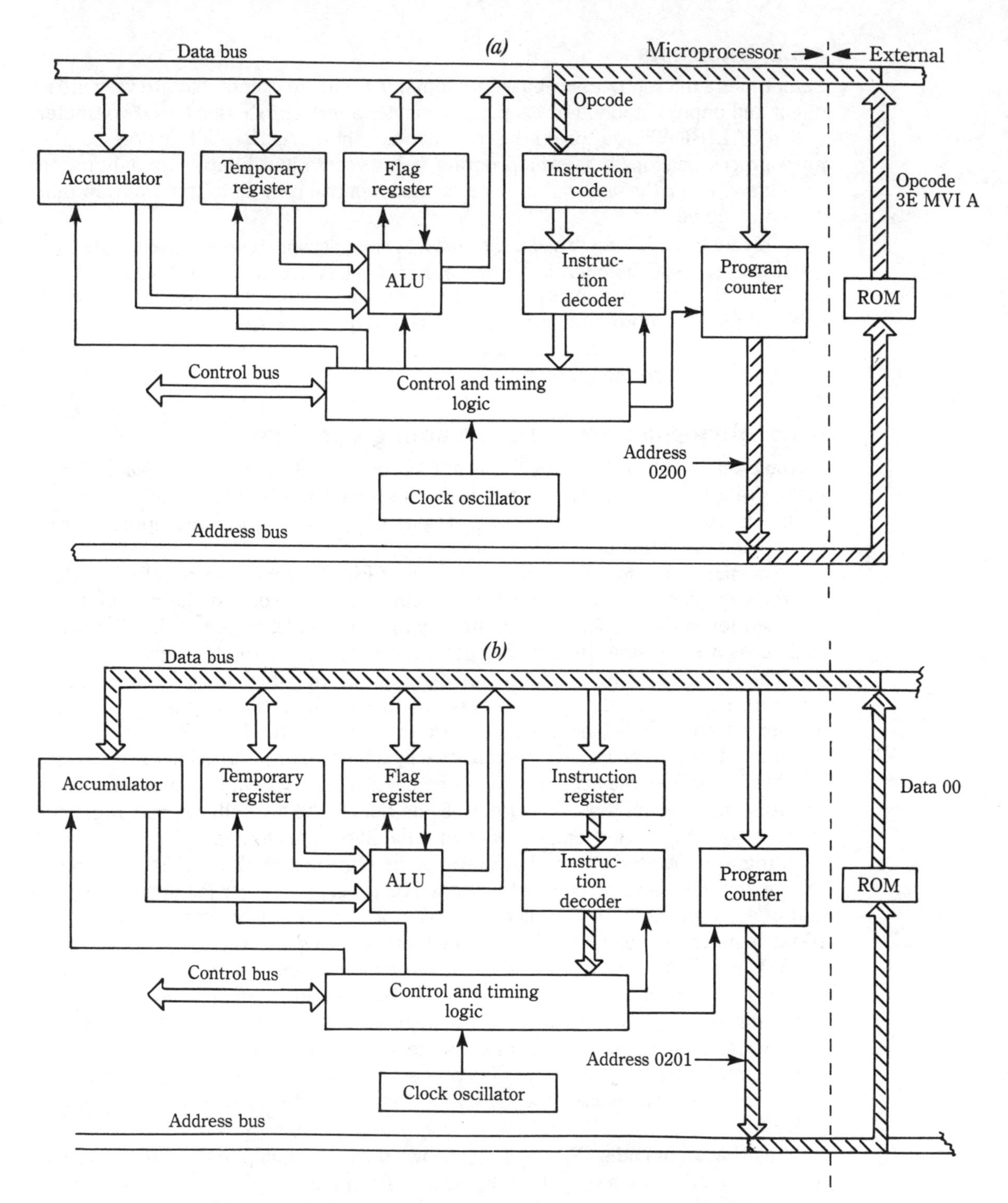

Fig. 4.12 Reading opcodes and data from an external ROM.

The ALU performs all data manipulation, such as incrementing a number or adding two numbers.

Feeding the other input of the ALU, the *temporary register* is invisible to the programmer and is controlled automatically by the microprocessor control circuits.

The flag register is a collection of flip-flops that indicate certain characteristics of the results produced by the most recent operations performed in the ALU. For example, the *zero flag* is set if the result of an operation is zero. The zero flag is tested by the JZ instruction.

The *instruction register, instruction decoder, program counter*, and *control/timing logic* are used for *fetching* instructions and directing the execution of the instructions.

Typical instruction cycle As an example of a typical instruction cycle, suppose that an instruction is about to be read from location 0200. First, the opcode is read from memory; this is the *instruction fetch*. The program counter, which contains the address 0200, is output to the address bus and causes memory location 0200 to be selected. The ROM then places the contents of location 0200 (presumably an opcode) on the data bus, and the microprocessor stores the opcode in the instruction register.

The instruction register feeds the instruction decoder; the decoder recognizes the opcode and provides control signals to the control/timing logic. The control/timing circuits are somewhat like a microprocessor within a microprocessor.

A ROM within the microprocessor (not the external ROM) contains a *microcode* (or *microprogram*) that tells the microprocessor exactly how to execute each machine-language instruction. The microcode, which is part of microprocessor design and generally cannot be changed, defines the microprocessor machine language. Writing microcode (which is usually done by the microprocessor manufacturer) is called *microprogramming* and should not be confused with writing programs to be executed by the microprocessor.

Reading data from memory Figure 4.12*b* shows the relationship of the internal registers to the external ROM when reading data from memory for an MVI A instruction. In this step, the control/timing logic first reads the opcode, 3E, and then increments the address in the program counter. The instruction decoder determines that this opcode is followed by a byte of data, so the contents of the memory location pointed to by the program counter are read into the accumulator.

The microprogram then indicates to the control/timing logic that the instruction is completed. The program counter is incremented, and the next byte of program (the next opcode) is read into the instruction register. The execution of this instruction then begins. This repetitive sequence performed by the microprocessor is generally called the *fetch-execute cycle*.

Execute phase The real work is done during the execute phase of the instruction. There are four basic types of operations that can be performed by the microprocessor shown in Figs. 4.5 and 4.12.

1. Read data from memory or an input port.
2. Write data to memory or an output port.

3. Do an operation internal to the microprocessor.
4. Transfer control to another memory location.

The first two types of operations are self-explanatory. The third, internal operations involves manipulation of the registers (such as the accumulator) without accessing the memory or I/O ports. For example, the contents of one register can be moved to another register, or the contents of a register can be incremented or decremented. The fourth group of operations includes such instructions as JMP, CALL, and RET.

In a typical computer application, the microprocessor keeps reading in sequence through the memory, one location after another, performing the indicated operations. Exceptions to this occur when a jump, call, or return instruction is executed. Another exception is an interrupt. Any of these events causes the microprocessor to interrupt the normal in-sequence flow and begin executing instructions from another address, as discussed in Sec. 4.5.2.

Machine-cycles The fetching and execution of instructions is divided into machine cycles. (The timing of these machine cycles is discussed in Sec. 4.6.) A machine cycle consists of setting the address on the address bus and then transferring information over the data bus.

Opcode fetch The first machine cycle of every instruction is the opcode fetch. An additional machine cycle is then required for each memory or I/O reference to provide time for the data transfer.

Most operations inside the microprocessor (such as incrementing the accumulator) are completed in the same cycle as the opcode fetch. A simple instruction such as INR A requires only one cycle, but STA requires four cycles (three to read the instruction and one to write the accumulator data into memory).

Note that opcodes and data are intermixed in memory. One address might contain an opcode, the next two a jump address, the next an opcode, and the next a byte of data. It is the programmer's responsibility to be sure that memory contains a valid sequence of opcodes and data. The microprocessor can distinguish between the information only by context.

The opcodes, jump addresses, and data are all bit patterns stored in memory. All such information is read in exactly the same way and all travel over the same data bus. The microprocessor must keep track of whether an opcode or data is being read and treat each appropriately.

The microprocessor assumes that the first location read contains an opcode and goes from there. If the opcode requires a byte of data, the microprocessor "knows" (from the instruction decoder) that the next byte is data and treats that byte accordingly. The microprocessor then assumes that the byte following the data is the next opcode. If a data byte is misinterpreted as an opcode, the system usually goes completely out of control (*crashes*).

4.5.4 DMA

Some, but not all computer systems have *direct memory access*, or DMA. This function, which is built into the microprocessor, allows data to be transferred between peripherals and memory without interference to the microprocessor.

DMA is often referred to as a method for speeding up data movement between elements of the computer, and allows fast peripherals (Sec. 4.7) access to the memory without taking some microprocessor program time. How much time is required depends on the DMA system used.

In a typical microprocessor with DMA, two DMA lines are provided for special types of byte transfer between memory I/O devices or peripherals. For example, activating a DMA-IN line causes an input byte to be immediately stored in a memory location without intervention by the program. Activating a DMA-OUT line causes a byte to be immediately transferred from memory to the requesting peripheral or I/O circuit. A register within the microprocessor is used as a DMA pointer.

Peripherals and I/O devices can cause data transfer by activating a flag line, an interrupt line or a DMA line. The flag lines must be sampled by the program to determine when flags become active (a request for service by the peripheral or I/O) and are used for signals that change relatively slowly. Activating the interrupt line causes an immediate microprocessor response, regardless of the program currently in progress, suspending operation of that program and allowing real-time access. Use of DMA provides the quickest response time with the least disturbance of the program.

4.6 Program timing and synchronization

All the computer functions described in Sec. 4.5 must be synchronized. For example, if a data byte is to be entered into memory at a particular address, that (and only that) data byte must be on the data bus when the desired address byte is on the address bus. If the two bytes (data and address) are not synchronized exactly, the data byte is entered at the wrong address (or at no address).

To overcome this problem, registers within the microprocessor are opened and closed at exact time intervals within the microprocessor. The time relationships and synchronization are shown by *timing diagrams*, such as Fig. 4.13.

Clock pulses are shown at the top of the timing diagram. In this case, there are two sets of eight clock pulses, with each pulse representing one bit in an eight-bit byte. As discussed in Sec. 4.5, some functions within the microprocessor are done within the first byte, and other functions require two or more bytes. Note that there is no reference to operating speed on the clock pulse, but that one clock cycle is shown as a time interval 1T. This is known as the *clock period*.

From a troubleshooting standpoint, microprocessor timing can be crucial. For example, as discussed in Sec. 4.8, one of the standard troubleshooting techniques for microprocessor-based systems is to display the microprocessor signal pulses (as many as possible) on a scope or logic analyzer. In effect, the timing diagram of Fig. 4.13 (or a significant portion of the diagram) is displayed, and the time relationships are compared.

As an example, note that the timing pulses TPA and TPB (sent from the microprocessor to control peripherals and I/O circuits) occur at eight-bit (one-byte) intervals. However, TPB occurs 5.5 bits (or 5.5 cycles) after TPA. If either TPA or TPB is absent or abnormal (wrong time relationships, say three bits apart instead

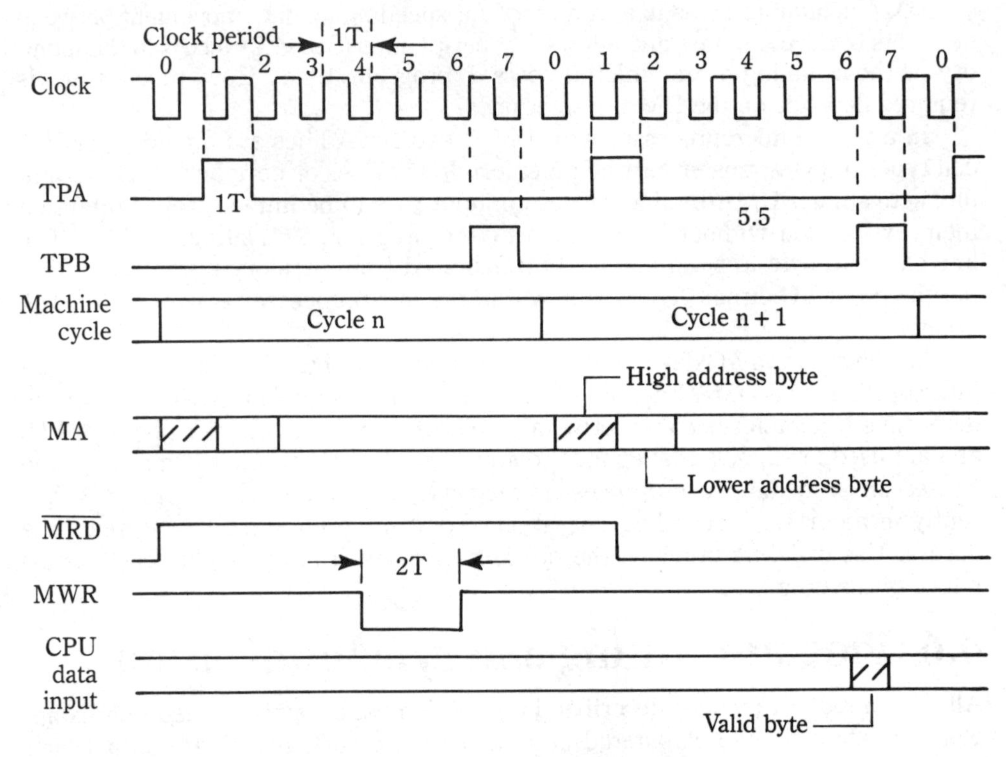

Fig. 4.13 Typical timing diagram.

of 5.5 bits), this pinpoints a fault in the computer system. From this, you can see that it is essential that you understand time relationships of a microprocessor to properly troubleshoot microprocessor-based systems. Fortunately, most computer service literature contains the necessary diagrams.

4.7 Peripherals

This section describes a few peripherals common to microprocessor-based computer systems. These peripherals include keyboards, video terminals, disk drives, CD-ROMs and modems. The primary function of peripheral equipment is to translate human instructions and data into microprocessor language (binary data bytes) and from microprocessor language into a form suitable for readout. In effect, the peripheral equipment (in conjunction with I/O ICs, Sec. 4.4.10 through 4.4.14) reconciles the outer world with the computer.

The most commonly used peripherals in real-time computers (such as the familiar personal computer) are keyboards and video terminals. These are described in Sec. 4.7.1 and 4.7.2, respectively. When the computer does not include any internal magnetic storage, an external disk drive provides for such storage. The drive can be hard disk or floppy disk. The basics of floppy disks are described in Sec. 4.7.3; CD-ROMs are described in Sec. 4.7.4. CD-ROM is becoming more popular as an external storage device. If the computer must communicate with other computers over telephone lines, some form of data communications is required. The modem discussed in Sec. 4.7.5 is now the standard for telephone line data communications.

4.7.1 Keyboards

Figure 4.14 shows the input relationship between a keyboard and a typical computer. With any system, there must be circuits between the keyboard and the computer input that (1) convert data instructions into bytes compatible with the microprocessor circuits and (2) store the bytes until they can be entered into the computer circuits without disrupting normal operation.

The conversion procedures (sometimes called *service routines*) are done by a decoder circuit, and the storage and timing is done by registers and buffers. The present trend is to include the registers and buffer functions in the computer I/O IC, with the decoder function in the keyboard.

Note that the data bytes are transmitted in both parallel and serial form between the keyboard and computer. The parallel method is faster and requires the simplest I/O circuit in the computer (for example, a PPI as shown in Fig. 4.9*b*). However it is also possible to communicate in serial form between the keyboard (or any peripheral) and the computer when a suitable I/O is used (for example the UART of Fig. 4.9*c*).

Serial input/output The serial input and output of many peripherals (including keyboards and video terminals) can be jumper wired for one or more standard data-transmission configurations. (The subject of digital data transmission is discussed further in Sec. 4.7.5.) Three common serial I/O configurations found on peripherals are TTL (Sec. 3.1.4), EIA (Electronic Industries Association), and TTY (Teletypewriter). TTY is also known as *electric typewriter* or *20-mA current loop*.

The two logic states of a serial input or output are called a *mark* and a *space*. For TTL, a mark is defined as a EIA logic 1. For TTL, a mark is a minus voltage. For TTY, a mark is 20 mA flowing in the current loop between the peripheral and the I/O. For TTL, a space is defined as a TTL logic 0. For EIA, a space is a positive voltage. For TTY, a space is zero current in the 20-mA current loop. The idle state of both the input and output is defined as a mark.

Figure 4.15 shows the circuit connections for a jumper-wired digital-data transmission system. These connections are between the UART and the data-transmission line, and (in this case) are located on a video terminal that includes a keyboard.

TTL input/output As shown in Fig. 4.15, a TTL input signal is coupled to pins 9 and 10 of IC_{606}. The output of IC_{606} drives IC_{609D} which, in turn, drives the serial input of the UART at pin 31. The TTL output signal from pin 30 of the UART is applied directly to the data-transmission line through IC_{609A}, IC_{609B}, and terminal B.

EIA input/output In the idle state, the EIA input is a negative voltage that keeps Q_{603} biased off. Because the collector of Q_{603} is 5 V, a logic 1 is placed at input pin 31 of the UART. When the EIA input goes high, Q_{603} turns on, and the

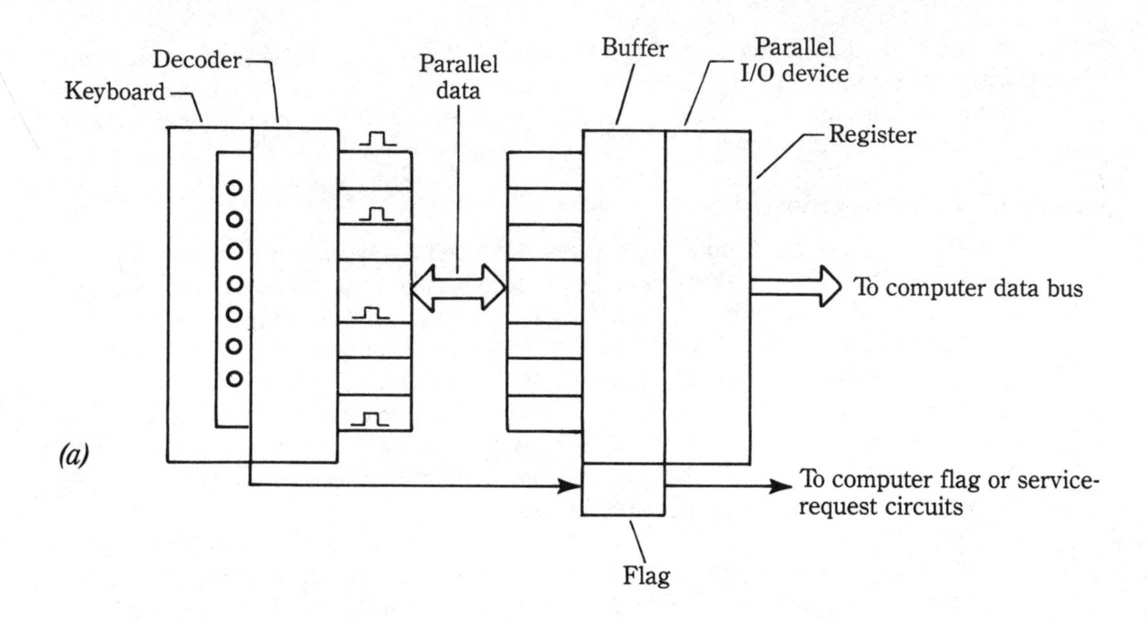

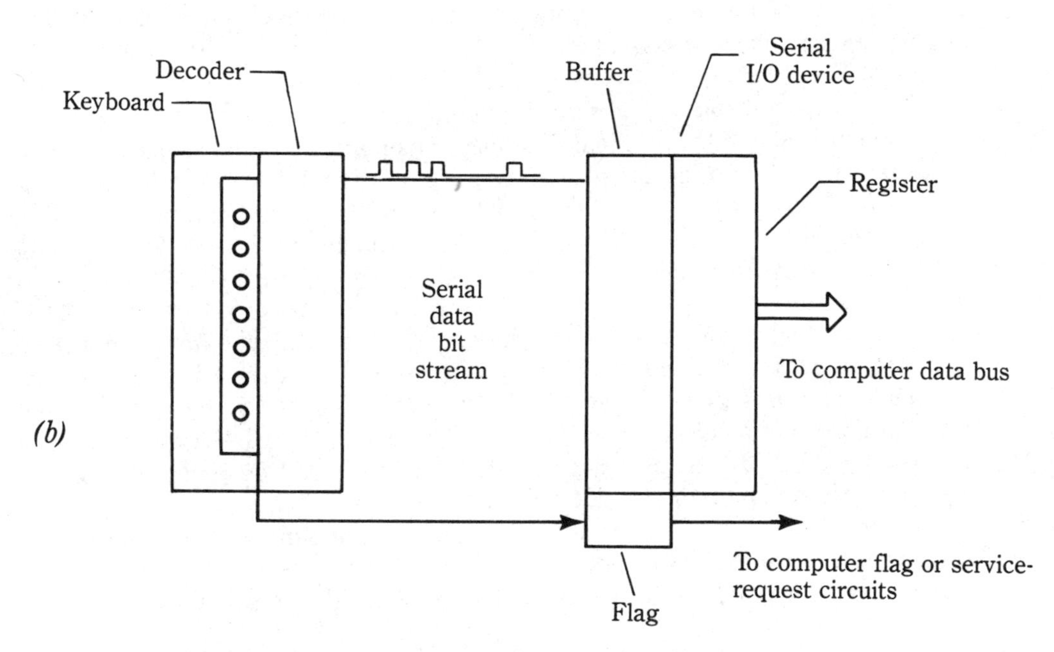

Fig. 4.14 Basic relationship between keyboard and computer.

collector voltage goes low. This places a space (logic 0) at the input of the UART. The serial output from the UART at pin 30 drives IC_{609A} which in turn drives IC_{609B}. The output of IC_{609B} drives optocoupler IC_{604} (Sec. 3.5.6).

When the output is a mark, the phototransistor in IC_{604} turns off and Q_{602} turns Q_{601} on. The current flows from the +12 V supply through R_{631} and Q_{601} to the

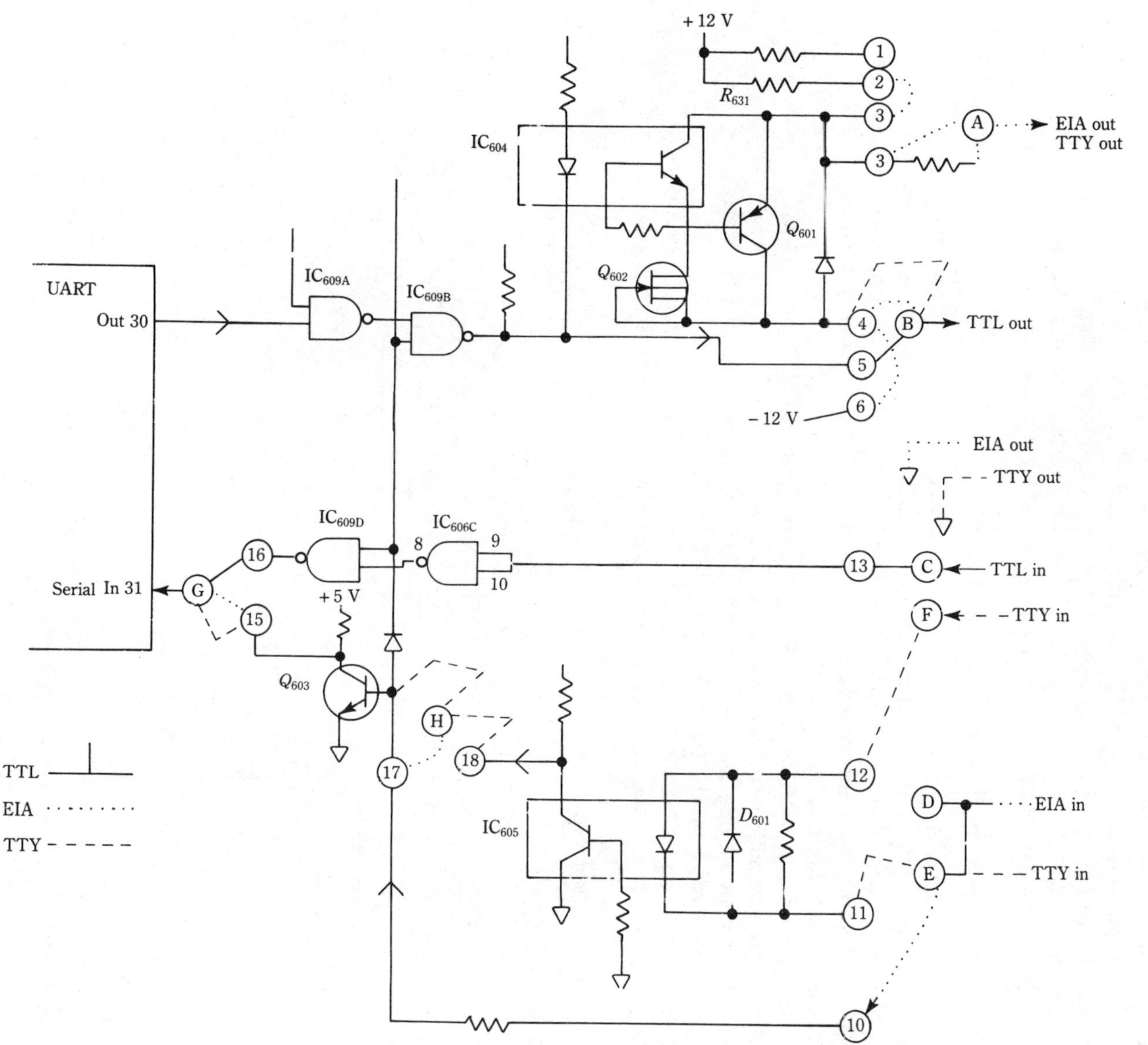

Fig. 4.15 TTL, EIA, and TTY (20-mA current loop) input/output connections.

– 12 V supply. This makes the EIA output approximately – 10 to – 12 V. As the output of IC_{609B} goes low, the LED in IC_{604} turns on the phototransistor, which turns Q_{601} off. This causes the EIA output to go to about + 12 V.

TTY input/output When a TTY mark is applied, a loop current passes through D_{601} and optocoupler IC_{605}, which turns Q_{603} off. A logic 1 then appears at the serial input of the UART (pin 31). The current-loop output circuit, composed of IC_{604}, Q_{601}, and Q_{602}, operates essentially the same as for EIA output.

4.7.2 Video terminal (keyboard/CRT)

The video terminal is used as an input/output device for computers and many other digital systems. Video terminals are particularly effective in inquiry/response situations where information is required for immediate use. Unless the CRT (cathode-ray tube) portion of the video terminal is connected to a printer, no permanent record of displayed data is kept.

As shown in Fig. 4.16, the video terminal is essentially a CRT (complete with vertical- and horizontal-sweep circuits similar to those used in TV sets) plus a keyboard. Information is displayed on the face of the CRT immediately after the information is received from the computer or entered from the keyboard.

In a video terminal, the keyboard output is converted into data bytes (in machine language suitable for input to the computer) by a decoder (which is gener-

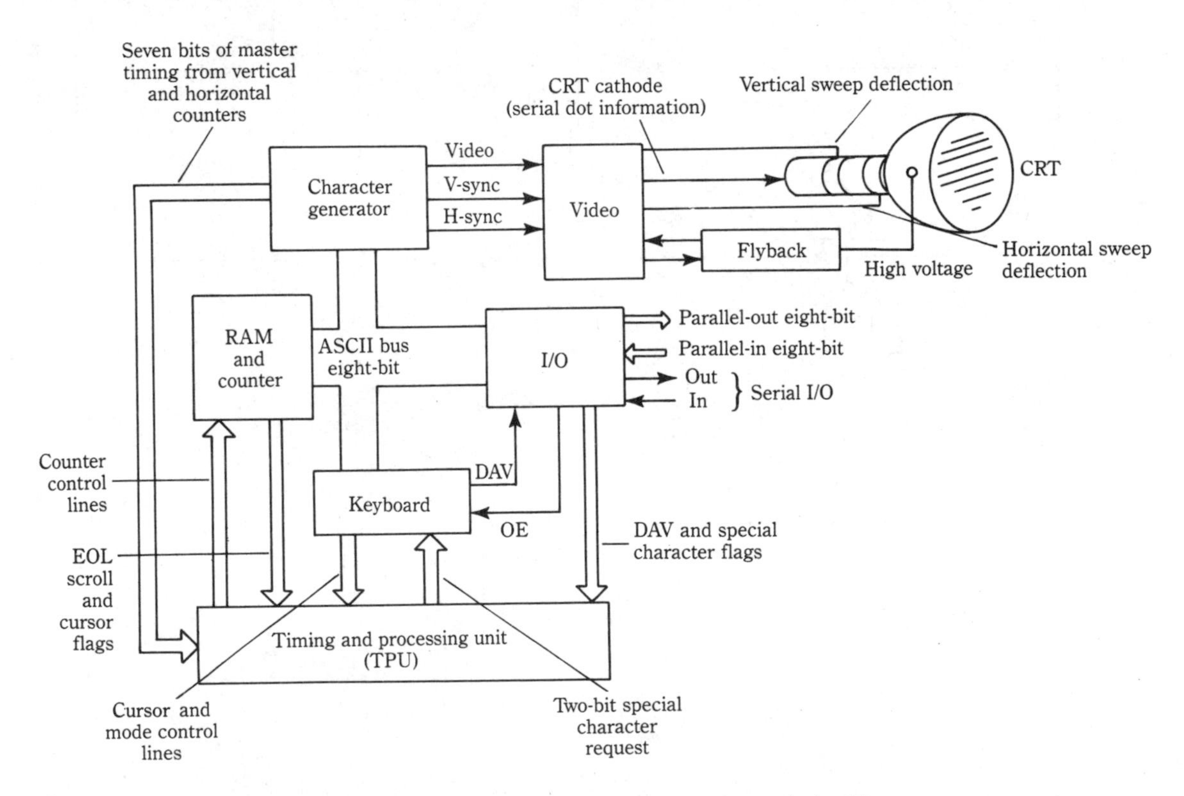

Fig. 4.16 Basic video terminal functions and circuits.

ally part of the video terminal). The output function (display of data from the computer) is performed by the CRT. Standard electromagnetic sweep deflection (similar to a TV set) is used for both the horizontal and vertical CRT deflection.

The electron beam is swept across the CRT face several hundred (or thousand) times per second. Typically the deflecting circuits sweep out 23 to 40 rows or horizontal lines of characters (letters, numbers, or symbols), with each line divided into 80 spaces (for 80 characters).

In all video-terminal systems, the persistence of the CRT screen is fairly short. This is necessary so that the readout is removed in a reasonable period of time (to permit new data bytes to be read in). Some video terminals contain storage registers that permit the displayed characters to be retained (or refreshed) on the screen until the operator clears the readout.

Many systems are used to produce the characters. Most video terminals use some type of *character generator*, usually in IC form. Such a generator is a form of decoder. The generator converts binary data bytes into electrical voltages that alter or manipulate the CRT electron beam to trace out characters on the screen.

Typical video-terminal circuits Figure 4.16 is the overall block diagram of a typical video terminal. This terminal can display information coming from computer circuits, or information typed in from the keyboard. The information is displayed on a 12-in (diagonal) CRT capable of displaying 960 characters at one time (12 rows with 80 characters per row).

A 67-key ASCII keyboard, which permits you to compose and edit directly on the CRT, has a capability of 64 different characters and 24 different functions. The message or program from the keyboard can be transmitted through serial or parallel I/O devices.

Figure 4.16 shows how the terminal can be divided into a series of functional blocks that coordinate the movement of data on an eight-bit ASCII bus. These blocks, which include the character generator, RAM and counter, keyboard, and the I/O, are interconnected directly to the ASCII bus.

Data bits are transferred from one block to another and to the external devices over the ASCII bus under control of the timing and processing unit (TPU). Other functional blocks include the video circuits, a line-operated power supply, and a high-voltage power supply or flyback system (similar to a TV set), which is operated from the video circuits.

The character generator has a master oscillator and dividers that generate the timing signals for all functional modules within the terminal. Basically the character generator takes data bytes from the RAM and converts them from eight-bit parallel ASCII information to *serial-dot information*. The serial-dot information is then used to modulate the display on the CRT screen.

The video module contains a video amplifier that converts the dot data from the character generator into voltage levels that drive the cathode of the CRT. The video circuits also have a vertical oscillator and amplifier that generate the vertical sweep and drive the vertical portion of the CRT yoke. The video circuits contain the horizontal oscillator and driver that drive the horizontal flyback and sweep systems. The vertical and horizontal oscillators are synchronized by sync pulses generated by the character generator.

The RAM counter contains the memory that stores the information to be displayed on the screen. Everything that appears on the face of the CRT is stored in the RAM. The RAM and counter circuits also contain the RAM address counter. The keyboard contains a standard ASCII keyboard and a series of other special-function keys that control operation of the video terminal. The keyboard also includes a special character generator that puts selected ASCII data on the bus whenever requested.

The I/O contains latches and buffers (instead of a PPI) to control parallel data transfers between the video terminal and any external device. The I/O also includes a UART that controls serial data transfers between the terminal and external devices. The TPU controls all other functional blocks. The TPU is the decision-making module within the video terminal. Data bits are moved, written, or transmitted under control of the TPU as a function of the front-panel switches and keyboard.

4.7.3 Diskette (floppy disk) recording basics

The floppy disk is a commonly used magnetic-storage device for microprocessor-based systems (such as personal computers). The floppy disk drive can be built into the computer or can be a peripheral. Most present-day software is in the form of disks (either hard or floppy). Full details of disk drives are not presented here. Instead, a basic peripheral floppy-disk drive is covered. If you understand this information, you should have no difficulty in understanding the various disk-drive systems now in use.

Figure 4.17*a* shows a basic floppy-disk or diskette drive. Such drives are peripherals that perform the electro-mechanical and read/write functions necessary to record and recover data on the diskette. As shown, the diskette is similar to the old 45-rpm phonograph record. However, the diskette rotates at a speed of 360 rpm (typical), and the data bytes are stored magnetically on concentric tracks over the face of the diskette. Data bytes are recorded serially.

Because of the high serial-data rates, special circuits are required for serial/parallel conversion, data recovery, and data error checking when interfacing a diskette drive and a computer. This hardware that performs this function is often called a *formatter*. The formatter also serves as a buffer between the computer and diskette, as shown in Fig. 4.17*b*.

The formatter combines with the computer to control the diskette drive, and sometimes the combination is called a *floppy-disk* or *diskette controller*. As used here, the term *controller* includes not only the system hardware, but also those microprocessor programs that directly or indirectly control the diskette drive. The program routines for diskettes are often called *drivers* or *control modules*.

Interfacing diskettes There are several considerations that must be made when interfacing diskettes with computers and microprocessors. They are not discussed here, because they are essentially the problem of the system designer. However, the following information should be of interest to both the student and troubleshooter.

Some diskette drives are designed for *daisy-chain interfacing*, where some of the interconnecting lines are shared and some are dedicated. Other drives use the

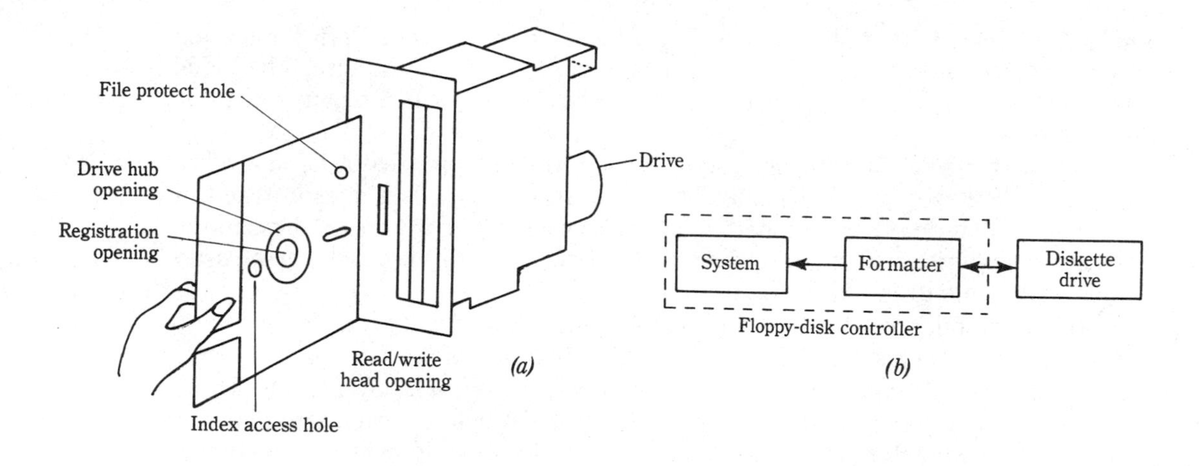

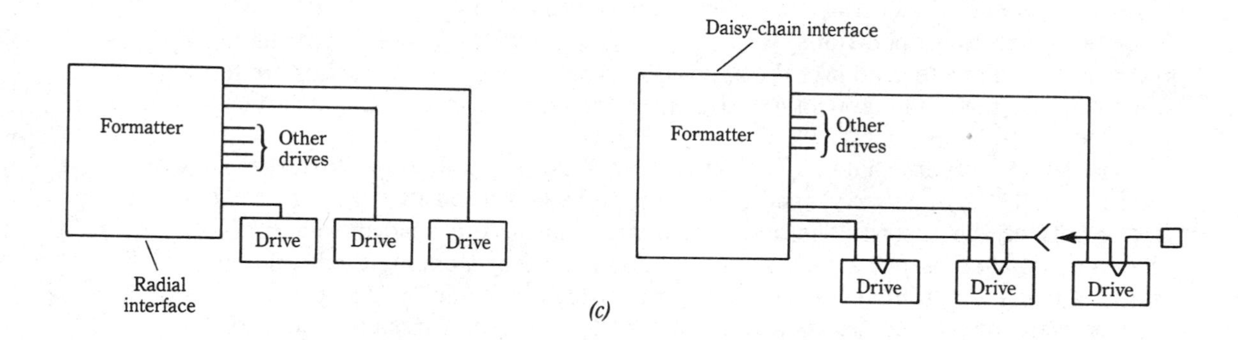

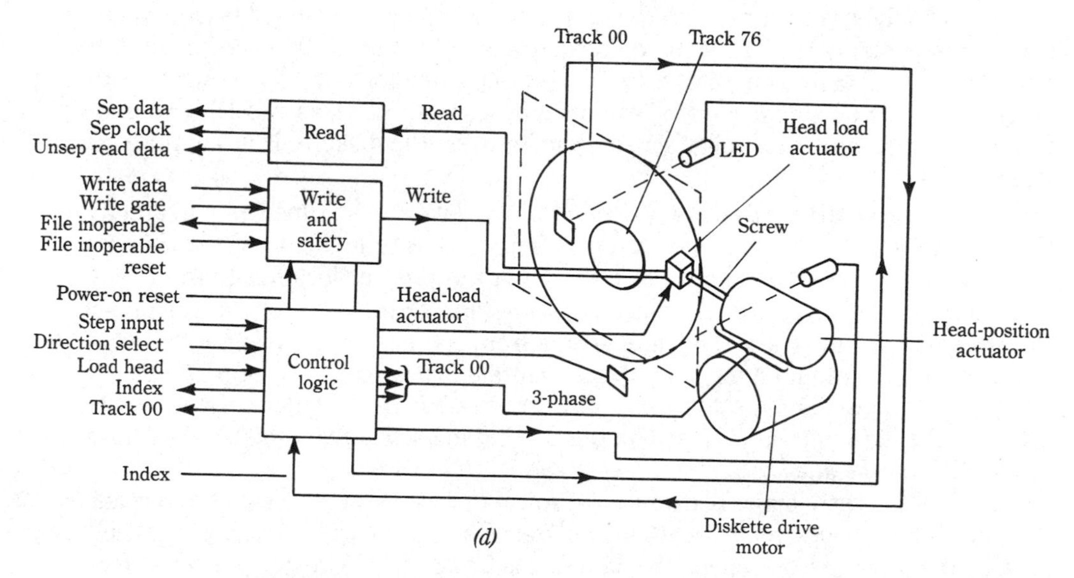

Fig. 4.17 Basic floppy-disk or diskette drive.

radial interface, where all interconnecting lines are dedicated. Both interfacing techniques are shown in Fig. 4.17*c*. Each technique has its advantages. The radial interface isolates (or buffers) each drive, whereas the daisy-chain requires less system hardware.

No matter what interfacing system is used, the microprocessor is busy all of the time when used with a diskette. This is because of high data rates involved. It also means that no other peripherals can be serviced while the computer is in the diskette read/write operation. As discussed in Sec. 4.5.4, this problem can be overcome with some form of DMA.

When the microprocessor includes DMA, there is a register (within the microprocessor) that serves as a DMA pointer. This built-in pointer is used to indicate the memory location for the DMA cycles. The program initially sets the DMA pointer to a beginning memory location. Each DMA-byte transfer automatically increments the pointer to the next-higher memory location. Repeated activation of a DMA line can cause the transfer of any number of consecutive bytes to and from memory, independent of concurrent program execution.

Seek-and-restore operations One of the functions of the diskette drive is to position the read/write head over the appropriate track where the data bytes are to be recorded or read. This function is sometimes referred to as a seek-and-restore operation.

The diskette records data on 77 circular tracks, numbers 00 to 76. To access a certain record, the read/write head must first be locked in position at the track which contains the record. The operation that performs the head movement is called a *seek operation*. For a diskette, a seek is executed by stopping the head one track at a time. The timing between stops is controlled from an interval timer.

The main difference between seek and restore is that a restore operation always moves the read/write head to track 00. After the seek operation is completed, the only way to verify that the proper track has been accessed is to read the *track address* magnetically recorded on the track. When track 00 is accessed, the diskette drive usually generates a *track 00* status signal or flag. As shown in Fig. 4.17*d*, the track 00 signal is often developed by a sensor (such as an LED and photocell). In some units, a similar sensor detects when the diskette is at an index or start position.

Data access time Diskette drives improve data access time (over magnetic tape and most magnetic storage systems). This access time includes *seek time,* or the time for the read/write head to move from the present location to the newly specified location (typically 10 ms/track); *settle time*, or the time for the positioner to settle onto a new track (typically 10 ms from the last stop pulse); and *latency time*, or the time required for the diskette to rotate to the desired position (typically 83.3 ms, average). The diskette spins at a fixed rate of 167 ms per revolution. On average, the data will be 1/2 a revolution, or 83.3 ms, away from the head. This is known as *average latency time*.

Typical diskette specifications A typical diskette system records and reads data at 250 kbits per second or 4 μs per bit. When used with a typical eight-bit parallel microprocessor system, the data rate is 250/8 = 31.25 kbytes per second, or 32 μs per byte. A single (typical) diskette has a capacity of 2,050,048 bits, or

256,256 bytes on the 77 tracks. There are 26 sectors on each track and 128 bytes on each sector.

Functional description of a typical diskette drive The driver shown in Fig. 4.17*d* consists of read/write and control electronics, drive mechanism, read/write head, track-positioning mechanism, and removable diskette. These components perform the following functions: interpret and generate control signals, move read/write head to the selected track, and read/write data.

General operation The *head-position actuator* positions the read/write head to the desired track on the diskette. The *head-load actuator* loads the diskette against the read/write head, and data bits can then be read from the diskette. The head-position actuator, which consists of an electrical stepping motor and screw, positions the screw clockwise or counterclockwise in 15° increments.

A 15° rotation of the screw moves the read/write head one track position. The microprocessor increments the stepping motor to the desired track. The diskette drive motor rotates the spindle at 360 rpm through a belt-drive system.

Signal interface The signal interface between the diskette drive and the microprocessor consists of lines required to control the drive and to transfer data to and from the drive. All lines in the signal interface are digital (pulse form) and either provide signals to the drive (input) or provide signals to the microprocessor (output).

Input signals There are six input signal lines as shown in Fig. 4.17*d*.

1. *Direction select*, which defines the direction of motion for the read/write head when the step line is pulsed. A binary 1 (open circuit) defines the direction as out, and if a pulse is applied to the step line, the read/write head moves away from the center of the diskette (toward track 00). Conversely, if the direction-select input is shorted to ground, or binary 0, the direction is defined as in, and if a pulse is applied to the step line, the read/write head moves toward the center of the diskette (toward track 76).
2. *Step input*, which is a control signal that causes the read/write head to move in the direction defined by the direction-select line. The access motion is initiated on each binary-0 to binary-1 transition of the step input signal.
3. *Load head*, which is a control signal to an actuator that allows the diskette to be moved into contact with the read/write head. A binary 1 deactivates the head-load actuator, and causes a bail to lift the pressure pad from the diskette. This removes the load from the diskette and read/write head. A binary-0 level on the load-head line activates the head-load actuator and allows the pressure pad to bring the diskette into contact with the read/write head (with the proper contact pressure).
4. *File inoperable reset*, which provides a direct reset for the file inoperable output signal. The file inoperable condition is reset when a binary 0 is applied to the file inoperable reset line.
5. *Write gate*, which controls the writing of data on the diskette. A binary 1 on the write-gate line turns off the write function. A binary 0 enables the write function and disables the stopping circuits.

6. *Write data*, which provides the data to be written on the diskette.

Output signals There are six output signal lines as shown in Fig. 4.17*d*.

1. *Track 00*, which indicates when the read/write head is positioned at track zero (the outermost data track).
2. *File inoperable*, which is the output of the data-safety circuits, and is a binary 0 level when a condition that jeopardizes data integrity has occurred.
3. *Index*, which is a signal provided by the diskette once each revolution (166.67 ms) to indicate the beginning of the track. Normally, the index signal is at binary 1 and makes the transition to the 0 level for a period of 1.7 ms once each revolution.
4. *Separated data*, which is the line over which read data bits are sent to the microprocessor. The signals written on the diskette (using a frequency-modulation system) are demodulated by the drive electronics, and are converted to data pulses. The data pulses (representing data bytes on the diskette) are sent over the separated-data line. Normally, the separated-data signal is at binary 1, and each data bit recorded on the diskette causes the signal to make the transition to a 0 level for 200 ns.
5. *Separated clock*, which provides the microprocessor with the clock bits recorded on the diskette. The levels and timing are identical to the separated-data line, except that a clock pulse occurs each 4 μs.
6. *Unseparated read-data*, which provides raw data (clock and data bits together) to the microprocessor.

4.7.4 CD-ROM

CD-ROM technology was derived from audio CD players. As a result, many of the circuits are identical, so this section concentrates primarily on the differences between audio CD and CD-ROM. (Read *Lenk's Laser Handbook*, McGraw-Hill, 1992, if you want a thorough understanding of CD, CD-ROM and CDV or Laservision technology.)

Instead of audio, computer information (software) is stored on a CD-ROM disc as variable-length pits in a continuous spiral track. (Note that the spelling *disc* is used when laser technology and compact disc or CDs are involved.) The pits are burned into a master disc by a laser beam and then used to press holes and flat surfaces into each copy disc.

As in the case of audio, a CD-ROM disc is read by a scanning laser beam that reflects from the silvered areas. The information is stored in the continuous spiral track and is read by the drive at CLV (constant linear velocity, the disk spins at a speed inversely proportional to the radius being read). Since each section of data must be read in the same amount of time, the disc spins faster on inner tracks and slower on the outer tracks (as does an audio CD).

A CD-ROM drive is similar to an audio CD player except that the CD-ROM handles computer data instead of audio. The CD-ROM drive is operated through a computer and software (as is a hard or floppy disk drive, Sec. 4.7.3), rather than by direct operation using controls (front panel or remote). Some CD-ROMs also provide for audio operation (using special software), but this is not the typical case.

CD-ROM is used for very large databases and for reference materials, such as medical, legal, or government libraries. CD-ROM products include maps, catalogs, training courses, and educational books. Any data that can be represented digitally can be stored on CD-ROM discs. This includes text, music, pictures, and computer graphics.

One CD-ROM disc can store about 552 megabytes of digital data when 60 minutes of disc space is used for mode-1 data. When mode-2 is used for data storage, 60 minutes holds about 630 megabytes. It takes the equivalent of a 1,533,360-kilobyte, 5.25-inch floppy disk to store the same amount of information. Approximately 270,000 typed pages of text can be reduced to just one CD-ROM disc. At the present time, CD-ROM discs and drives are read-only.

Major differences between CD-ROM and CD audio The following is a summary of the major differences between CD-ROM drives and audio CD players:

1. *Minimal user controls and displays:* A typical CD-ROM drive has a power switch and LED and possibly some additional LEDs to show that functions are being performed in the CD-ROM system microprocessor (such as disc-tray open and close, "busy" functions where the computer and drive are accessing data from the disc, etc.). A drive-select switch is also found on some CD-ROMs, particularly when more than one CD-ROM is used or when other drives (floppy, hard, etc.) are used. This makes it possible for the computer to select drives in a given order or to omit drives, as necessary.
2. *A bidirectional data and communications port*, and control lines, between the host computer and the CD-ROM system microprocessor. Figure 4.18 shows the basic connections.
3. There are *five layers of internal error correction* for audio data and CD-ROM data, compared to four layers for a typical audio-only CD player.
4. There are *minimal audiophile circuit features* (usually none at all) in a CD-ROM drive.
5. A CD-ROM drive provides *much greater programmability* through use of the host-computer resources and programming.
6. CD-ROM drive has the *ability to return Q-code, drive-status, or CD-ROM data* to the host computer. Figure 4.19 shows a comparision of CD-ROM and audio CD formats, both in sector format and Q-code format.
7. A CD-ROM drive has the *ability to transfer 153,600 bytes of data per second* over 60 minutes from a CD-ROM disc.
8. A CD-ROM provides *very fast access* to the inner and outer tracks of a disc. Some CD-ROM drives have a front-panel control (often called FG-IN or some similar term) that drives the pickup at high speed to inner or outer tracks (similar to the fast-forward or fast-reverse functions of an audio CD player).

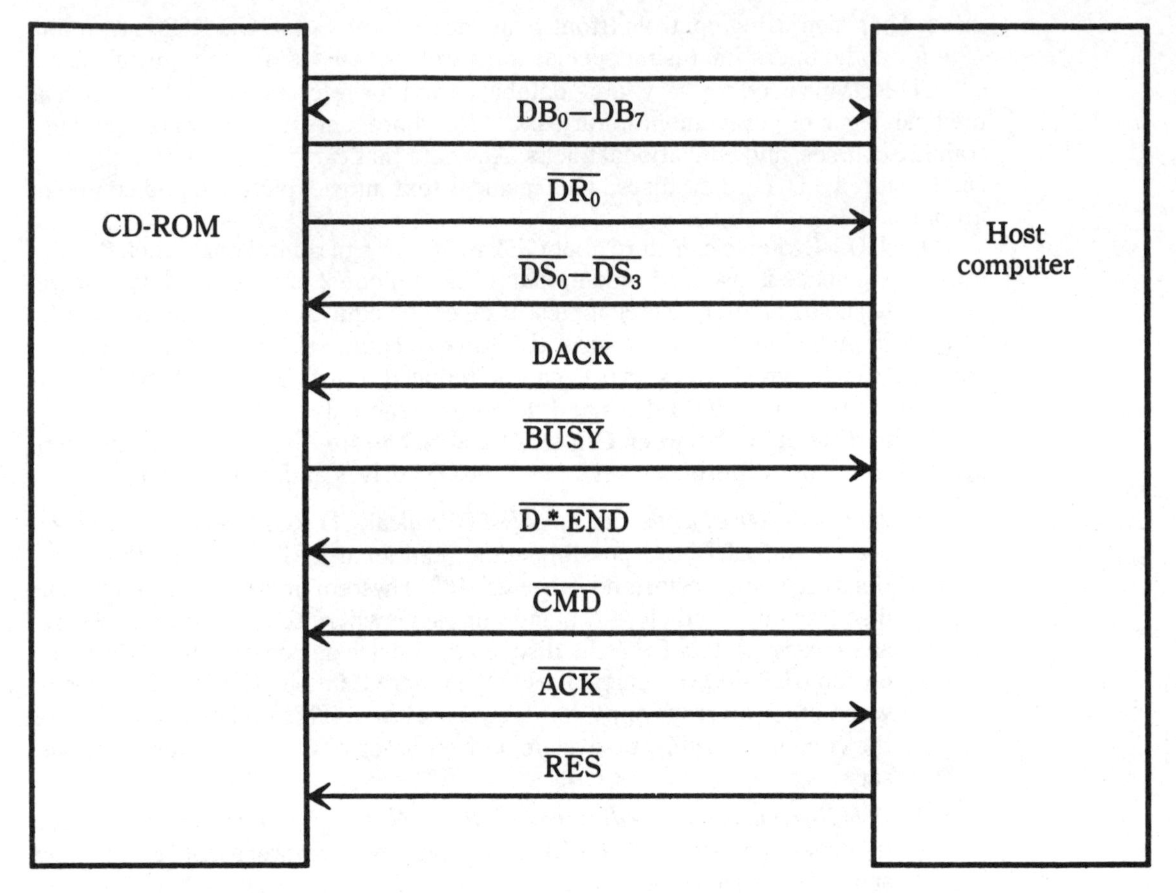

Fig. 4.18 Basic connections between CD-ROM and host computer.

9. A CD-ROM drive *converts serial data* to eight-bit parallel-data using a CD-ROM conversion IC and a RAM.

10. Most CD-ROM drives *use direct-drive (DD) disc motors*.

11. The major *circuit differences* found in a CD-ROM are shown in Fig. 4.20. These include an IC used for error correction and data conversion (IC_{302}), additional RAM (IC_{303}) for the CD-ROM data and control circuits, and I/O line drivers and interface circuits (IC_{304} and IC_{305}). Also note that there are differences in the CD microprocessor IC_{104}, such as an FG-IN input, as well as outputs to LEDs that show open and close, and busy or access functions.

Typical CD-ROM/computer interface The following discussion assumes that the host computer is an IBM PC-XT/AT, PS/2 Model 30, or any direct hardware-compatible computer running MS-DOS.

External drive The hardware installation is fairly simple with external CD-ROM drives. You simply remove the computer cover, plug one of the compatible PC-bus I/F cards (a CDIF14A for example) into a vacant eight-bit slot (note that AT-type computers have 16-bit and eight-bit slots). Then connect the I/F card-compatible external CD-ROM drive (such as the Hitachi CDR-1502S) with a matching cable (such as a CDCBL). No internal switches in the computer need be set.

Internal drive Internal drives (such as the Hitachi CDR-2500 or CDR-3500) require more effort than external drives. First, there must be room inside the computer to install the drive. For example, front-panel clearance and full-height slot is necessary for the Hitachi CDR-2500. A half-height slot is needed for a Hitachi CDR-3500.

A full-height slot is the same size as a full-height floppy disk drive, and a half-height slot is the same as a half-height floppy. Thus, two half-height devices can be mounted in the same space used by a single full-height floppy disk drive. Special mounting hardware is required for internal installation. (Mounting kits are available from many computer stores.)

The internal drive requires power from the host computer and generally uses the same power connector as a floppy-disk drive. Obviously, if there is no empty power connector, a power connector must be provided.

Recognizing and accessing the CD-ROM drive When the computer is first turned on, the microprocessor jumps to a particular address in memory, and starts executing machine-language instructions in ROM. The computer runs various tests and prepares to run some type of user software. Then the computer checks for a disk drive and boots up DOS software from a floppy or hard disk. Some computers need not boot to DOS software from a disk (when the computer has the disk-handling software in ROM).

When the DOS code is loaded, the device-handler code (such as config.sys in MS-DOS) installs the device name and handler code into the device table. (If you are interested, read the MS-DOS manual for details of this file.)

The computer then looks for a file of autorun software (called autoexec.bat in MS-DOS) and executes the code found in that batch file. When these programs are finished running, the computer turns control over to the user at the DOS prompt.

After the device handler is installed, the CD-ROM drive is accessed as the next drive in the system (except the drive is read-only). Files cannot be saved in a CD-ROM disc or drive.

If the device-handler code for a Hitachi CD-ROM drive is on the boot disk at boot-up, the CD-ROM drive can be recognized by the operating system. Generally, the CD-ROM drive becomes the next drive after A and B and any other drive handlers installed by the device-handler drive. Typically, the CD-ROM becomes drive C if there is no hard drive installed, or drive D if a 20-megabyte hard drive is installed.

CD-ROM information layout Note that information organization is similar, but not identical, for audio CD and CD-ROM. In both cases, the bit stream has sync signals to allow the bit data to be separated for use in maintaining a constant-speed data stream.

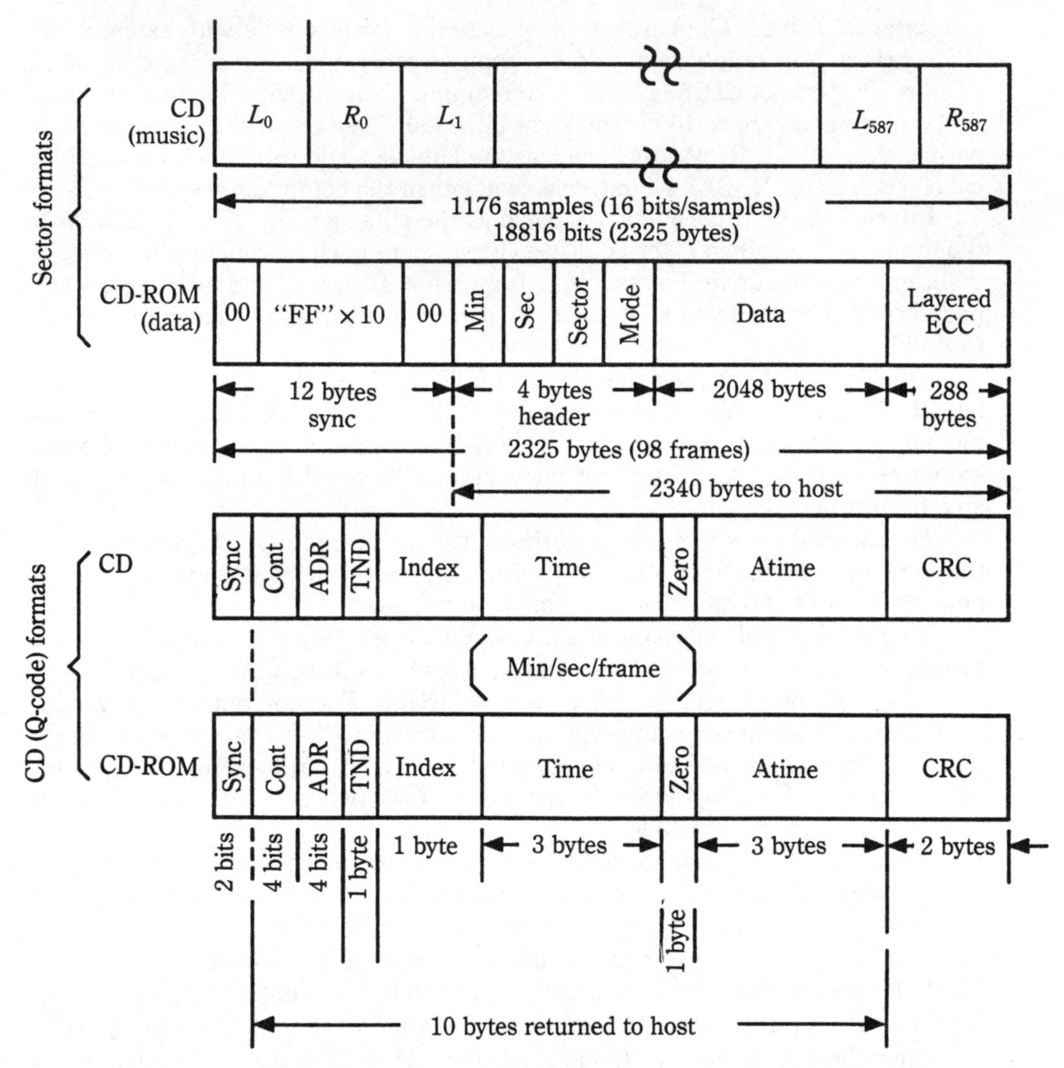

Fig. 4.19 Comparison of CD-ROM and audio CD formats.

During the disc-recording process, data bytes are *interleaved* (scrambled in a fixed order), error-correction information is added, and the data bits are converted from 8 to 14 bits and recorded in a non-return-to-zero format. This is reversed in play by a custom CD-decoding chip set and 2-kbyte static RAM. The chip set (IC_{301} through IC_{305}) for a typical CD-ROM provides five layers of internal error correction, compared to four layers for audio CD.

Frame information As shown in Fig. 4.19, the low-level data bits are set into *bit frames* (similar to a line of video) and *frames* (a group of 98-bit frames, similar to a frame of video). The frame rate is 75 Hz, so one frame occurs in 1/75 second.

Each frame contains 98 bytes of control information (channels P to W) and 2353 bytes of data.

Control code channels Each bit of control information is considered a *channel*, P, Q, R, S, T, O, V, and W. Channels P and Q are generally used in CD-ROM. Ten bytes (80 bits) of Q-channel or Q-code information can be returned to the host computer. Q-code contains information about the type of disc data, and timing information for positioning. There are Q-code data bits in the table of contents (TOC) area and in the program area.

The TOC of a disc should not be confused with the directory of files on the disc. Any directory of files on a disc is contained in individual sectors according to the data format (such as High Sierra, IS09660, etc.) or custom formats. The program area of the disc is identified in the Q-code by absolute time in minutes (0 – 59 +), seconds (0 – 59), and frames (0 – 74). CD-ROM discs are usually rated for 60 minutes of mode-1 data. The maximum possible upper limit is 99 minutes for BCD-format data but is limited to 89 minutes with small-computer (SCSI) command formats.

Frame data The 2353 bytes of data contain 1176 words of audio, or 1 sector of CD-ROM data. When the data bits are in a CD-ROM, the bits are passed to the custom CD-ROM controller IC_{302} and the 6-kilobyte RAM IC_{303} for serial-to-parallel conversion, and to other processing in IC_{301}. After conversion, four bytes of header, 2048 bytes of data, and 288 bytes of L-ECC are returned to the computer.

The CD-ROM header contains the CD-ROM block in Min, Sec, Block, and a data Mode byte. The mode byte identifies whether the 288 bytes are L-ECC (mode 1) or other (mode 2). The CD-ROM block is not necessarily the same as the Q-code frame number.

CD-ROM electrical and data interface Figure 4.20 shows a typical interface between a CD-ROM and a host computer. Note that the interface can be divided into two sections: the eight-bit parallel data and the control signals. The data bus transmits all information (the *data read* from the CD-ROM, the *control-command code*, and the *status information* of the CD-ROM system). The control signals control transmission of this information.

Interface lines and buses The following describes the function of each interface line and bus between the CD-ROM and host computer (Figs. 4.18 and 4.20).

DBO-7 (*data bus*): DB is a bidirectional eight-bit parallel bus. Data bits from the CD-ROM, control-command code to the CD-ROM, and status information of the CD-ROM system are transmitted through this eight-bit data bus. All eight bits are passed to and from the CD-ROM system-control microprocessor IC_{104} through interface connector JK_{301} and bus transceiver IC_{304}.

System-control microprocessor IC_{104} is connected to the CD-ROM error-correction microprocessor IC_{302} through a bus and control lines. Note that IC_{302} is not to be confused with the CD signal microprocessor IC_{301}, which has functions similar to the signal processors of audio CD players. Both IC_{302} and RAM IC_{303} are unique to CD-ROM players and provide the five layers of error correction necessary for use with a computer.

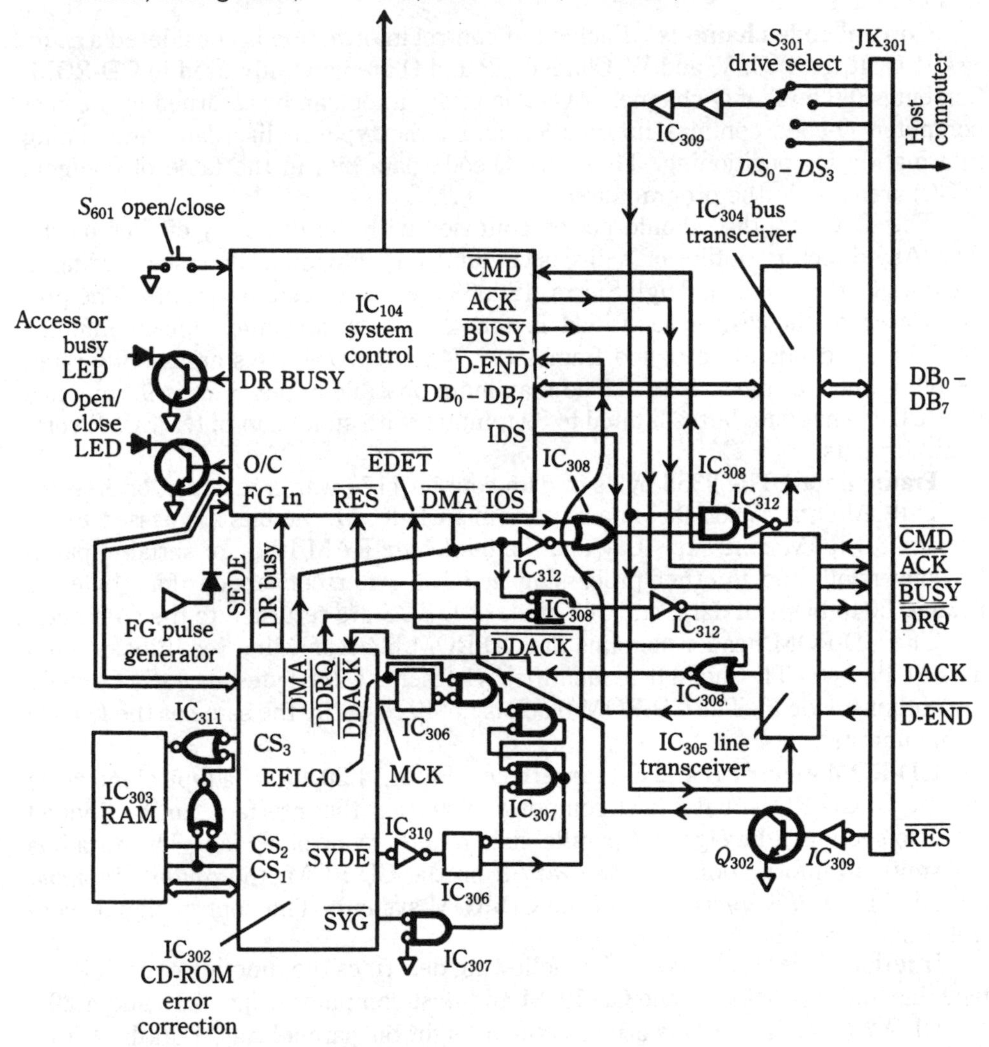

Fig. 4.20 Major circuit differences between CD-ROM and audio CD players.

$\overline{DS_0}$ through $\overline{DS_3}$ (*drive select*): $\overline{DS}$ is the signal input that enables the selected CD-ROM drive. The CD-ROM drive is selected when one of $\overline{DS_0}$ through $\overline{DS_3}$ goes low. When $\overline{DS_0}$ through $\overline{DS_3}$ are high, all signal lines are in the high-impedance state. The four DS bits are used to control both IC_{304} and IC_{305} through the front-panel drive-select switch S_{301}, IC_{309}, IC_{308} and IC_{312}.

$\overline{DRQ}$ (*data request*): $\overline{DRQ}$ is the signal output that controls data communications. When $\overline{DRQ}$ is active (low), the CD-ROM data on the eight-bit data bus is valid (and only the $\overline{DRQ}$ is low). $\overline{DRQ}$ shakes hands with DACK (data acknowledge) and transfers the data from the CD-ROM to the host computer. $\overline{DRQ}$ is a combination of

DMA from IC_{104} and DDRQ from IC_{302}, and it is applied to the host computer through IC_{308}, IC_{305}, IC_{312} and JK_{301}.

DACK (*data acknowledge*): DACK is the signal input that controls data communications. When the host computer had read data from the CD-ROM, the computer makes DACK active (high). Then the CD-ROM makes $\overline{\text{DRQ}}$ inactive (high) after confirming that DACK will become active (high). The CD-ROM outputs the next data on the data bus and makes NRC active (low) when DACK becomes inactive. DACK is applied to IC_{302} (as DDACK) through JK_{301}, IC_{305} and IC_{308}.

$\overline{\text{BUSY}}$ (*bus busy*): $\overline{\text{BUSY}}$ is the signal output that informs the condition (accessing or not accessing) of the data bus. While IC_{104} is in the data-transmission mode, a command code cannot be accepted from the host computer. During the data-transmission mode, IC_{104} makes $\overline{\text{BUSY}}$ active (low). When IC_{104} receives a $\overline{\text{D-END}}$ signal from the host computer, IC_{104} makes $\overline{\text{BUSY}}$ inactive (high).

$\overline{\text{D-END}}$ (*data transfer end*): $\overline{\text{D-END}}$ is the signal input which indicates the end of the data-transmission mode. IC_{104} changes from the data-transmission mode to the command mode and makes $\overline{\text{BUSY}}$ inactive (high) after receipt of a $\overline{\text{D-END}}$ signal from the host computer. $\overline{\text{D-END}}$ is applied to IC_{104} through JK_{301} and IC_{305}.

$\overline{\text{CMD}}$ (*command pulse*): $\overline{\text{CMD}}$ is the signal input that controls transmission of command codes and status information. IC_{104} transfer status information or receives command codes, by shaking hands with the host computer through $\overline{\text{CMD}}$ and $\overline{\text{ACK}}$ signals. When BUSY is active, IC_{104} does not accept a CMD signal. $\overline{\text{CMD}}$ is applied to IC_{104} through JK_{301} and IC_{305}.

$\overline{\text{ACK}}$ (*acknowledge*): $\overline{\text{ACK}}$ is the signal output that controls transmission of command codes and status information. ACK becomes active (low) after receipt of command codes or when status information is being output. $\overline{\text{ACK}}$ shakes hands with the $\overline{\text{CMD}}$ signal, and is applied to the host computer through IC_{305} and JK_{301}.

$\overline{\text{RES}}$ (*reset*): $\overline{\text{RES}}$ is input from the host computer. If $\overline{\text{RES}}$ is active (low), IC_{104} and the CD-ROM are reset. $\overline{\text{RES}}$ is applied to IC_{104} through JK_{301} and IC_{305}.

4.7.5 Digital data communications peripherals

Digital or data communications is the transmission of digital data (bytes and bits) from one point to another. As discussed, bytes and bits are transferred between components within a digital device (computer, etc.) by buses under the supervision of control lines. When a digital device must communicate with another digital device, such communication takes one of two forms, depending on the distances between the devices.

Digital communications over short distances When digital devices (computers, video terminals, etc.) communicate with each other over short distances (typically within the same building) they generally do so using *buses* (in the form of electrical cables) and interface devices. The buses are often extensions of the buses found in a typical computer (data buses, control buses, etc.).

The interface devices (sometimes peripherals and sometimes built in) have two basic functions. First, they match the I/O characteristics of the various equipment for proper signal levels, impedances, and so on. More important, the interface devices function to supervise or control communications between the various digital equipments connected to the system or network. When several devices are interconnected over short distances, the system is often called a *local area network*, or LAN (although this is not necessarily a true definition).

Standardized digital communications (IEEE-488) At one time, there was little or no standardization for the buses and interfaces used in digital communications. Most bus and interconnecting cable systems were custom designed. It was almost impossible to connect the digital device of one manufacturer to another without special adapters, cables, patch cords, and so on. Sometimes, manufacturers even had difficulty in connecting all of their own digital equipment into a single, compatible system.

Today, the bus system used in digital communications, is governed by IEEE-488. Figure 4.21 shows the basic arrangement of the IEEE-488 system, in both *linear* and *star* configurations, as well as a typical GPIB (general-purpose interface bus) connector outline. The basic hardware, characteristics and interface functions are as follows:

16-line bus The IEEE-488 system uses a 16-line bus to interconnect up to 15 digital instruments. Each instrument on the bus is connected parallel to the 16 lines of the bus. All active interface circuits are contained within the various instruments, and the interconnecting cable (containing 16 signal lines) is entirely passive. The cable simply interconnects all devices in parallel, whereby any one device can transfer data to one or more other participating devices.

Eight of the 16 lines in the cable are used to transmit data. The remaining lines are used for data byte transfer control (or handshake) and general interface management (bus management). Data bytes are transferred by an *interlocked handshake*, permitting asynchronous communications over a wide range of data rates.

Talker, listener, controller Every participating digital device must be able to perform at least one of three roles: talker, listener, or controller.

A *talker* can transmit data to other devices via the bus, and a *listener* can receive data from other devices on the bus. Some devices can perform both functions. For example, a programmable digital voltmeter can listen to receive instructions from the controller, and can talk to send the voltage measurement to another device.

A *controller* manages operation of the bus system by designating which devices are to send and receive data. The controller may also command specific actions within other devices.

A minimum IEEE-488 system consists of one talker and one listener, without a controller. In this configuration, data transfer is limited to direct transfer between one device manually set to talk only. For example, a measuring instrument can be set manually to talk to a printer.

Connections and structure As shown in Fig. 4.21, the IEEE-488 system has a *party-line* structure, where all devices on the bus are connected in parallel. The 16 signal lines within the passive interconnecting cable are grouped into three clusters according to function as follows:

- Data bus (eight lines)
- General interface management bus (control) (five lines)
- Data byte transfer control (handshake) (three lines).

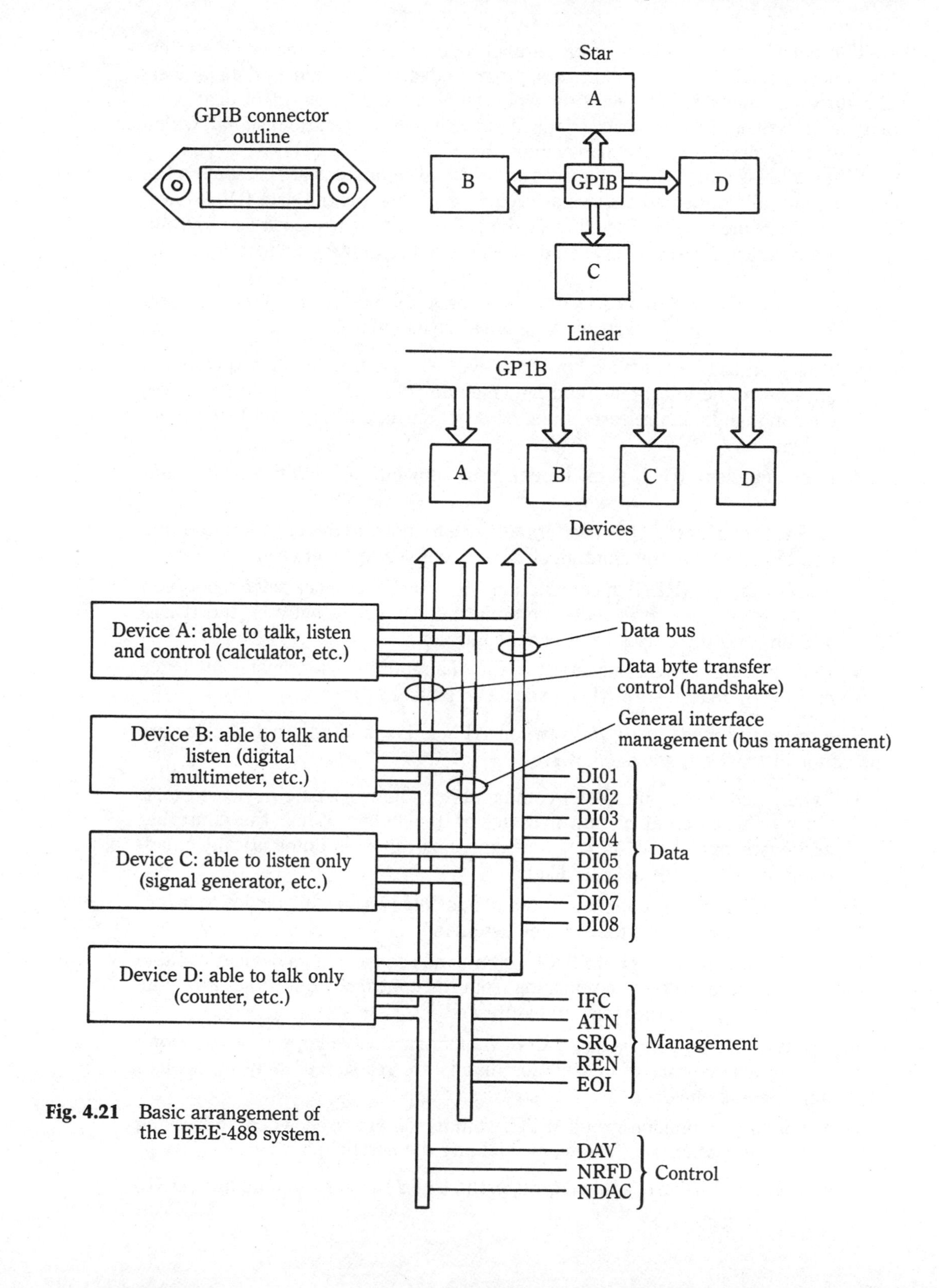

Fig. 4.21 Basic arrangement of the IEEE-488 system.

The data bus carries data in bit-parallel, byte-serial format across the interface. These signal lines carry addresses, program data, measurement data, universal commands, and status bytes to and from the network devices. Identification of the type of data on the data lines is indicated by the ATN (attention) signal, which is one of the general interface management lines.

When ATN is true, either addresses or universal commands are present on the data bus, and all connected devices are required to monitor the data (DIO) lines. When the ATN message is false, device-dependent data bytes (such as programming data bytes) are carried between devices previously addressed to talk and listen.

The *general interface management lines* manage the bus to produce an orderly flow of messages. The functions of these lines are as follows:

1. Attention (ATN) specifies how data bytes on the data input/output (DIO) lines are to be interpreted and by which devices. ATN is pulled low (true) for commands and released (high or false) for data transferred by the controller.
2. Interface clear (IFC) puts the entire system into a predefined quiescent state.
3. Service request (SRQ) is used by a device to indicate the need for attention and to request an interruption of the current sequence of events.
4. Remote enable (REN) in conjunction with other messages selects between alternate sources device-programming data (typically between the IEEE-488 bus and the device front-panel controls).
5. End or identify (EOI) indicates the end of a multiple byte transfer sequence or, in conjunction with ATN, executes a polling sequence.

The *command mode instructions* (when ATN is low and all other devices are waiting for instruction) are as follows:

1. Talker address group (TAG) commands enable a specific device to talk. Only one device at a time may act as the talker. When the controller addresses one device to talk, the previous talker is automatically unaddressed and ceases to be a talker.
2. Listener address group (LAG) commands enable a specific device to listen. Up to 14 devices at a time can be listeners.
3. Universal command group (UCG) commands cause all bus devices capable of responding to these commands from the controller to do so at any time regardless of whether they are addressed.
4. Addressed command group (ACG) commands are similar to universal commands except that addressed commands are recognized only by devices addressed as listeners.
5. Secondary command group (SCG) commands are used when addressing extended listener and talkers, or enabling the parallel poll.

In addition to the five command groups, the Unlisten address command (UNL) unaddresses all listeners that have been previously addressed to listen, and the Untalk address command (UNT) unaddresses any talker that had been previously addressed to talk.

Data byte transfer control (handshake) Transfer of each byte on the data bus is done through three handshake signal lines:

1. Data valid (DAV) indicates the availability and validity of information on the DIO lines.
2. Not ready for data (NRFD) indicates the state of readiness of devices to accept data.
3. Not data accepted (NDAC) indicates the condition of acceptance of data by devices.

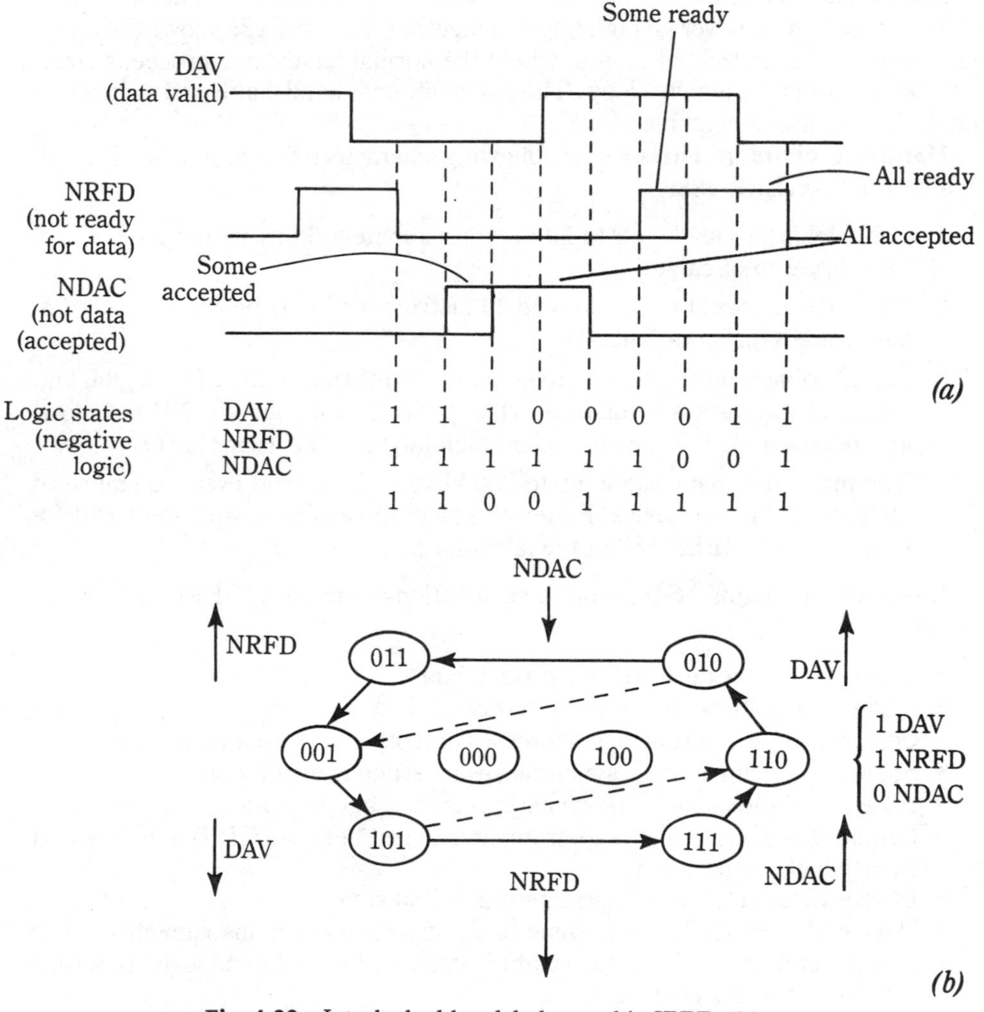

Fig. 4.22 Interlocked handshake used in IEEE-488.

The DAV, NRFD, and NDAC signal lines operate in a three-wire interlocked handshake process to transfer each data byte across the interface, as shown by the timing sequences of Fig. 4.22*a*.

A handshake sequence is entered with the listener-controlled NRFD and NDAC both low. The DAV line is high. As each listener is ready to accept data, the listener releases its NRFD line. When all listeners have released their respective NRFD lines, pull-up resistors on the line pull NRFD high.

The talker signals new data valid by pulling the DAV line low. Listeners respond by pulling their NRFD outputs low. During the period that listeners accept data, the listeners release the NDAC line. When data bits have been accepted by all listeners, the NDAC line goes high. Acknowledgment by the talker releases the DAV line, and the handshake is completed by the listeners by pulling the NDAC line low.

A legal handshake must proceed as shown in Fig. 4.22*a*. Note that the NRFD and NDAC lines may never go high (logic 0) together. Figure 4.22*b* shows the handshake sequence as a cycle of events, where the normal handshake proceeds counterclockwise around the outer loop. The two additional handshake paths shown in Fig. 4.22*b* are also acceptable.

Hardware characteristics The following characteristics apply to all hardware covered by IEEE-488:

1. Cable lengths may be up to 20 m (approximately 66 ft) with a device load for every 2 m of cable.
2. Up to 15 devices (1 controller and 14 instruments) may be connected in linear or star configurations.
3. Signal voltage on the buses are generally TTL-compatible. Bus signal and data lines are asserted (or true) when pulled low (about +0.8 V or below) and released (or false) when pulled high (about +2.0 V or above).
4. The maximum data rate is up to 250 kbytes per second over a distance of 20 m, with 2 m per device. Faster data rates are available with some restrictions. Refer to IEEE-488 of the latest issue.

Interface functions Some interface functions defined by IEEE-488 are as follows:

- Talker (T): allows instrument to send data.
- Listener (L): allows instrument to receive data.
- Controller (C): sends device addresses and other interface messages.
- Source Handshake (SH): synchronizes message transmission.
- Acceptor Handshake (AS): synchronizes message reception.
- Remote-Local (RL): allows instrument to select between GPIB interface and front-panel programming
- Device Clear (DC): puts instrument in initial state
- Device Trigger (DT): starts some basic operation of the instrument
- Parallel Poll (PP): allows up to eight instruments (simulataneously) to return a status bit to the controller.

Digital communications over long distances When digital devices communicate with each other over long distances (to other buildings, to other cities across the country, or perhaps to cities in other countries) they do so using the telephone line already available. In some cases, a central computer serves several users on a time-sharing basis, with each user at some different remote location. Figure 4.23 summarizes the digital data-transmission systems in common use.

Data set, modem, acoustic coupler Although teletype lines were used extensively at one time for digital communications, the present trend is to use the telephone lines (public, leased, private, satellite, etc.) to transmit data from one digital device to another. This requires the translation of digital data (usually in the form of logic pulses) into a form suitable for transmission across telephone lines.

The term *data set* can be applied to any device that converts digital data into a form suitable for transmission and vice versa. A data set usually contains a *modem*, (*mod*ulator-*dem*odulator) or possibly an *acoustic coupler* (on older data sets). The data set might also include a standard telephone, but this is not the typical case in present-day digital communications.

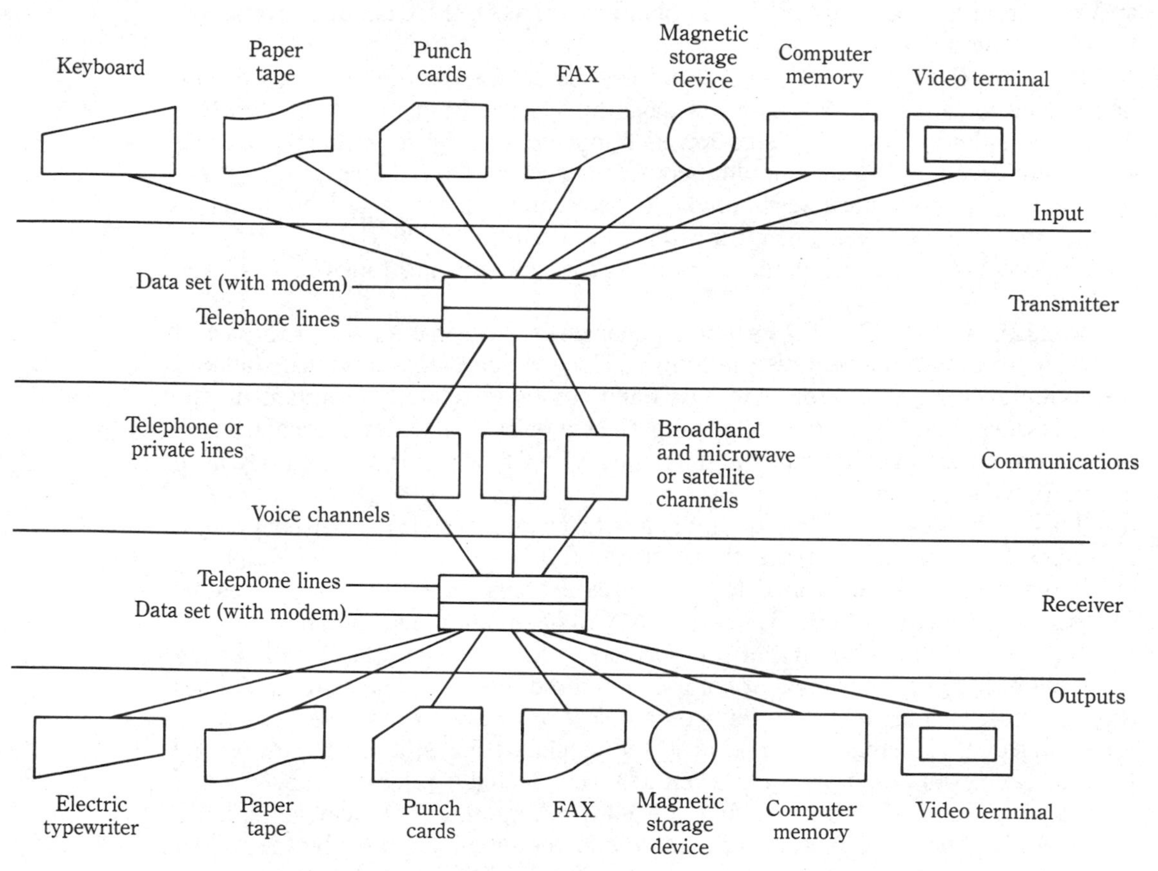

Fig. 4.23 Digital data-transmission (communications) systems in common use.

The difference between a modem and an acoustic coupler is that the modem must be connected directly into the telephone lines, usually through a modular telephone jack. Some computers include a modem. An acoustic coupler produces audio tones that can be used with a standard telephone, making it unnecessary to connect into the telephone lines or wiring. The modem has all but replaced the coupler in present-day digital communications.

There are several systems for transmitting digital data over telephone lines, such as *frequency-shift keying* (FSK) and *differential phase-shift keying* (DPSK). In the most common system, a mark (or logic 1) is converted to a tone of one frequency, and a space (or logic 0) is converted to a tone of another frequency.

Basic modem system Figure 4.24 shows a typical modem system, including the standard frequency assignments. As shown, a modem fills the need in a data-communications network to provide interface between a telephone network, which marries analog information, and a computer system that operates on digital information.

Note that both the transmitting (modulation) and receiving (demodulation) functions are contained in each modem, permitting *duplex* operation (transmission and reception at both ends of the telephone lines, as in the case of conventional telephone operation).

Basically, the modem converts 1 and 0 levels into audio-frequency tones and back again to 1s and 0s. These tones have the specific frequencies listed in Fig. 4.24. Two pairs of tones are listed for each modem, one set for transmitting and one set for receiving, so that simultaneous two-way (*full-duplex*) operation is possible over a single transmission channel. The modem that places a call is referred to as the *originate modem* (transmitting tones of 1070 and 1270 Hz), whereas the modem receiving this call is the *answer modem* (transmitting tones of 2025 and 2225 Hz).

RS232C and CCITT V.24 The frequencies shown in Fig. 4.24 are part of EIA (Electronic Industries Association) RS232C, which is the universal standard for communications between computers and computer-related equipment in the United States. RS232C is generally compatible with CCITT (International Consultative Committee for Telegraphy and Telephony) V.24, which is a commonly used international standard.

This book does not attempt to duplicate the standards in full detail. Instead, it describes the essential features of the standard as they relate to the equipment being discussed. The complete RS standards are available from the Electronic Industries Association, 2001 Eye Street, N.W., Washington, DC 20006.

Figure 4.25 shows the pin or line format for RS232C. A total of 25 lines are available between the terminal or computer and the modem. Note that all 25 lines are not necessarily used in all data-communications networks. However, when a line is used, the format or function is the same for all equipment covered by RS232C. For example, if pin 8 of a computer is connected to pin 8 of a modem (as it is in most systems), the signal on that line tells the computer that a carrier signal is being received from the answering modem over the telephone lines by the calling modem.

DTE and DCE RS232C defines all equipment as either *data terminal equip-*

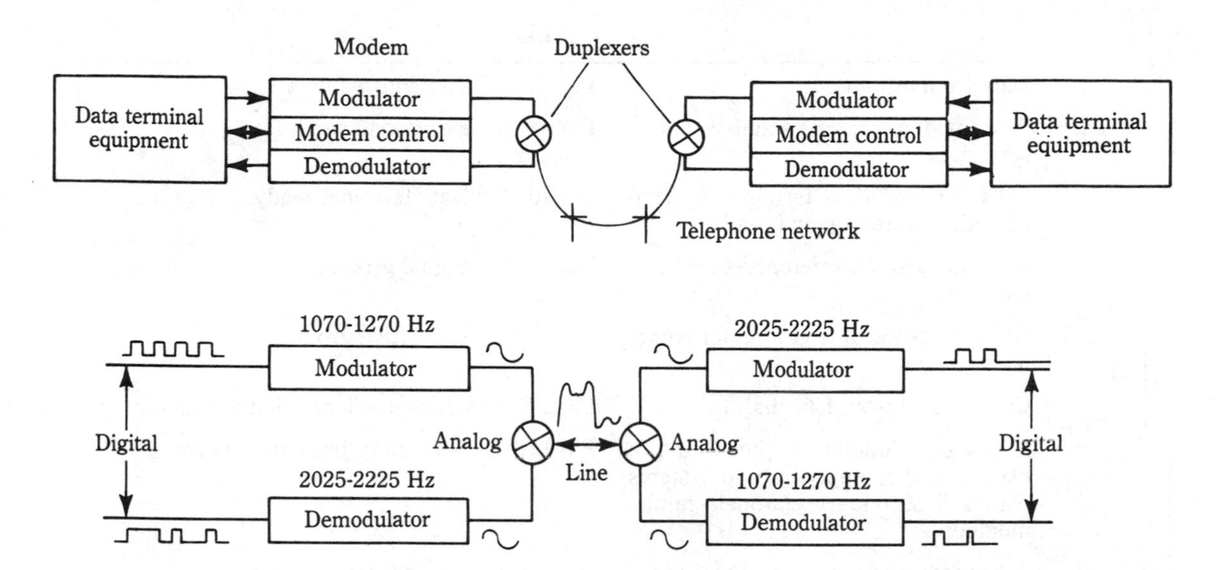

Mode	Originate (Hz)		Receive (Hz)	
	Send	Receive	Send	Receive
Mark	1270	2225	2225	1270
Space	1070	2025	2025	1070

Fig. 4.24 Basic modem system.

ment (DTE) or *data communications equipment* (DCE). The video terminal (Sec. 4.7.2) is one of the most popular DTE devices. In a typical application, a time-sharing user can have a video terminal (with built-in modem or peripheral data set) at an office or home location, connected by telephone lines to a computer (or possibly to another video terminal) at some remote point. The input and output of the video terminal are connected through the modem to the telephone lines. With this arrangement, the video terminal (in effect) connects directly to the computer (or to the other remote video terminal). In personal computers, the two computers (including the video terminals) are connected through their respective modems and the telephone lines.

Typical serial data transmission system The following paragraphs provide a brief description of a typical serial data transmission system. Two devices are involved, an ACIA (asynchronous communications interface adapter) and a digital modem.

ACIA Figure 4.26 shows the interfacing relationship to the system, and basic block diagram, of an ACIA. As shown, the ACIA provides the data formatting and control to interface serial data between the computer and modem. Parallel data of the computer bus system is serially transmitted and received (full duplex) by the AICA, with proper formatting and error checking. (The term *full duplex*

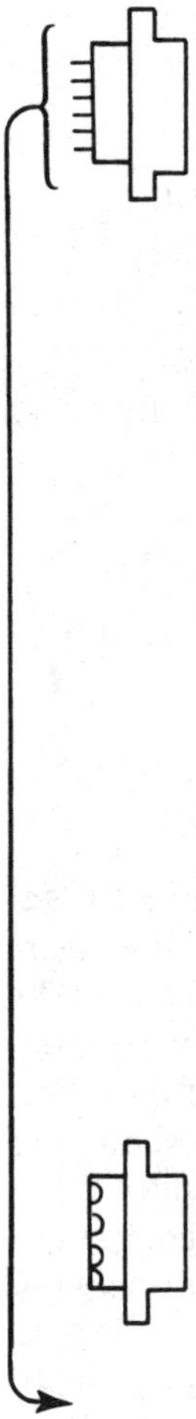

DTE		
Data from terminal	Pin 2	Transmitted data
Tells modem that terminal wants to send data	Pin 4	Request to send
Tells modem that terminal is connected, powered up and ready	Pin 20	Data terminal ready
Common ground reference or all signal lines	Pin 7	Signal ground
Safety or power-line ground for equipment	Pin 1	Protective ground
Clock signal from terminal	Pin 24	Transmit signal element timing
Identical in function to pins 3 and 5, except as they apply only to systems with full secondary channels implemented	Pin 14	Secondary transmitted data
Tells modem to turn on the secondary channel carrier; used for supervisory operation	Pin 19	Secondary request to send
Used by modem/terminals with programmable data-rate selection	Pin 23	Data-rate signal selector
Pins receiving signals from modem: 1, 3, 5, 6, 7, 8, 12, 13, 15, 16, 17, 21, 23		

DCE (Modem)		
Received data	Pin 3	Data from CPU
Clear to send	Pin 5	Tells terminal that it may place data on transmit data line (pin 2)
Data set ready	Pin 6	Tells terminal that modem is connected, powered up, and ready
Signal ground	Pin 7	Common ground reference for all signal lines
Protective ground	Pin 1	Safety or power line ground
Received line signal detector	Pin 8	Tells terminal that carrier is being received from answering (or remote) modem
Transmission signal timing Receive signal timing	Pin 15 Pin 17	Clock signal from modem (used only with synchronous modems)
Secondary received data Secondary clear to send	Pin 16 Pin 13	Identical to pin 2, except applies only to systems with full secondary channel implemented

Fig. 4.25 RS232C pin or line format.

Secondary received line signal detect	Pin 12	Tells terminal that carrier is present on secondary channel
Signal quality detector	Pin 21	Used by some modems which have sign-evalulating circuits
Data rate signal selector	Pin 23	Used by modem/terminals with programmable data-rate selection

Pins receiving signals from terminal: 1, 2, 4, 7, 14, 19, 20, 23, 24

Fig. 4.25 Continued

applied here means that serial data bits can be transmitted and received simultaneously on the same line or channel.)

The functional configuration of the ACIA is programmed via the data bus during system initialization. A programmable *control register* provides variable word lengths, clock division ratios, transmit control, receive control, and interrupt control. Three I/O lines are provided to control external peripherals or modems. A *status register* is available to the computer and reflects the current status of the transmitter and receiver.

Modem Figure 4.27 shows the relationship to the system, and a block diagram, for the modem. In most cases, the modem is actually a subsystem designed to be integrated into a wide range of equipment using serial data communications.

The modem provides the necessary modulation, demodulation, and supervisory-control functions to implement a serial data-communications link, over a voice-grade channel (telephone line), using FSK at bit rates up to 600 bits per second. The modem can be implemented into a wide range of digital devices (video terminal, computer, storage device, etc.) or it can be a peripheral device.

4.8 Basic troubleshooting approaches

Details of how basic solid-state digital troubleshooting techniques can be applied to microprocessors and related equipment are discussed throughout the remainder of this book. In this section, you review the basic approaches for troubleshooting both control-type microprocessors (Chapter 9) and those microprocessors used in computer-type applications. As you will see, the basic troubleshooting approaches for the two types of microprocessors are quite different.

4.8.1 Troubleshooting control-type microprocessors

When troubleshooting control-type microprocessors (where programs, as such, are not involved), the logic probe and logic pulser are generally sufficient to locate most troubles. In fact, you can probably get by with only a scope in most cases. (Logic probes and pulsers, as well as a variety of digital test instruments, are discussed in Chapter 5.)

The basic approach for control-type microprocessors (such as those used in VCRs, videodisc players, TV tuners, CD players, camcorders, etc.) is to monitor the inputs and outputs of the microprocessor with a logic probe or scope. This is

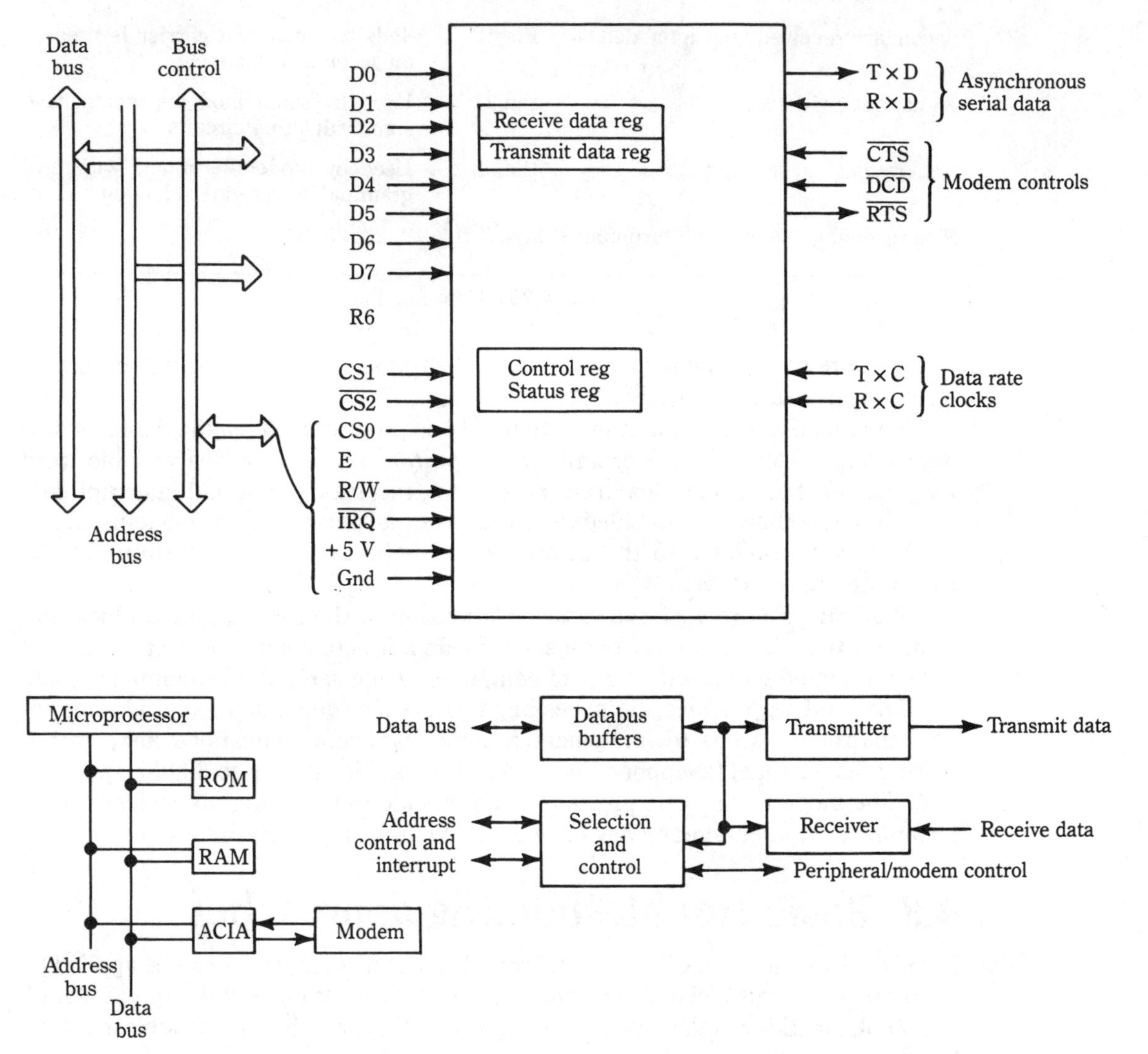

Fig. 4.26 Typical ACIA circuits and functions.

relatively easy when compared to troubleshooting programmed equipment (computers, etc.). However, you still must study the troubleshooting literature to determine the relationship between inputs and outputs of the microprocessor.

An oversimplified example is shown in Fig. 4.28. Assume that a control-type microprocessor is used to control operation of the cylinder motor in a VCR. During search operation, tape speed is increased by five (5X). Because of this increase in tape speed, the cylinder rotation speed must also be changed slightly to maintain proper horizontal sync in the playback video signal. To get this control, the slow and search control signals from the system-control microprocessor are applied to search-correction microprocessor IC_3, at pin 6.

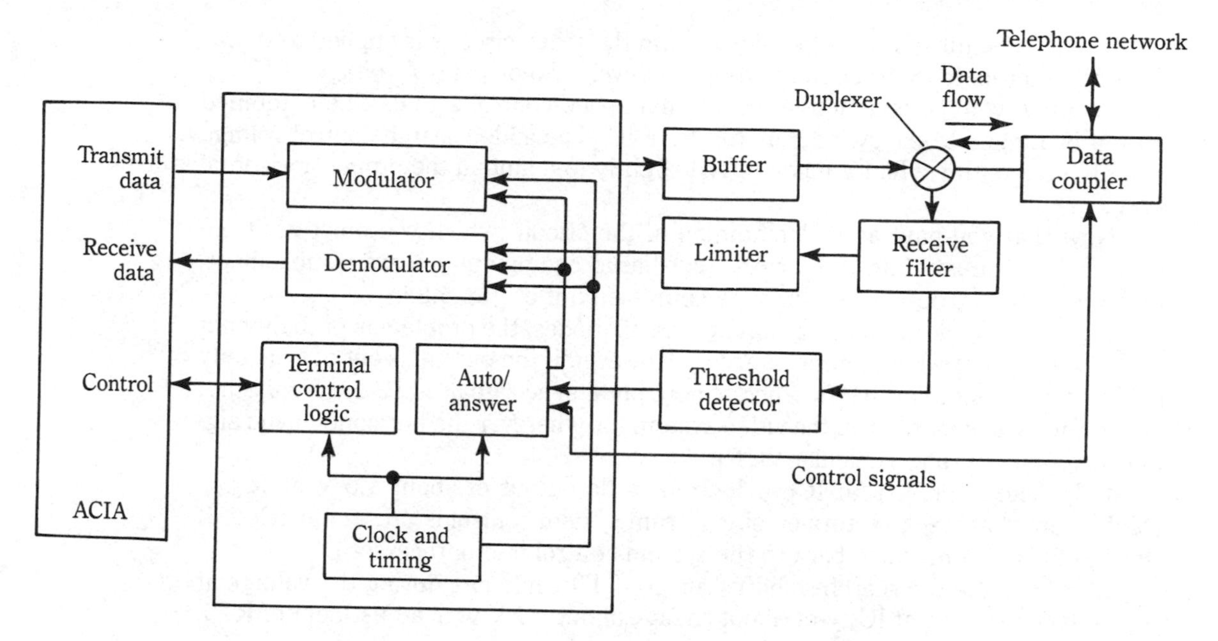

Fig. 4.27 Typical modem circuits and functions.

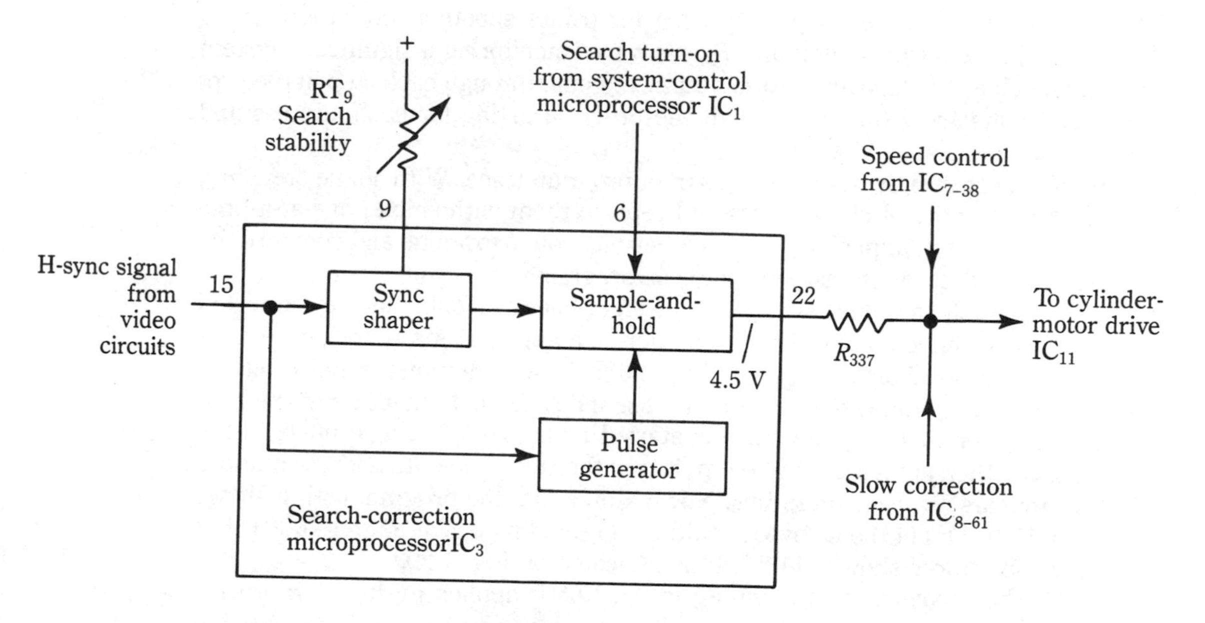

Fig. 4.28 Example of troubleshooting for control-type microprocessors.

The horizontal sync (H-sync) signal from the video circuits is applied to IC_{3-15}. A sample-and-hold (S/H) circuit within IC_3 develops a dc control voltage at IC_{3-22}. This control voltage is added to the normal speed-control signal. The combined output is applied to the cylinder motor drive IC. The added search-control voltage from IC_3 shifts the cylinder motor speed slightly to maintain the proper horizontal sync.

Now that you have an understanding of the circuit (which you must get from the service manual of any microprocessor-based equipment you are troubleshooting), see how to troubleshoot the very simple circuit of Fig. 4.28.

If the picture is out of horizontal sync in all modes, the problem is probably not in the search-control microprocessor IC_3. However, if the picture is out of sync only during search, look for the presence of horizontal-sync signals at IC_{3-15}. If missing, check the sync separator in the video system (or wherever the horizontal-sync signals originate in that particular VCR).

If H-sync is present at IC_{3-15}, look for a dc output of about 4.5 V at IC_{3-22}. (Make sure that a search turn-on signal from system control is present at IC_{3-6}. If the signal is absent, trace back to the system-control microprocessor.)

Slowly rotate the search-stability control RT9 while monitoring the voltage at IC_{3-22}. If the voltage at IC_{3-22} does not change about ± 1 V with adjustment of RT9, suspect IC_3 and/or RT9.

Admittedly, this is an oversimplified example and is based on an exact knowledge of the circuit function. However, the example is quite typical of troubleshooting in the circuits of control-type microprocessors.

4.8.2 Troubleshooting computer-type microprocessors

The *program trace* is the classic approach for troubleshooting any programmed device. In simple terms, a program trace involves monitoring a significant system function (such as the data and address busses), going through each step in the program, and comparing the results with the program listing for each address and step.

Single stepping is the most basic form of program trace. With single stepping, you remove the normal clock pulses and replace them with single, one-at-a-time pulses from a switch or pushbutton. This permits you to examine and compare the data at each address with that shown in the program.

As an example, assume that each of the eight lines on a data bus are connected to a multitrace scope so that the bit on each line appears as a pulse on a corresponding trace, as shown in Fig. 4.29*a*. Under these conditions, a pulse on the trace indicates a 1, and the absence of a pulse indicates 0, on that line of the bus.

A single-stepping program trace is started by applying a single pulse to the reset line and then applying sufficient pulses to the clock line until address number 1 (0001) appears on the address bus. Now assume that the program listing shows that hex 7F (0111 1111) data byte should be on the data bus at address 0001, but that the scope traces show 0011 1111, as indicated on Fig. 4.29*a*.

Under these conditions, the wrong instruction is applied to the microprocessor, and the system fails. This failure can be caused by a broken (or grounded) line on the data bus (say line 6 is grounded), a defect in memory, or absence of a mem-

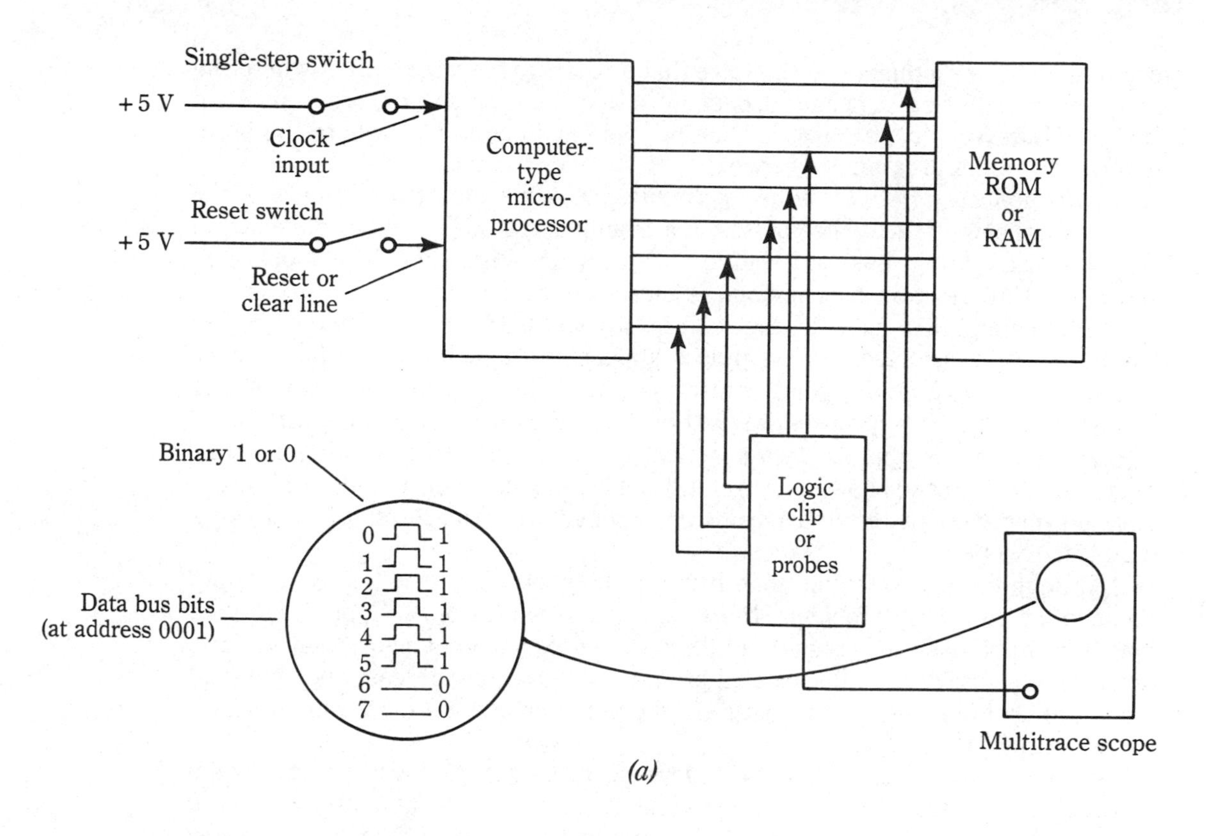

(a)

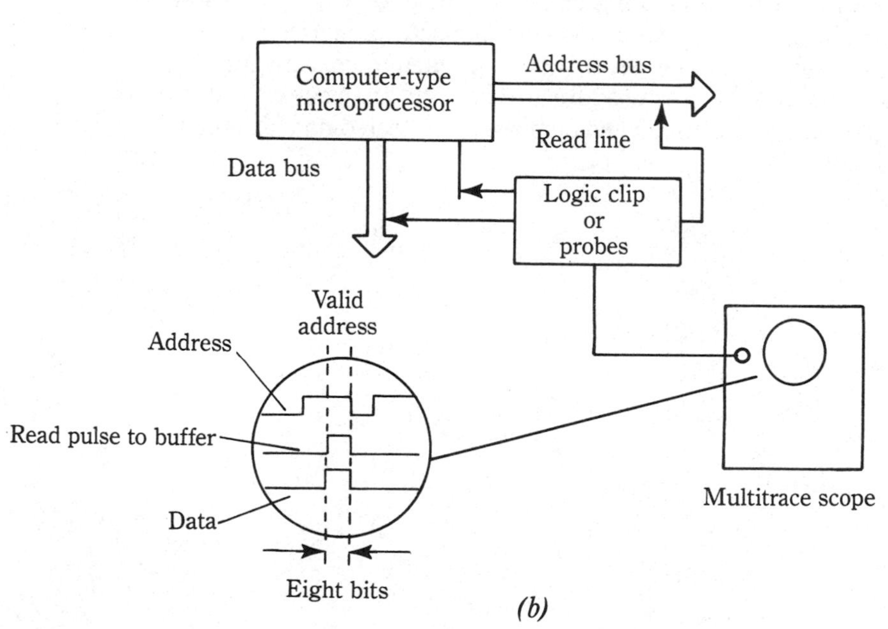

(b)

Fig. 4.29 Example of troubleshooting for computer-type microprocessors.

ory-read pulse (or a memory-read pulse that appears at the wrong time, where the memory data buffer is opened too soon or too late), or several other possible causes. However, you have isolated the problem and determined where the problem occurs in the program sequence.

If the problem appears to be one of timing, the scope can be used to check the time relationship of the related pulses. For example, the scope can be connected to the data bus, address bus, and read line, as shown in Fig. 4.29*b*. The scope then shows the time relationships among the pulses on these lines.

In this simplified example, the read pulse must hold the memory data buffer closed until the selected address pulses appear on the address bus (sometimes known as the *valid address* point), must hold the buffer open just long enough for all eight data bits to appear on the data bus, and then must close the buffer until the next address is applied. In a practical case, the entire timing diagram (discussed in Sec. 4.6 and shown in Fig. 4.13) can be duplicated on a multitrace scope, or important sections of the diagram can be monitored (two lines at a time) on a dual-trace scope.

Although single stepping, together with a check of system timing, can pinpoint many computer problems, there is an obvious drawback. Typically, a data byte is eight bits and thus requires eight clock pulses or eight one-at-a-time pushes of the single-step button. Because all program steps require at least one byte (and often two or three bytes), you must push the button many times if the malfunction occurs far into the program.

In any event, all forms of single stepping cause you to spend endless hours comparing program listings against machine (binary) readouts at addresses. If you are already familiar with the troubleshooting of programmed devices, you know that the most time-consuming part of the task is making such comparisons.

The *logic analyzer* (Chapter 5) overcomes this basic problem by permitting you to select for display all data at a particular address. The logic analyzer then runs through the program at near-normal speed and displays the selected data at breakpoints in the program.

5 Digital test equipment

It is assumed that you already know how to use basic electronic test equipment (meters, scopes, generators, etc.), including all of the precautions for general safety (your safety as well as that of the equipment). If not, and you plan to troubleshoot digital equipment, consider reading the author's books on test equipment and test procedures. Troubleshooting for the simplest of digital devices requires a thorough knowledge of test equipment and test procedures.

This chapter is devoted to test equipment used specifically to troubleshoot digital equipment. This is particularly important for microprocessors used in computer applications, since digital networks can be an electronic nightmare when hundreds or thousands of circuits are interconnected (as they are in the simplest of computer and in many control-type microprocessor applications).

Digital system faults are located best by analyzing test results (response to input signals, presence or absence of pulses and signal levels, step-by-step tracing of instructions and data through each step of the program, etc.).

The chapter includes a variety of test equipment for digital troubleshooting and covers operating principles and/or characteristics. The discussions describe how features found on present-day equipment relate to specific problems in troubleshooting.

5.1 Meters

The meters for digital troubleshooting should have the same characteristics as those for other solid-state troubleshooting. The ohmmeter portion of the meter should have the usual high-resistance ranges. Many of the troubles in the most sophisticated and complex computers are caused by such common problems as cold-solder joints, breaks in PC wiring that result in high resistance, or shorts and partial shorts between wiring or traces (producing an undesired high-resistance condition). The internal battery voltage of the ohmmeter should not exceed any of the voltages used in the circuit being checked.

Typically, accuracy should be ± 2% or better for dc voltage and ± 3 to ± 5% accuracy for the ac scales. Note that the ac scales are generally used only in checking power-supply or input-power functions, because most digital signals are in pulse form (requiring a scope for display and measurement).

Microprocessor operating voltages are low (generally less than 24 V), and digital pulses and logic levels are typically 5 V. This means that you must have good resolution on the low-voltage ranges (as is the case with most digital meters). For example, assume that a 0 is represented by 0 V and a 1 is represented by anything over 3 V. This means that if an input of 3 V or greater is applied to an OR gate, the gate output is a 1 (true or high). An input of something less than 3 V, say 2 V, produces a 0 (false or low) output.

Under these conditions, if the voltmeter is not capable of an accurate readout between 2 and 3 V on some scale, you can easily arrive at a false conclusion. If the OR gate in question is tested by applying a supposed 3 V, which is actually 2.8 V, the OR gate might or might not operate to indicate the desired 1 output. As I discuss in Chapter 6, a high-resolution voltmeter is often useful in locating abnormal PC-board currents (shorts, etc.).

5.2 Scopes

Scopes with a *dual-trace* or *multi-trace* feature are essential for digital troubleshooting. Many troubleshooting procedures are based on monitoring two time-related pulses (for example, an input pulse and output pulse or a clock and a readout pulse). With dual trace, the two pulses can be observed simultaneously—one above the other or superimposed.

In some digital tests, the clock pulse is used to trigger both horizontal sweeps. This permits a *three-way time-relationship measurement* (clock pulse and two circuit pulses, such as one input and one output). A few scopes have multi-trace capabilities. However, this is not standard. Usually, such a scope is provided with plug-in options that increase the number of horizontal sweeps.

A scope for digital troubleshooting should also have a *triggered horizontal sweep*. Preferably, the delay introduced by the trigger should be very short. Often, the pulse to be monitored occurs shortly after an available trigger. In other cases, the horizontal sweep must be triggered by the pulse to be monitored.

In some scopes, a delay is introduced between the input and the vertical-deflection circuits. This permits the horizontal sweep to be triggered before the vertical signal is applied, thus assuring that the complete pulse is displayed.

Because the pulses used in most digital circuits are about 5 V or less, the sensitivity of both the vertical and horizontal channels should be such that full-scale deflection can be obtained (without overdriving or distortion) with less than a 5-V signal applied. Although *storage scopes* (for display of transient pulses) and *sampling scopes* (for display and measurement of very short pulses) are recommended in some service literature (primarily for lab work), most routine digital troubleshooting can be done with a conventional scope.

5.3 Generators

Some manufacturers recommend a *pulse-generator* for troubleshooting digital circuits. A pulse generator should be capable of duplicating any pulse present in the circuits being tested. So the generator output should be continuously variable (or at least adjustable by steps) in amplitude, pulse duration (or width), and frequency (or repetition rate) over the same range as the circuit pulses.

Typically, repetition rates (or clock frequencies for microprocessors) are less than 10 MHz. The pulses are rarely longer than 1 s, or shorter than 1 ns, although there are exceptions. Again, most digital-circuit pulses are 5 V or less in amplitude. In rare cases, the pulses can be from 10 to 15 V.

For routine digital troubleshooting, the logic pulser described in Sec. 5.5 can supply all pulses necessary. However, there might be special cases where pulse generators with special features must be used. For example, some tests require two output pulses from the generator, with a *variable delay* between the pulses. Other tests may require an output pulse that can be triggered from an external source.

5.4 Logic probe

The logic probe is sometimes called the *digital screwdriver* because the probe is as commonly used by troubleshooters of digital circuits as the screwdriver is by mechanics. By means of a simple lamp indicator, a logic probe tells you the logic state of a digital signal and allows very brief pulses to be detected.

A typical logic probe is shaped like a pencil and has a lamp near the probe tip. When the probe is touched to the circuit or line, the lamp gives an immediate indication of the logic states, either static or dynamic, existing in the circuit.

Logic probes detect and indicate a high and low (1 and 0) level, as well as intermediate or "bad" logic levels (including open circuits) on a single line of a digital circuit. For example, the probe can be connected to one of the address or data-bus lines (or the clock line, chip-select line, etc.), and will indicate the state (0, 1, or high-impedance) of that line. A logic probe can also check each output of a control-type microprocessor (Chapter 9) as well as the corresponding inputs.

Figure 5.1 shows how the logic-level lamp responds to voltage levels and pulses found in typical digital circuits. The lamp can give any of four indications: (1) off, (2) dim (about half brilliance), (3) bright (full brilliance) or (4) flashing on and off.

The lamp is normally in the dim state and must be driven to one of the other three states by voltage levels at the probe tip. The lamp is bright for inputs at or above the 1 state, and off for inputs at or below 0. The lamp is dim for voltages between the 1 and 0 states and for open circuits. Pulsating inputs cause the lamp to

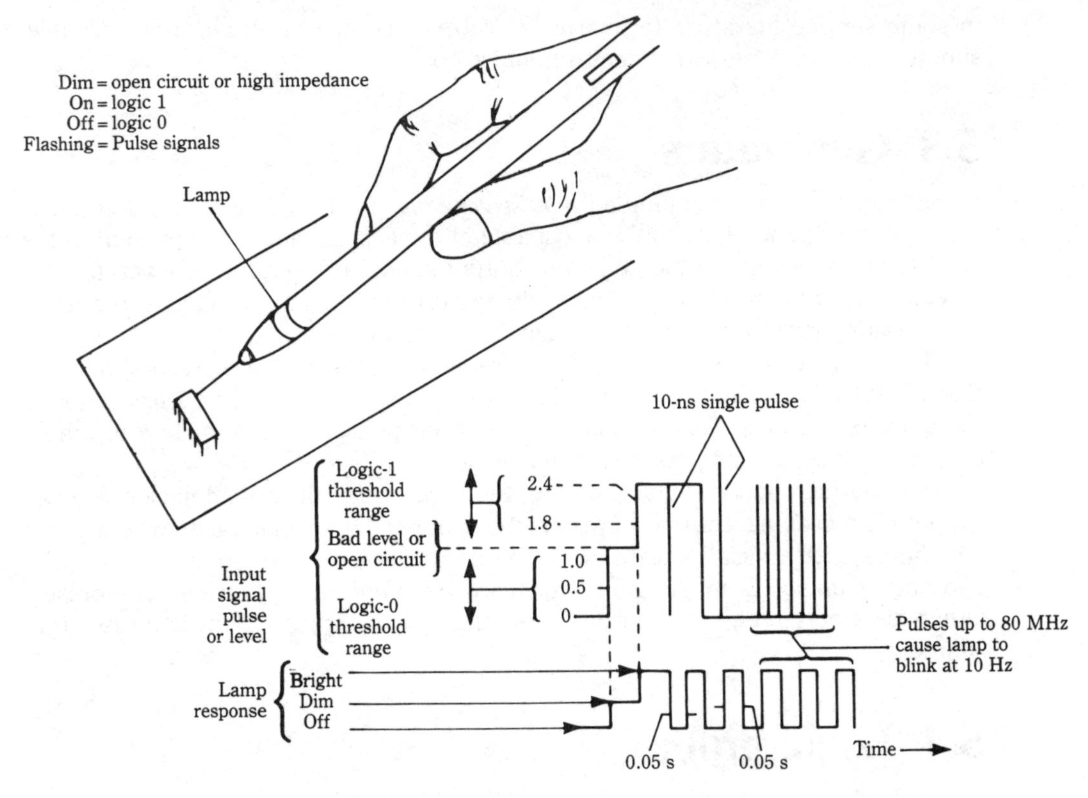

Fig. 5.1 Logic-probe lamp response.

flash at about a 10-Hz rate (regardless of the actual pulse rate on the line being probed).

In some digital troubleshooting situations, the probe is used alone to detect the presence of pulse activity on lines that interconnect components or to check the level on that particular line under a given set of conditions. In other cases, the logic probe is used with the logic pulser (Sec. 5.5).

For example, the pulser can be used to simulate the input to an element in a digital circuit, while the probe senses activity of the element (at the element output). The logic probe is probably the most nearly universal of all digital test instruments, because the probe can be used to verify continuity; signal flow; address decoding; or clock, switch, and bus activity (both with and without a pulser).

The following paragraphs provide brief descriptions of how you can use the logic probe in troubleshooting of digital equipment. More detailed procedures and approaches are given in the remaining sections of this chapter, and in Chapters 7 and 8.

5.4.1 Logic-probe power supply

The logic probe can be powered from the digital-equipment power supply or from a regulated dc power supply. If a separate power supply is used, the supply and digital-equipment grounds should be connected together. A ground wire (provided with most probes) can be used for the connection. The ground wire is a convenient means of connecting grounds when the external supply is used, and (according to some manufacturers) improves both pulse-width sensitivity and noise immunity. However, the use of a ground wire is optional (on most probes). The typical power-supply voltage is 4.5 V to 15 V.

5.4.2 Pulse detection characteristics

Typically, positive pulses of about 10 ns or greater trigger the indicator on for at least 50 ms. Negative pulses cause the indicator to go off momentarily. Note that the logic probe is ideal for detecting short-duration and low-repetition-rate pulses that are difficult to observe on a scope.

5.4.3 Monitoring three-state logic outputs

The bad-level feature of a logic probe is useful for testing three-state outputs. The 1 and 0 states produce lamp-on and lamp-off indications, respectively. The third (or high-impedance) state is detected as an open-circuit (or bad level) condition and produces a lamp-dim indication. The bad-level indication is also useful for detecting floating or disconnected inputs that look like a bad level to the probe.

5.4.4 Basic probe techniques

One very useful technique when using the probe to troubleshoot digital circuits is to monitor all significant control signals (such as reset, halt, memory read, flag, clock, and so on) with the clock running at normal speed. This quickly answers such questions as "Is there a clock signal?" or "Is there a memory-read signal?" or "Is there a reset ?"

If the probe lamp flashes on and off, indicating pulse activity, it is reasonable to assume that all control signals are normal. (Of course, timing can be off, but this requires comparison of scope displays against timing diagrams found in the service literature.)

This initial check of control signals with the probe can prove quite helpful. For example, if there is no flashing indication when the probe is connected to the clock line, there is no clock pulse. Likewise, if the probe does not flash when connected to a memory-read line in a computer, the microprocessor is probably stuck somewhere in the program (because the memory-read line is usually pulsed at each step of the program to open the RAM or ROM at each address).

Once you are certain that the control signals are all present, replace the clock pulse with a very slow pulse from a logic pulse and monitor all remaining control signals on a real-time basis. Real-time analysis, coupled with the ability to inject pulses anywhere in the circuit with a logic pulser, makes for rapid troubleshooting and fault finding in the control lines. Unfortunately, one of the drawbacks to a logic probe is its inability to monitor buses or other multiple lines simultaneously. That is one reason why you need a logic analyzer (Sec. 5.9) to troubleshoot bus-structured devices (such as a computer).

5.5 Logic pulser

Figure 5.2 shows operation of a logic pulser. As shown, the pulser is the ideal troubleshooting companion to the logic probe. When used in this application, the logic pulser is an in-circuit stimulus device that automatically outputs pulses of the required logic polarity, amplitude, current, and width to drive lines and other points (*nodes*) high and low.

A typical pulser has several pulse modes available. In operation, pulsers are hand-held logic generators used for injecting controlled pulses into digital circuits. The pulses are compatible with most digital equipment. Pulse amplitude depends on the supply voltage (typically 3 to 18 V) used as the supply for the pulser. The frequency and number of pulses generated are controlled by a push-slide switch on the pulser. A flashing LED located on the tip indicates the output mode. Pulse current and width depend on the load presented by the circuit under test.

The pulser is programmed by pushing the switch once for each single-pulse output or a specific number of times for continuous pulse streams or pulse bursts

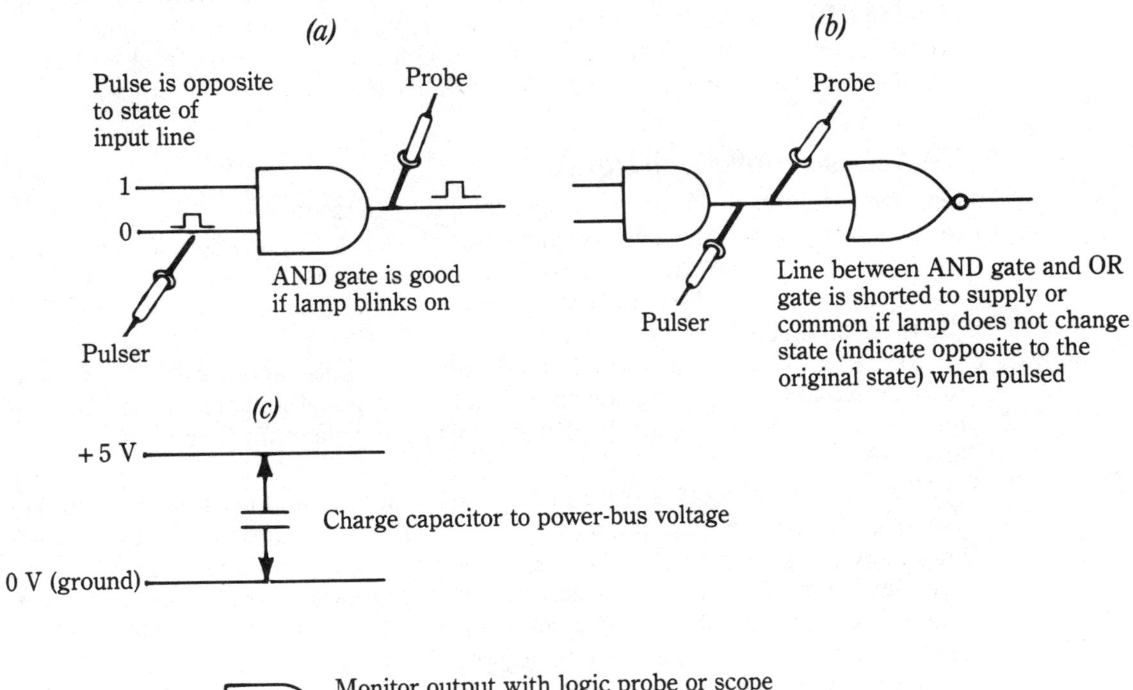

Fig. 5.2 Logic-pulser operation.

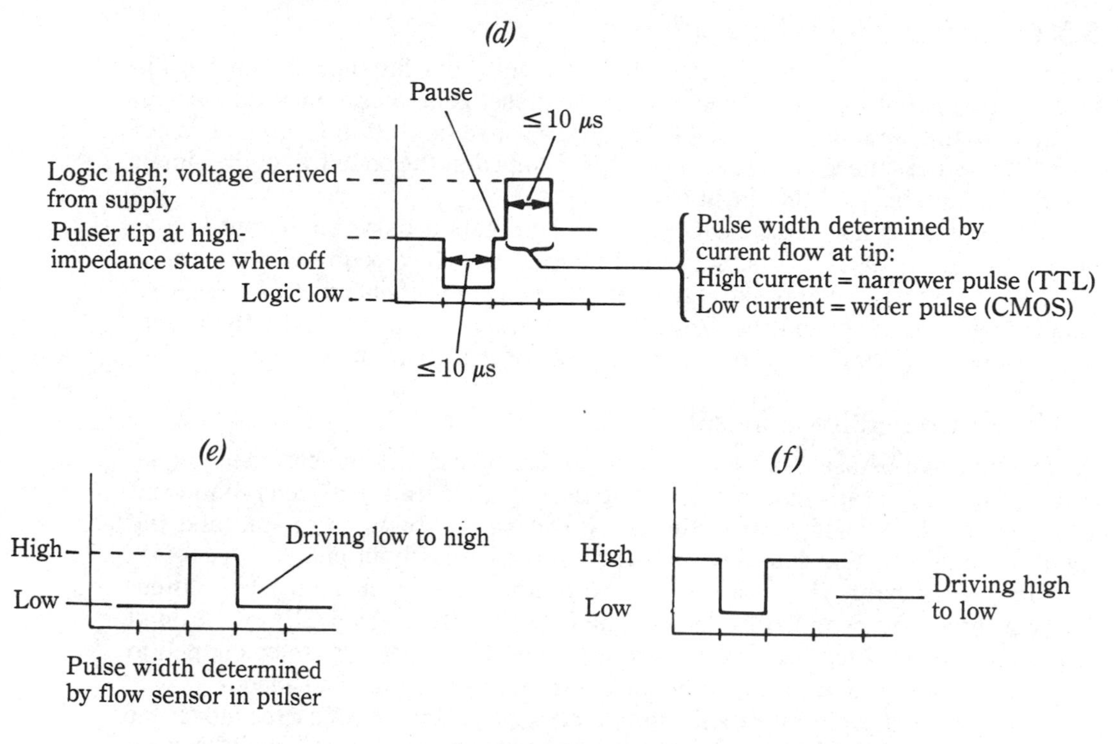

Fig. 5.2 Continued

at selected frequencies. The pulses are also applied to the LED on the pulser tip. (The pulses applied to the LED are slowed down for visibility.) Pressing the switch automatically drives a line (usually connected to an IC input or output) from low to high (0 to 1) or vice versa. The high source and sink current of the pulser can override IC output points that were originally in either the 0 or 1 state.

Pulses can be injected while the circuit is operating, and no disconnects are needed. Some probes have a multi-pin kit or accessory that provides up to four pulses simultaneously (say to simulate four bits of an eight-bit word on a data or address bus). The pulser output has three states. In the off state, the high output impedance of the pulser ensures circuit operation is unaffected by probing until the pulser switch is pressed. Pulses can be injected while the circuit is operating.

In the most common troubleshooting application, the digital circuit inputs are pulsed, while the effects of the corresponding outputs are monitored with a probe. However, the pulser can also be used to force ICs on or off as a substitute for the clock pulse. Likewise, the pulser can force overriding pulses into logic circuits (either single pulses or pulse streams).

The following paragraphs provide brief descriptions of how the logic pulser can be used in troubleshooting basic digital equipment. More detailed procedures and approaches are given in the remaining sections of this chapter, and in Chapters 7 and 8.

5.5.1 Testing gates with a pulser

Figure 5.2*a* shows how an AND gate can be tested by pulsing the gate input while monitoring the output with a logic probe. The pulser generates a pulse opposite to the state of the input line and should change the gate output. If the output does not change, it is possible that the gate output is clamped in the state by another input (such as a high on the other input of an OR gate).

If the pulse is not detected at the AND-gate output, pulse the output line as shown in Fig. 5.2*b*. If the output is not shorted to +3 V or ground, the probe should indicate a pulse opposite to the original indication. If not, check for external shorts (solder bridges and the like) before removing the IC. Shorts at the inputs and outputs are best located by the current tracer described in Sec. 5.6.

5.5.2 Simulated logic pulser

A capacitor can be used as a logic pulser. Figure 5.2*c* shows the basic technique. Simply charge the capacitor by connecting the capacitor leads between ground and a power-bus line. Then connect the charged capacitor between ground and the input to be tested. The capacitor discharges, creating an input pulse.

Be sure to charge the capacitor to the correct voltage level and polarity. Often, it is convenient to connect one lead of the capacitor (typically 1 μF) to a ground clip, with the other lead connected to a test prod. The capacitor can be clipped to ground and then the prod tip can be moved from bus to input as needed.

Although a capacitor is no substitute for a logic pulser, use of a capacitor is far superior to connecting the line to be tested directly to a power bus. This is because some circuits can be damaged by prolonged application of the voltage. Equally important, the results of such a test are inconclusive. For example, many faulty circuits operate normally when a fixed bus voltage is applied (even momentarily) but do not respond to a pulse.

5.5.3 Pulser output waveforms

Figures 5.2*d* through 5.2*f* show the output-pulse sequence from a typical logic pulser. When the button is pressed, the pulser outputs a single dual-polarity pulse. The pulse first goes low, pauses, and then goes high. When the pulser tip is pulsed into an open circuit, the pulse appears as shown in Fig. 5.2*d*.

The output pulse shown in Fig. 5.2*d* is at the maximum width (for a typical logic pulser). When pulsing into a circuit, the current flow through the pulser tip is sensed by an *output-sensing circuit* that shuts the pulser off. This built-in circuit turns the pulser off faster for TTL than CMOS, thus keeping the total energy low so as to eliminate any damage to the circuit being tested.

Note that pulse height, or amplitude, is derived from the power supply to which the pulser is connected. For this reason, the pulser should always be powered from the digital equipment under test or a power supply of the same voltage.

When driving a point in the low state to a high state, as shown in Fig. 5.2*e*, the pulse has no effect until the circuit goes high (because the point is already low). When the pulser goes high, there is sufficient output drive to take any normal circuit or bus high momentarily. Total energy is limited (low duty cycle), however, to exclude the possibility of circuit damage.

When driving a point in the high state to a low state, as shown in Fig. 5.2*f*, the circuit can be taken low by the pulser (because the circuit starts high). However, the circuit cannot be driven higher if already in the high state. As a result, the high portion of the pulser output has no effect on the circuit.

5.6 Current tracer

Current tracers are hand-held probes that permit precise localization of low-impedance faults (shorts, etc.) in many electrical systems, particularly in digital PC wiring. The current tracer is one of the most sophisticated test instruments for troubleshooting digital circuits. Unfortunately, because of the cost, current tracers are not usually found in most consumer-electronics service shops.

The current tracer detects current activity on PC traces (or other circuit points) with an inductive pickup at the probe tip. By adjusting the sensitivity control and observing the intensity of the tracer lamp when placed on a pulsating trace or line, you can identify current paths and relative magnitudes to locate a bad device.

A current tracer senses the magnetic field generated by a pulsing current internal to the circuit, or by current pulses supplied from an external stimulus such as a logic pulser (Sec. 5.5). Indications of current pulses is provided by the tracer lamp. Adjustment of tracer sensitivity over the typical range of 1 mA to 1 A is provided by a sensitivity control near the lamp. The tracer is self-contained and requires less than 75 mA at 4.5 to 18 V from any convenient source.

In a typical digital troubleshooting situation, the current tracer is used to track down the cause of stuck lines or IC terminals. The tracer can tell approximately how much pulse current is present, and (more important) what path the current takes. When a logic pulser is used to inject current into a circuit without pulse activity, the impedance and general nature of the problem (line shorts, gate shorts, etc.) can be estimated. Then, the actual low-impedance point can be found by tracing the path of the current from the logic pulser to that point in the circuit. The current either goes someplace it should not (a shorted line), or the current enters a component (probably an IC) that is stuck, shorted or turned on.

The following paragraphs provide brief descriptions of how the current tracer can be used to troubleshoot basic digital equipment. More detailed procedures and approaches are given in the remaining sections of this chapter and in Chapters 7 and 8.

The current tracer operates on the principle that whatever is driving a low-impedance point must be delivering the majority of the current. Tracing the path of this current leads directly to the fault. Problems compatible with this method follow:

- Shorted inputs of ICs
- Solder bridges on PC boards

- Shorted conductors in cables
- Shorts in voltage distribution networks (such as VCC-to-ground shorts)
- Stuck data or address buses including three-state buses
- Stuck wired-AND circuits

5.6.1 Basic current-tracer troubleshooting

The current tracer is most effective when used to locate the specific source of excessive current if you suspect a short or low-impedance condition. In typical operation, you align the mark on the tracer tip along the length of the PC trace (at the driver end) and adjust the sensitivity control until the lamp just turns on. You then move the tracer along the PC track, or place the tracer tip directly on the terminal (usually IC pins), while observing the lamp.

This method of following the current path leads directly to the fault responsible for the abnormal current flow. If the driving point does not provide pulse stimulation, the terminal can be driven externally by a logic pulser at the driving point. The following paragraphs describe troubleshooting techniques for some of the more common problems.

5.6.2 Current-tracer sensitivity

As shown in Fig. 5.3*a*, current-tracer sensitivity can be varied over a range from about 1 mA to 1 A. It is critical to troubleshooting success to set the sensitivity correctly using the following steps.

1. Select the circuit, PC trace, or IC terminal to be traced.
2. Place the tracer tip at the driver output. This can be the output from the logic pulser tip (if used) or from an IC that drives a line or PC trace.
3. Align the tracer tip. As shown in Fig. 5.3*b*, the tracer tip is directional. That is, current paths oriented 90° out of phase with the pickup coil tend to null out. So, proper tip orientation helps eliminate crosstalk from traces on different layers or at different angles.

(a)

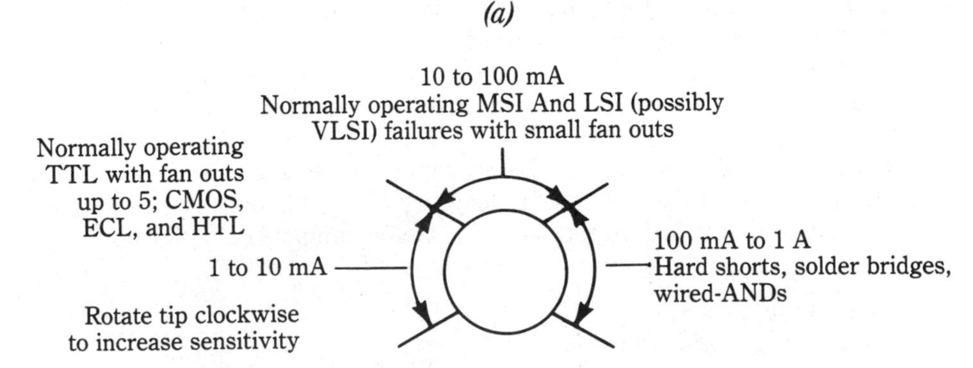

Fig. 5.3 Current-tracer sensitivity range and orientation.

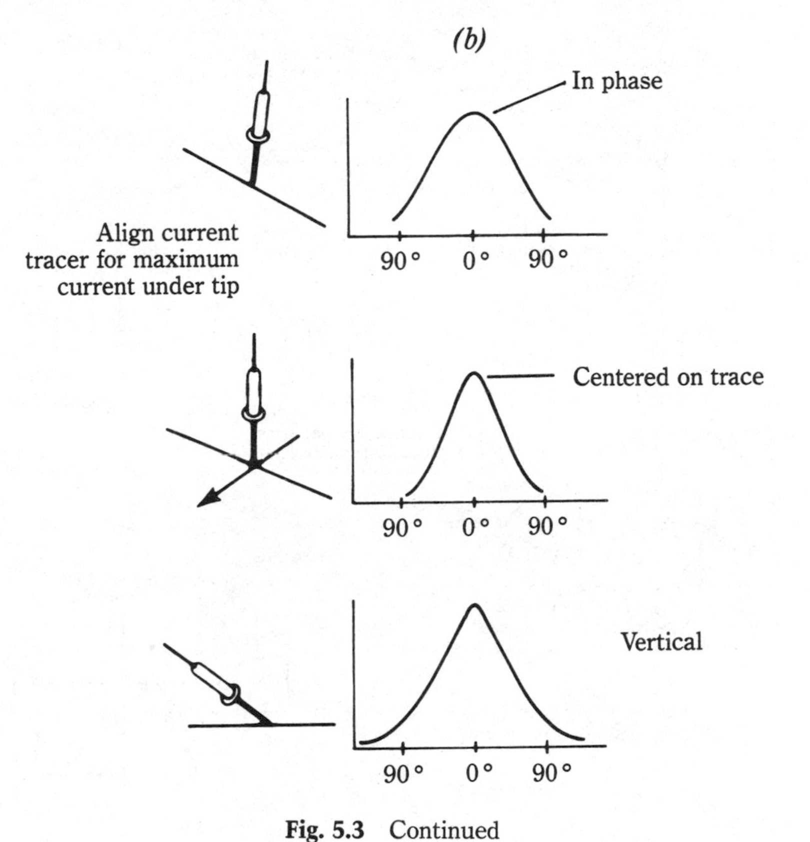

Fig. 5.3 Continued

4. Set the sensitivity control for half-brilliance on the tracer indicator lamp.
5. Leave the sensitivity control at the same setting until the fault is located or until test conditions change.

5.6.3 Wired-AND troubleshooting

One of the most difficult problems in troubleshooting digital ICs is a stuck wired-AND circuit. Typically, one of the open-collector gates connected in the wired-AND mode can continue sinking current after the gate has been turned off. The current tracer provides an easy method of identifying the faulty gate. (Of course, if the gate is located in an IC, the entire IC must be replaced. However, you have still located the problem and can take whatever action is best suited to the situation.)

To locate a wired-AND problem with a current tracer, place the tracer tip on the gate side of the pull-up resistor, as shown in Fig. 5.4*a*.

Align the mark on the tracer tip along the length of the PC trace, and adjust sensitivity until the tracer lamp is just fully on. If the lamp does not go on, use a logic pulser to excite the line as shown. Place the tracer tip on the output pin of each gate. Only the faulty gate (or pin) causes the tracer lamp to go on.

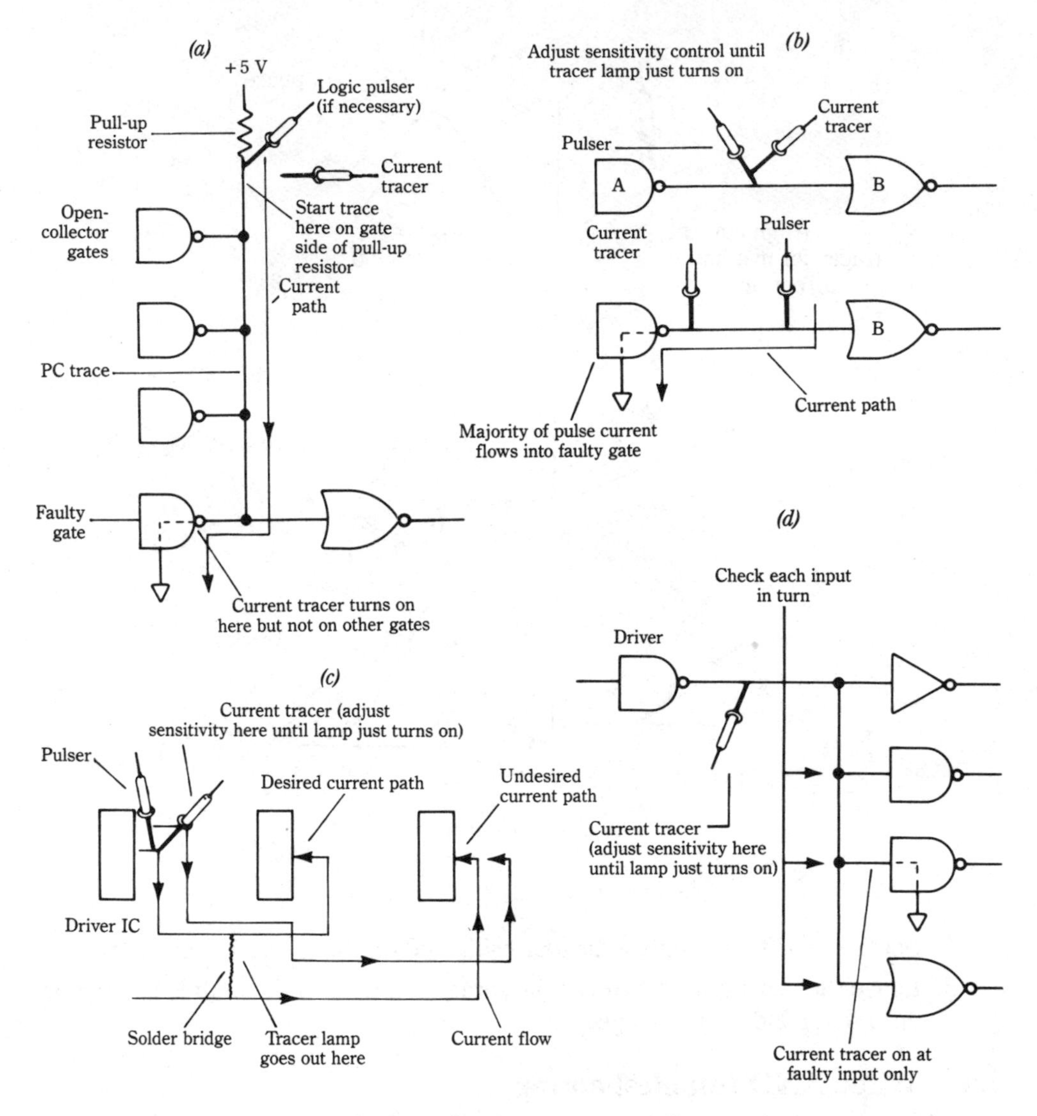

Fig. 5.4 Current-tracer operation.

5.6.4 Gate-to-gate troubleshooting

The current tracer and logic pulser can be combined to pinpoint the defect quickly when a low-impedance fault (short, full, or partial) exists between two gates. An example is shown in Fig. 5.4*b* where the output of gate A is shorted to ground.

Place the pulser midway between the two gates and place the current tracer tip on the pulser pin. Pulse the line and adjust the tracer sensitivity control until the tracer lamp just turns on. Now place the tracer tip next to gate A, and then gate B, while continuing to pulse the line.

The tracer should turn on only at the gate-A side, because gate A (the defect in this example) is sinking the majority of the current. If the tracer does not turn on when placed between the pulser and gate A, look for a short on the line between the pulser and gate B (or within gate B).

5.6.5 Solder bridge problems

Figure 5.4*c* shows an example of an incorrect current path because of a solder bridge between two traces on a PC board. To locate such a problem with the current tracer, start the tracer at the driver and follow the trace.

As the tracer moves along the trace, the lamp remains on until the tracer passes the bridge and then goes off. This indicates that the current has found some path other than the trace.

Remember that the tracer lamp goes off (as the tracer is moved past the bridge along the trace) only if properly adjusted at the driver. As shown in Fig. 5.4*c*, you must adjust sensitivity so that the lamp just goes on at the driver end.

5.6.6 Multiple-gate input problems

Figure 5.4*d* shows how to check the trace between a single output, or driver, and multiple inputs. In this case, place the tracer tip on the output pin of the driver and adjust the sensitivity control until the tracer lamp just turns on. Then check the input pins of the remaining gates. If one of the input pins is shorted, that pin will be the only one to turn the lamp on.

If it is impossible to turn the lamp on when the tracer is at the driver output (with full sensitivity), the problem probably exists in the driver. To confirm this, use the tracer as described for gate-to-gate faults in Sec. 5.6.4.

5.6.7 Current-tracer troubleshooting tips

The preceding paragraphs describe basic operation of the current tracer. Now consider some techniques that simplify use of a current tracer. (Use of the tracer to solve actual troubleshooting problems is discussed in Chapters 7 and 8.)

Trace spacing The tracer current-sensing coil is very small (about 0.010-in diameter), although the protective plastic cover makes the coil appear larger. This means that PC-board traces can be quite close together, but current tracing is still possible. As shown in Fig. 5.5*a*, two traces carrying identical current can be located 0.020-in apart (edge to edge), and current flow between the two can be differentiated by a properly adjusted current tracer. However, if the two traces have a 10-to-1 current differential between them, spacing should exceed 0.075-in for the tracer to be effective.

CMOS current tracing At first glance, it might seem that a current tracer should not be capable of sensing CMOS current flow (because the input to CMOS devices is essentially a "capacitor" that blocks direct-current flow). However, the tracer easily "sees" CMOS current for the following reasons.

CMOS inputs charge similar to capacitors. The charging current is greater than the threshold sensitivity of the tracer. CMOS IC manufacturers specify the direct current drawn by their devices, not the input charging current.

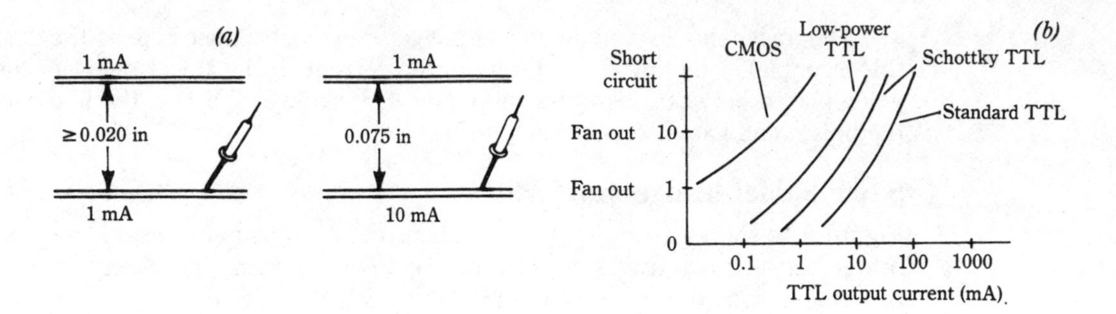

Logic family	Input capacitance (pF)	Stray capacitance (pF)	Rise time (ns)	ac current (mA)	dc current change	Total current change (ma)
TTL	5	5	10	5 (10 ns)	2 mA	7
CMOS	5	5	50	1 (50 ns)	40 pA	1

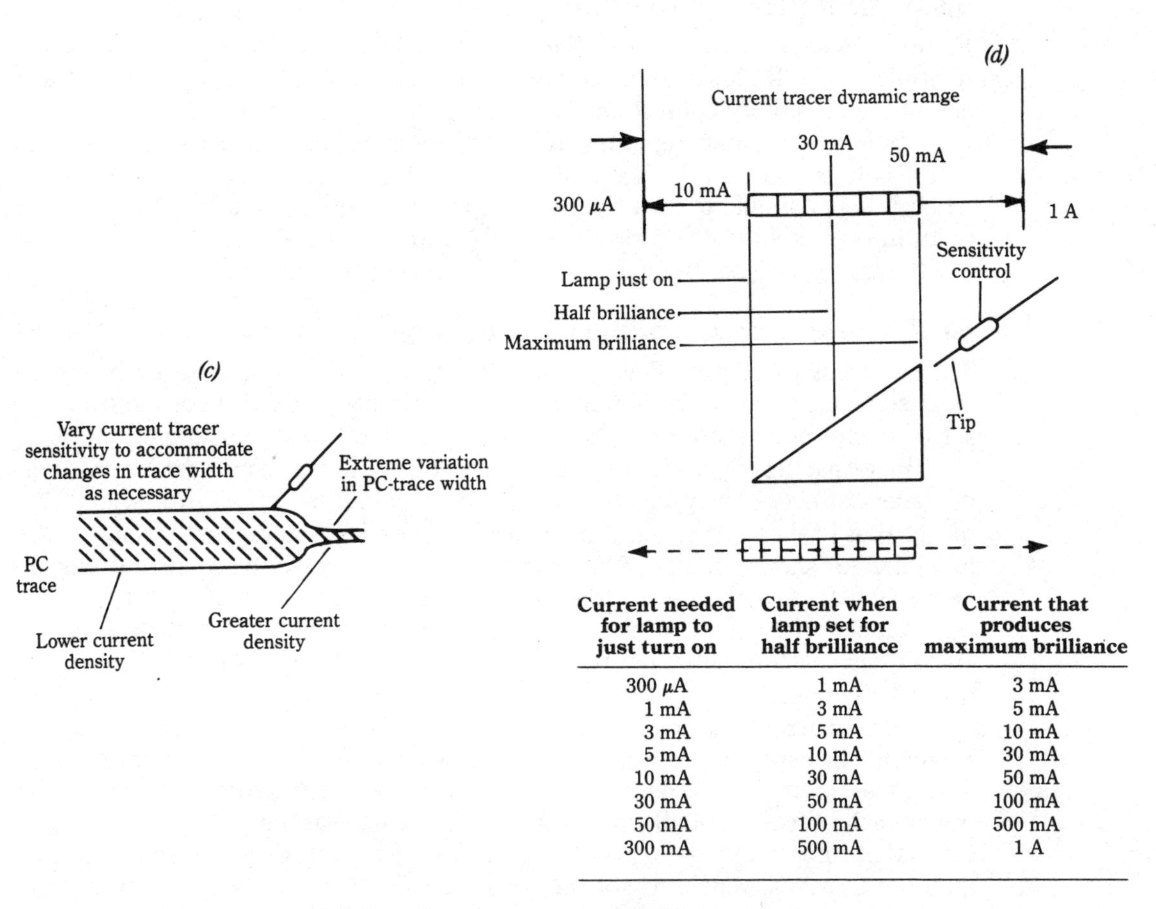

Current needed for lamp to just turn on	Current when lamp set for half brilliance	Current that produces maximum brilliance
300 μA	1 mA	3 mA
1 mA	3 mA	5 mA
3 mA	5 mA	10 mA
5 mA	10 mA	30 mA
10 mA	30 mA	50 mA
30 mA	50 mA	100 mA
50 mA	100 mA	500 mA
300 mA	500 mA	1 A

Fig. 5.5 Current-tracer troubleshooting tips.

Current-tracer sensitivity exceeds the typical 1 mA current specified for CMOS. More important, the current tracer is used to "look" at faults where current is considerably higher than normal. Also, the current tracer is often used with a logic pulser that raises the in-circuit current to a much higher level and thus allows easy-to-follow current tracing.

Remember that the current tracer is not voltage sensitive and responds only to current changes. So whether the circuit is TTL or CMOS, the tracer can sense any ac in the circuit. Figure 5.5*b* shows the range of currents usually present in both TTL and CMOS.

Trace width PC traces that vary greatly in width cause flux density changes under the tracer tip, as shown in Fig. 5.5*c*. This can be important when tracing supply-to-ground shorts. It might be necessary to vary the tracer sensitivity to accommodate changes in trace width.

Reference setting Setting the reference (tracer sensitivity) on a circuit point you suspect as faulty is critical to proper trace operation. However, the setting for that particular point has little or nothing to do with other points (because of fan out or other circuit variations). Typically, proper setting of the sensitivity control permits you to "see" current as low as 300 μA with virtually no upper limit.

Figure 5.5*d* shows typical current ranges for various settings of the sensitivity control. For example, if the sensitivity control is set so that 10 mA barely turns on the tracer lamp, 30 mA produces half brilliance, and 50 mA is the point where the lamp reaches full brilliance. Current in excess of 50 mA also produces full brilliance. As another example, when sensitivity is set so that 10 mA produces half brilliance, the lamp barely turns on at 5 mA and has full brilliance at 30 mA (or greater).

To get the maximum current range, set the sensitivity for half brilliance with the tracer at the suspected faulty point. Remember that this setting applies only to that point.

When a PC board has many traces side by side, all carrying substantial current (an LED driver IC is such an example), you can move away from these areas to trace current between components as follows. Set a reference current directly on the driver IC output pin. Then simply go from pin to pin on the ICs instead of attempting to follow along the PC traces.

Getting rid of crosstalk The grounded metal case surrounding the tip, along with an internal ferrite core, allows the signals directly under the tracer tip to be coupled into the pickup coil, but signals not directly under the tip are open-ended and attenuated. Even with this design, crosstalk can be a problem in current tracing.

One way to minimize crosstalk is to use proper tip alignment (Fig. 5.3) and proper reference sensitivity. If neither of these eliminate crosstalk between adjacent PC traces, try using a logic pulser (Sec. 5.5) with the current tracer. This effectively boosts viewable current in the circuit. (Make sure to reset the current-tracer sensitivity when using the pulser to allow for the higher current flow.)

Tracing current on power-supply lines Supply and ground lines or traces tend to be very noisy (with respect to current spikes) on microprocessor circuits. These traces often produce the highest level of ac current on a PC board. Try to avoid tracing current on these lines. If this is not possible, use the logic pulser in conjunction with the current tracer.

5.7 Logic clip

Typically, logic clips accommodate only 16-pin ICs and are thus of marginal value for most present-day digital equipment. A typical microprocessor has at least 40 pins (often 64 pins), and 24 pins is typical for most ROMs, RAMs and I/Os. However, there might be larger logic clips in the future. Also, in some computers, there might be 16-pin IC that has access to the data and address buses (or some other significant group of interconnections).

If a logic clip can have access to both buses, the clip can be used for a program-trace function with single stepping. In this case, the clip serves to read out the data word at each address of the program. However the logic analyzer (Sec. 5.9) is the most practical instrument available today for making a rapid, thorough program trace.

Although the logic clip might not be the most effective tool for troubleshooting present-day digital equipment, you should have some knowledge of how the clip can be used. A typical logic clip is used to monitor the logic states at each pin of a 16-pin DIP IC. LEDs on top of the clip indicate the logic levels at each IC pin. (The LEDs are turned on to show a logic 1, and off to show a logic 0.)

The clip has no controls to set, needs no power connections, and requires practically no explanation as to how it is used. Because the clip has internal logic for locating the ground and 5-V pins, the clip works equally well upside down or right side up. Buffered inputs ensure that the circuit under test is not loaded by the clip. Simply clipping the unit onto an IC makes all logic states visible at a glance.

When the clock is slowed to about 1 Hz, or is triggered manually (real-time basis), timing relationships become especially apparent. The malfunctions of gates, FFs, counters, and so on then become readily visible because all inputs and outputs of an IC are seen in perspective. When pulses are involved, the logic clip is best used with the logic probe. Timing pulses are observed on the probe, while the associated logic-state change can be observed on the clip.

5.8 Logic comparator

As in the case of a logic clip, the logic comparator is not used extensively in troubleshooting of present-day digital equipment (because of the 16-pin limitation). However, operation of a logic comparator is interesting from a troubleshooting standpoint.

Such comparators clip onto 16-pin ICs and, through a comparison scheme, instantly display any logic-state differences between the test IC and a reference IC. Logic differences are identified to the specific pin or pins of the IC. A turned-on LED at the comparator corresponds to a logic difference at that pin.

To use the logic comparator, you insert a reference board with a good IC (of the same type to be tested) into the comparator. You get an immediate indication if the test IC operates differently than the reference IC. Even very brief dynamic errors are detected, stretched, and displayed.

Note that the use of a comparator is somewhat similar to *piggybacking* described in Sec. 6.22. (The comparator operates by connecting the test and reference ICs in parallel.)

In the comparator, the reference IC sees the same signals that are input to the test IC. The outputs of the two ICs are compared, and any differences in outputs greater than 200 ns in duration indicates a failure. A failure on an input pin, such as an internal short, appears as a failure on the IC driving the failed IC. So any failure indication pinpoints the malfunctioning pin.

5.9 Logic analyzer

Logic analyzers can be considered as very specialized digital scopes that let you examine specific portions of a program running through a computer or other program-based digital device. The analyzer captures a selected program sequence in *tabular form* (such as shown in Fig. 5.6*a*), stores the sequence in an internal memory, and can display the tabular sequence indefinitely. The analyzer also allows you to look at activity before or after the operations shown. In addition to a tabular display, most logic analyzers also have a *timing display*, and many have a *mapping display*.

A logic analyzer is similar to a scope in many ways. Both instruments have signal inputs, a trigger-timing circuit, and a CRT display. The analyzer signal inputs differ from those of a scope in that the analyzer has 16 to 32 threshold-sensitive inputs to detect logic levels (1 and 0). The trigger circuit of an analyzer is also much more sophisticated than that of a scope.

Many present-day analyzers can trigger: (1) on a particular address, (2) on the nth occurrence of that address (for software loops), (3) after n clock cycles past that address, or (4) after a specified sequence of trigger addresses (to detect a particular

Address	**Data (at corresponding address)**
A15 A0	**D7 D0**
0001 1000 0011 1111	0111 0011
0001 1000 0011 1110	1000 1000
0001 1011 0011 1111	0011 0011
0001 1011 0011 1111	0011 0011
0001 1000 0011 1111	0111 0011
0001 0000 0011 1111	0011 0111
0001 0000 0011 1101	1001 0011
0001 1001 0011 1111	0111 0111

(*a*)

Fig. 5.6 Logic-analyzer displays.

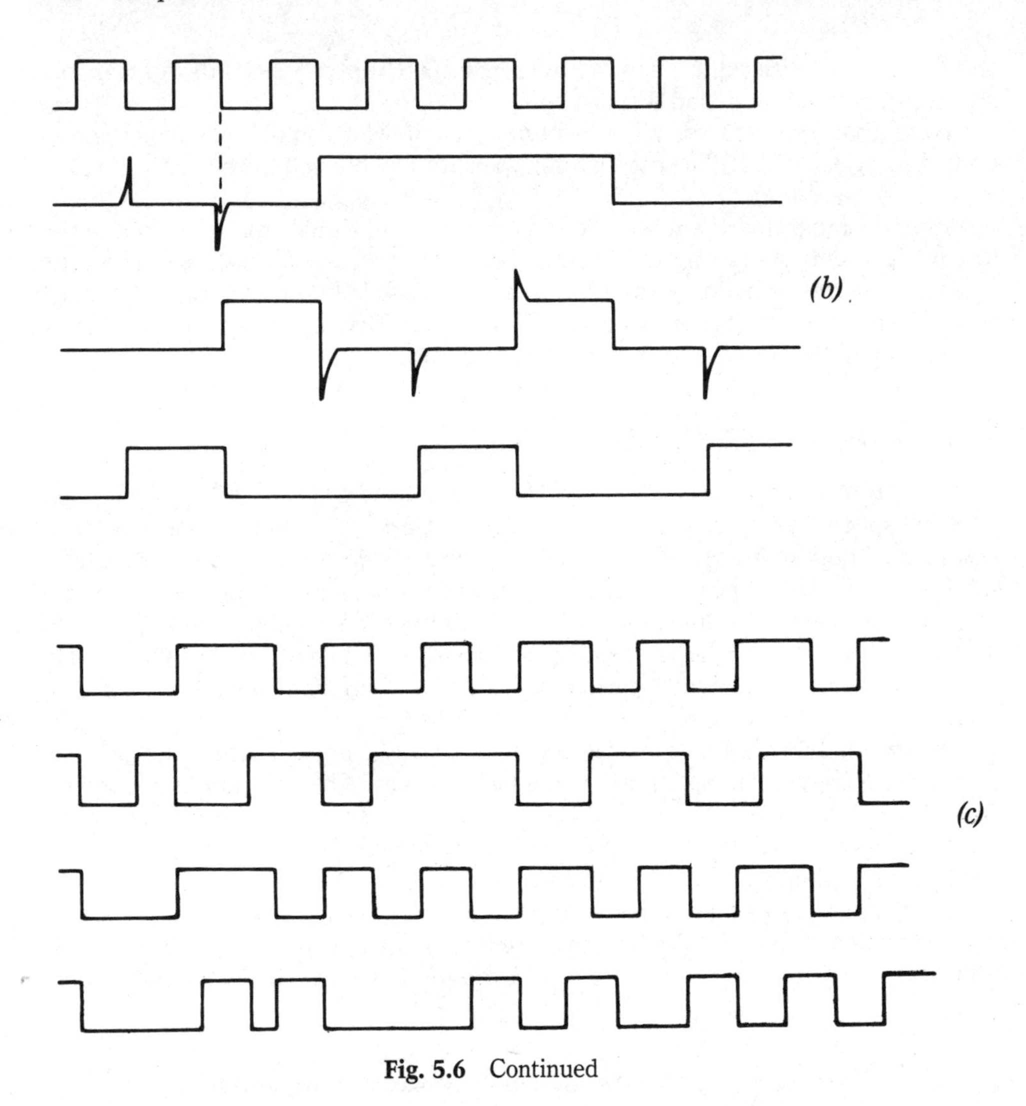

Fig. 5.6 Continued

program path). The trigger words need not be addresses but can be data, control, or any combination of logic signals.

Unlike scopes, which display data bits occurring after a trigger event, logic analyzers also can display data prior to the trigger. This capability, called *negative time recording*, can be used for troubleshooting any program-based equipment. You simply choose a faulty system operation (or point at which program fails) and let that point be the trigger. You can then observe the recorded events that lead up to the fault.

In most cases, logic analyzers are timed by the clock signal of the circuit under test. Typically, a new line of display is generated for each clock input. Time advances from top to bottom, one line per clock input. This overcomes the problem of tedious single-stepping and permits you to run through all (or any selected portion) of a program at normal speed.

5.9.1 Program trace

As discussed in Sec. 4.8, the basic troubleshooting approach for computer-type digital equipment is to monitor portions of the program while looking for abnormalities. The data-display format (tabular display) of the analyzer is used to display data bytes (as they appear on the data and/or address buses) in binary or hex (or possibly octal) form. In effect, the bytes appear as they would on paper (1s and 0s rather than the presence or absence of pulses).

Figure 5.6*a* shows that several data bytes can be displayed simultaneously (in binary in this case). This makes it possible to check the data words before and after the selected area of interest on the program. With some analyzers, the selected data word is indicated by extra brightness on the display. In Fig. 5.6*a*, the top data word is the selected word (the breakpoint starts at this word and continues for eight steps or words).

To use the tabular data format, you connect the analyzer to the data and/or address buses with patch cords or probes (supplied with the analyzer). Then you select a particular data word and/or address breakpoint from the program listing with the analyzer controls (typically keyboard or pushbuttons) and start the program. The computer then runs through the complete program, but only the desired portion of the program is displayed. In some cases, the computer runs to a certain point in the program and then stops. The analyzer lets you check each stop of the program up to the failure point. You then have a starting point for troubleshooting (or debugging, if there is a problem in the program, rather than in hardware).

Analyzer-display versus program listing Figure 5.7 shows a comparison of typical program listing versus a logic analyzer display (in binary). In this example, note that both the address and data bytes are shown for 10 steps of the program. However, only the first nine addresses are of interest, with address 0004 being of special interest.

Compare the display of Fig. 5.7 with that of the timing display in Fig. 5.6*b*. Even if you are not familiar with digital troubleshooting, you will note the advantages of a tabular display for troubleshooting where programs are involved. Remember that the program is operated at near normal speed and can be examined line by line, several lines at a time.

5.9.2 Timing displays

Timing is important for most digital equipment, and is critical for computer applications (or wherever a program is involved). The multitrace feature of a logic analyzer can be used to reproduce timing diagrams similar to those produced on a conventional multitrace scope. Figure 5.6*b* shows a typical timing display as produced on a logic analyzer.

Some logic analyzers have a feature called timing analysis, such as shown in Fig. 5.6*c*. The display is similar to that of a scope or analyzer timing diagram, except that the display of Fig. 5.6*c* is *digitized* (or rounded off) to discrete time and

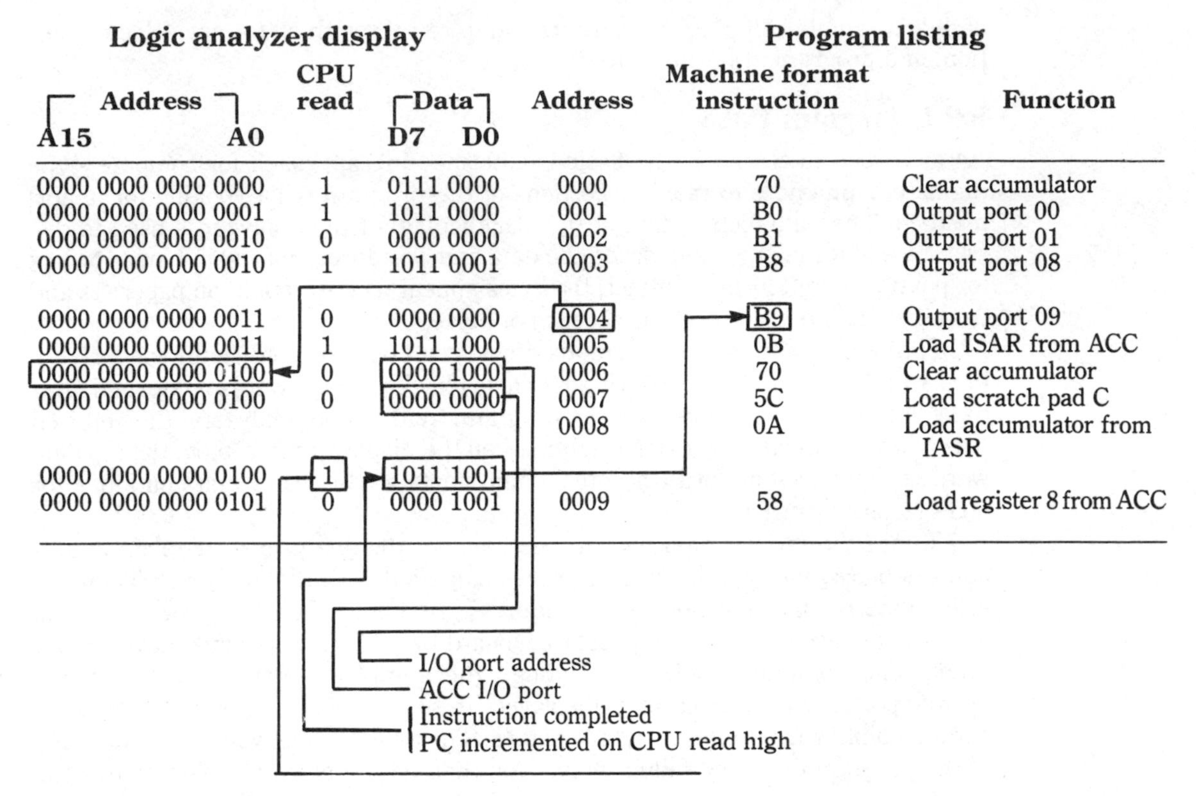

Logic analyzer display: Address (A15 … A0)	CPU read	Data (D7 … D0)	Program listing: Address	Machine format instruction	Function
0000 0000 0000 0000	1	0111 0000	0000	70	Clear accumulator
0000 0000 0000 0001	1	1011 0000	0001	B0	Output port 00
0000 0000 0000 0010	0	0000 0000	0002	B1	Output port 01
0000 0000 0000 0010	1	1011 0001	0003	B8	Output port 08
0000 0000 0000 0011	0	0000 0000	0004	B9	Output port 09
0000 0000 0000 0011	1	1011 1000	0005	0B	Load ISAR from ACC
0000 0000 0000 0100	0	0000 1000	0006	70	Clear accumulator
0000 0000 0000 0100	0	0000 0000	0007	5C	Load scratch pad C
			0008	0A	Load accumulator from IASR
0000 0000 0000 0100	1	1011 1001			
0000 0000 0000 0101	0	0000 1001	0009	58	Load register 8 from ACC

Fig. 5.7 Comparison of program listing versus logic-analyzer display.

voltage limits (to give a clear graphic display) without specific waveshape information.

When using the timing-analysis display, narrow pulses are detected by internal detection and stretching circuits. Many inputs can be observed at once. By taking full advantage of the triggering features, you can observe the timing of small, specific portions of a long program sequence. This elaborate triggering and digitizing is made possible by a RAM in the analyzer that stores waveform information in digital form.

5.9.3 Mapping display

The mapping function of a logic analyzer is most effective on computer-type digital equipment. A mapping display is formed by connecting the most significant bits (MSBs) of a data word to the vertical-deflection circuits of the analyzer display, and the LSBs are connected to the horizontal-deflection circuits. This produces a series of dots (or bright traces) in a format shown in Fig. 5.8*a*.

Each dot represents one possible combination of the 16 input lines. An input of all 0s (0000 in hex) is at the upper-left corner of the display, whereas an input of all 1s (FFFF in hex) is at the lower-right corner. The dots are interconnected so that the sequence of data changes can be observed.

In a mapping display, the interconnecting dot line gets brighter as the sequence moves toward a new point, thereby showing the direction of data flow. Some analyzers have a *cursor* to help locate specific points in the mapping display. The cursor is positioned over an area of interest on the mapping display, and the address or data word is read out on the analyzer controls.

The display assumes a unique pattern (sometimes called a *map signature*) when mapping is used to monitor a digital program. Figure 5.8*b* shows an 8085 microprocessor display in an expanded mapping-display mode. The program begins at address 0800. When the program gets to address 0808, the program stores a data byte at address 3000. At address 0810, the program jumps to address 0F20 and then continues. Address 0F20 contains an instruction that causes a data byte to be stored at address 3838. Finally, a jump instruction at 0F24 causes the program to go back to address 0800 and repeat.

The map function is particularly useful for checking a digital system that shows bus activity but is not functioning properly. (Bus activity is indicated by a flashing LED when a logic probe is connected to one or more of the bus lines.) The map display shows where the system is spending time (which subroutine or peripheral is using most of the time). When available, a map of a known-good product or system can be compared with the map of a malfunctioning system as a first step in troubleshooting.

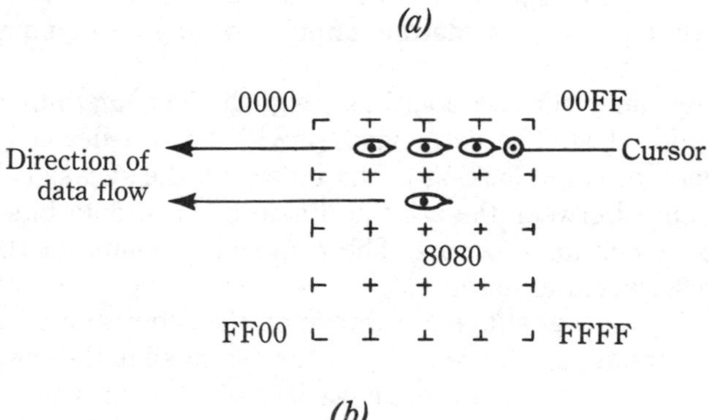

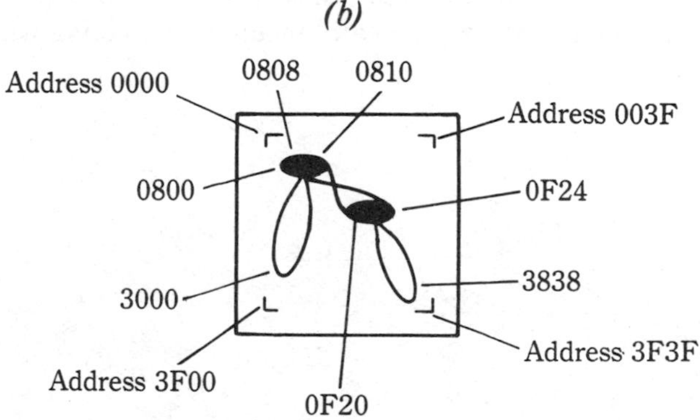

Fig. 5.8 Mapping displays.

5.10 Signature analyzer

The signature analyzer is most effective for troubleshooting digital equipment where there are *data streams* that pass through the circuits (typically in programmed devices). Signature analysis is an easy-to-use and highly accurate technique for identifying faulty digital circuits. The signature analyzer can convert the long, complex serial-data streams present in most digital circuits into four-digit *signatures*. These signatures indicate whether the circuit (or a particular point in the circuit) is acting properly.

Signature analyzers use 16 characters (0 to 9, A, C, F, H, P, and U) to monitor digital circuits. Points in the suspected circuit are probed until a signature is found that does not agree with the one documented in the service manual. The signal path is then traced back until a correct signature is found, thus localizing the fault (at the point between the good and bad signatures).

Once the fault is located, the bad component at that point usually becomes apparent (or can be located using the logic pulser and probe, current tracer, etc.). Signatures at points in digital circuits are used in much the same way as voltage and waveforms in analog-circuit service schematics.

The signature analyzer has one drawback. If there are no standard or correct signatures available in the service literature, you must make your own using known-good digital equipment! Note that there is no such thing as a signature being "almost right." The signature 8F37 is no more related to 8F38 than to PPPP. So make sure that you take signature samples on digital equipment in perfect working order.

The signature analyzer uses four basic signals. The *data* input receives data from the point under test. The *start* signal, provided by the device under test, tells the analyzer when to begin looking at the data, and the *stop* signal tells the analyzer when to stop. Between the start and stop signals, data bits are processed each time a new clock input occurs. The connection points for these inputs are specified in the service literature.

In some cases, the signatures are shown on the schematic or logic diagrams. In other cases, there are signature tables, often arranged in the same physical pattern as the ICs. You might find a great variety of ways in which signatures are listed in digital service literature (up to and including no listing whatsoever).

6
Digital troubleshooting

Digital systems are often bus-structured, and many of the devices on the bus are complex LSI (large-scale integration) or VLSI (very large-scale integration) ICs. This is true even of the nonprogrammed devices described in Chapter 9. The signal activity between the devices can be very complex, so it is often useful to break the data bus (the main feedback path) to isolate a fault that causes the entire equipment to malfunction. (Although useful, it is not always practical to break the data bus, or any bus.)

The basic digital troubleshooting techniques (such as the troubleshooting trees found in digital service literature) provide an orderly approach to locating system faults, but they are not always adequate. This chapter introduces the reader to many of the time-honored techniques, procedures, and tricks that can be effective in diagnosing, isolating, and locating faults in any digital equipment.

6.1 Basic troubleshooting problems (bus-structured)

All computers and many microprocessor-based devices are bus structured. These buses create testing problems. Unless you can remove each device on the bus, one at a time, it is difficult to determine which device is bad. (The current tracers described in Sec. 5.6 are helpful in finding a bad device on a line connected to many devices.)

One particular problem with any bus-structured system is that the bus acts as a *signal-feedback path* and tends to propagate errors through good circuits and then back to the fault source. The best way to overcome this problem is to open the feedback path when possible. The procedures for opening the feedback path are discussed in this chapter and in Chapter 7.

Another problem is that computer equipment is usually operated at high speeds, with the entire program being completed in seconds (or a fraction of a second). This makes it impossible to monitor any function (system or component) in real time, or to stop the program at a point where failure occurs, unless you use a logic analyzer.

Many devices in computers, particularly those connected on buses, cannot be tested using simple stimulus-response testing (where you apply a single input and look for a corresponding output, as you do for the control-type equipment of Chapter 9). In many cases, the outputs are not clearly defined or simply cannot be observed. The only course is to substitute a new component (which sounds simple on paper but is tedious when multi-pin ICs must be soldered and unsoldered).

The logic comparator (discussed in Sec. 5.8) can be helpful when substitution is dictated as a logical step in troubleshooting. It is also possible to piggyback ICs (as described in Sec. 6.22). However, the problem is usually solved only by physical substitution.

All computer program flow depends on a long sequence of instructions and events. If even a single bit of information is incorrect, the entire system fails. Noise glitches and bad memory bits are the most common sources of single-bit errors.

As an example, if a noise glitch is of the same approximate amplitude as the pulses on a data or address bus, the glitch can be confused with a logic 1. Similarly, if a logic-1 bit is stored in a memory, and the bit is degraded (say the bit amplitude is below the normal threshold of other pulses), the bit can appear as a logic 0. Either of these conditions causes the system to go to the wrong address, resulting in total failure (which is difficult to pinpoint because the entire system appears to be good, when only one bit is in error).

All of the problems discussed thus far are general in nature. The rest of the chapter looks at some specifics in troubleshooting digital equipment.

6.2 Clocks

A bad clock can cause a number of malfunctions. For example, clock problems can show up as a total failure to function (no activity on data or address buses when monitored with a probe, scope, or analyzer), the ability to function only open loop (free running), or when a meaningless and undefined program sequence occurs. Because these same symptoms can be caused by other malfunctions, your only course is to check for correct clock pulses (early in the troubleshooting process).

In some digital equipment, the clock is a single pulse that appears at a fixed repetition rate and is applied to all ICs via a clock line. Other systems have more than one clock, possibly two pulses that appear on the clock line with some fixed delay or phase relationship between the pulses. Some microprocessors have a built-in clock, whereas others require an external clock. Clock speed is controlled by a quartz-crystal circuit or by a simple RC (resistor/capacitor) circuit.

Because many digital systems run at the maximum rated speed of the microprocessor, the system can fail if clock speed is even slightly over the limit. Also, if the system runs too slowly, dynamic storage cells on the ICs in the system might fail. Both of these problems are more likely to occur when RC-clock circuits are used. However, crystals can sometimes break into the third-overtone oscillation, causing a much higher than expected clock rate.

When digital systems required multi-phase and nonoverlapping clocks, the timing/phase relationships are usually very stringent. Likewise, in a few cases, clock voltage levels are not necessarily compatible with all system components. If clock specifications are not found in the service literature, try looking up the microprocessor datasheet. Use a scope or frequency counter to measure clock speed and a scope to monitor amplitude, width, shape, etc., of the clock pulses.

6.3 Resets

Most microprocessor-based equipment has some form of reset function. Typically, a reset pulse is applied to the microprocessor when power is first applied, or under certain conditions, causing the program to go to the start (typically the program counter in the microprocessor goes to 0000 or possibly 0001). The program then starts when a certain condition occurs, or automatically, depending on the system. If the reset pulse is absent, too short, not of the correct amplitude, noisy, or too slow in transition (a sloping pulse instead of one with steep edges), the program can start at the wrong point, resulting in out-of-sequence, partial, or no-reset functions.

Problems can also occur in reset circuits that are susceptible to power-supply glitches. Slow edges on the reset pulses can cause timing problems from one device to another within some systems. This can cause some of the devices to power up before the others, resulting in erroneous behavior. A too rapid on-off-on power sequence fails to restart some systems, and it might be necessary to increase the off time to allow the power supplies and restart circuits to discharge.

Simply because a system starts and continues to run does not mean that there are no reset problems. Unless a reset is complete, some systems run part way through the program and then stop or lock up on a meaningless program loop or even perform most of the normal operations. Reset must be complete to ensure that all the test, control, and initialization operations are performed.

Because reset and power-up circuits are normally operative only when the system is initially turned on, you can monitor these functions at that time with storage scopes, logic analyzers, and (in some cases) with signature analyzers. When these instruments are not readily available (which is the case in a typical consumer-electronics service shop), you can override the reset circuits and control the circuit externally during test. For example, with most microprocessors, you can apply a fixed dc reset pulse and see if the system returns to program start.

6.4 Interrupts

Most computers are capable of responding to an interrupt signal or service request (say from an external terminal, disk, or printer). An interrupt request causes the control logic to interrupt program execution temporarily, jump to a special routine to service the interrupting device, and then automatically return to the main program. When interrupt lines are stuck or noisy, the system might work but does so very slowly or erratically.

If system changes occur sporadically, or if peripheral inputs and outputs take place at improper times, this indicates possible noise on the interrupt line.

If the system operates very slowly (with clock speed normal), or if the system spends time servicing a *phantom* interrupt, this indicates a stuck interrupt line.

When there are several interrupts, each interrupt is usually assigned a priority. This eliminates a conflict when two or more interrupts occur simultaneously. When the system does not respond to certain I/O interrupts, this might indicate that a higher-priority interrupt is disabling a lower-priority interrupt.

You can monitor interrupts with a logic probe, scope, or logic analyzer much as you would any other pulse or level at an IC pin. In some cases, you can control (enable or disable) the interrupt for test. For example, with many microprocessors, you can apply a fixed dc pulse or level and see if the system is interrupted.

6.5 Memories (patterns and checksums)

Most computers have both ROMs and RAMs (usually many of both). A defect in either a ROM or RAM can result in system failure. Some computers have self-test programs to check the ROMs and RAMs. (These self tests usually occur during the power-up/reset sequence.) The self test can locate a memory failure quickly (unless the defect prevents the self test from being completed).

RAMs are generally harder to test than ROMs and generally produce more disastrous results when a failure occurs. For example, when even a one-bit error occurs in that portion of a RAM reserved for the *stack* (an address where the program counter information is stored temporarily), the entire program crashes.

From a practical troubleshooting standpoint, it is generally necessary to substitute a known-good RAM for a suspected RAM. If the RAM is dynamic rather than static (where external data-refresh circuits are required for the dynamic RAM), testing and troubleshooting are even further complicated, because any failure in the refresh circuit can cause a good RAM to apparently fail.

ROM failures are generally not as disastrous, and are generally easier to locate, although ROMs fail as frequently as RAMs. A typical ROM failure is where the program runs normally until a defective address is reached in the ROM. (Of course, if this is at the beginning of the program, there is total failure thereafter.)

One technique for testing ROMs is to free-run the system and use a signature analyzer to verify documented signatures. Another technique is to compare the outputs of a suspected ROM with those of a ROM in a known-good system (if you are so fortunate as to have a known-good system!).

Some ROMs and RAMs have built-in self-test features and programs. The *checkerboard* memory pattern is the most common self test for a RAM. The *checksum* is a program found in some ROMs. The following are brief descriptions of both features.

6.5.1 Checkerboards

The checkerboard is one of the many different patterns used for RAM self tests. The RAM is tested by writing the checkerboard into memory, reading the checkerboard back (with an analyzer), and verifying if the checkerboard has or has not changed.

With the checkerboard, all the bits at each address in the RAM are set to alternating 1s and 0s. Once all memory locations have been tested, the pattern is repeated with each bit reversed (verifying that each bit of the RAM can store a 1 and a 0). Although the checkerboard shows that each bit can store either 1 or 0, this is no guarantee that the RAM is good, because some RAMs are *pattern sensitive*. For example, one location might correctly store 0101 0101 but not a pattern such as 0011 0111.

RAM test credibility is generally much lower than that of ROMs, because it takes a very long time to test every possible pattern sequence. As a troubleshooting guideline, if a system passes a checkerboard test, the RAM is probably good. If a system fails a checkerboard, something is wrong (probably the RAM).

6.5.2 Checksums

When a ROM with the checksum program is programmed, all of the ROM words are added together (ignoring any carries that result). This number is complemented and stored in the last (or sometimes the first) word of the ROM, so that when all the words are added together (including the checksum stored in the last byte), the result is zero. If the total is not zero at the end of the test sequence, something is wrong with the ROM.

In many systems, the checksum is calculated to make the total a specific number rather than zero. This number is then compared with a corresponding number during the power-up sequence. If the numbers do not compare, the system is halted at that point.

As in the case of the checkerboard, the checksum program is not guaranteed. This is because there are combinations of two or more errors that can cancel and produce the correct checksum, even though single-bit and multiple-bit errors produce an incorrect checksum. As an example, if one error causes the undesired bit to be omitted, the check appears to be correct. This makes the checksum a negative test.

As a troubleshooting guideline, if a system fails the checksum, something is wrong, probably (but not necessarily) the ROM. If the system passes the checksum, the ROM is probably good.

6.6 Signal degradation

Crosstalk and transmission-line problems can occur in digital equipment with long, parallel buses and control lines (especially on crucial lines such as clock and chip-enable). These problems often show up as glitches on adjacent signal lines (crosstalk) or ringing (overdriven pulses) on the driving line (causing multiple transitions through a logic threshold). Either of these situations can inject fault data or control signals that are very difficult to detect.

Signal degradation problems are most common when signals must pass over long lines. When extender cards are added to these systems, or when high-humidity conditions exist, the chance for failure because of degraded signals increases. Also, cross-coupling of lines on extender cards can be a problem when fast signal-transition lines (such as Schottky gate outputs) run alongside other signal lines (even when the lines are on opposite sides of a PC board).

6.7 Troubleshooting trees and charts

A troubleshooting tree is a graphic means of showing the sequence of checks performed on a device under test. These trees are often drawn as flowcharts in which the results of each test determine what step is taken next. The use of troubleshooting trees for digital equipment can save considerable time and effort if the trees are well planned. Theoretically, the tree should lead you to the fault by means of the actions taken and decisions made along the tree. Unfortunately, such is not always the case.

A perfect troubleshooting tree must consider all possible failures, an almost impossible task for the person writing the tree. Also, troubleshooting trees tend to be generalized, lacking the specifics desired for making tests and decisions. Few troubleshooting trees provide practical information about how a specific test or measurement relates to what the circuit does or is supposed to do. If the troubleshooting tree fails to direct you to the actual fault, you might be left at a dead end with no idea of where to go next. However, the troubleshooting tree is often your best (or only) guide (at least as a starting point).

Figure 6.1 shows a portion of a typical troubleshooting tree found in the service literature of microprocessor-based equipment (a CD-ROM drive in this case). Compare this tree with the troubleshooting diagrams shown in Chapter 9. Note that troubleshooting diagrams are not found in all service literature, only in well-written literature. Consider yourself lucky if you have troubleshooting diagrams for the digital equipment you are servicing.

6.8 Interface troubleshooting

Many digital systems interface with other systems through external communications lines (IEEE-488, RS232C, telephone modems, etc., as discussed in Chapter 4). These lines are often long and might be exposed to sources of electrical interference (produced by relays, motors, transformers, solenoids, and even lightning).

The electromagnetic interference (EMI) from these various sources can cause the transmission of faulty data, overstressing the interlace circuits and (especially in the case of lightning) cause component failure.

Generally, output line-driver circuits tend to have higher-than-average failure rates because of both *EMI stressing* and *high transition currents* caused by driving *capacitive interface circuits*.

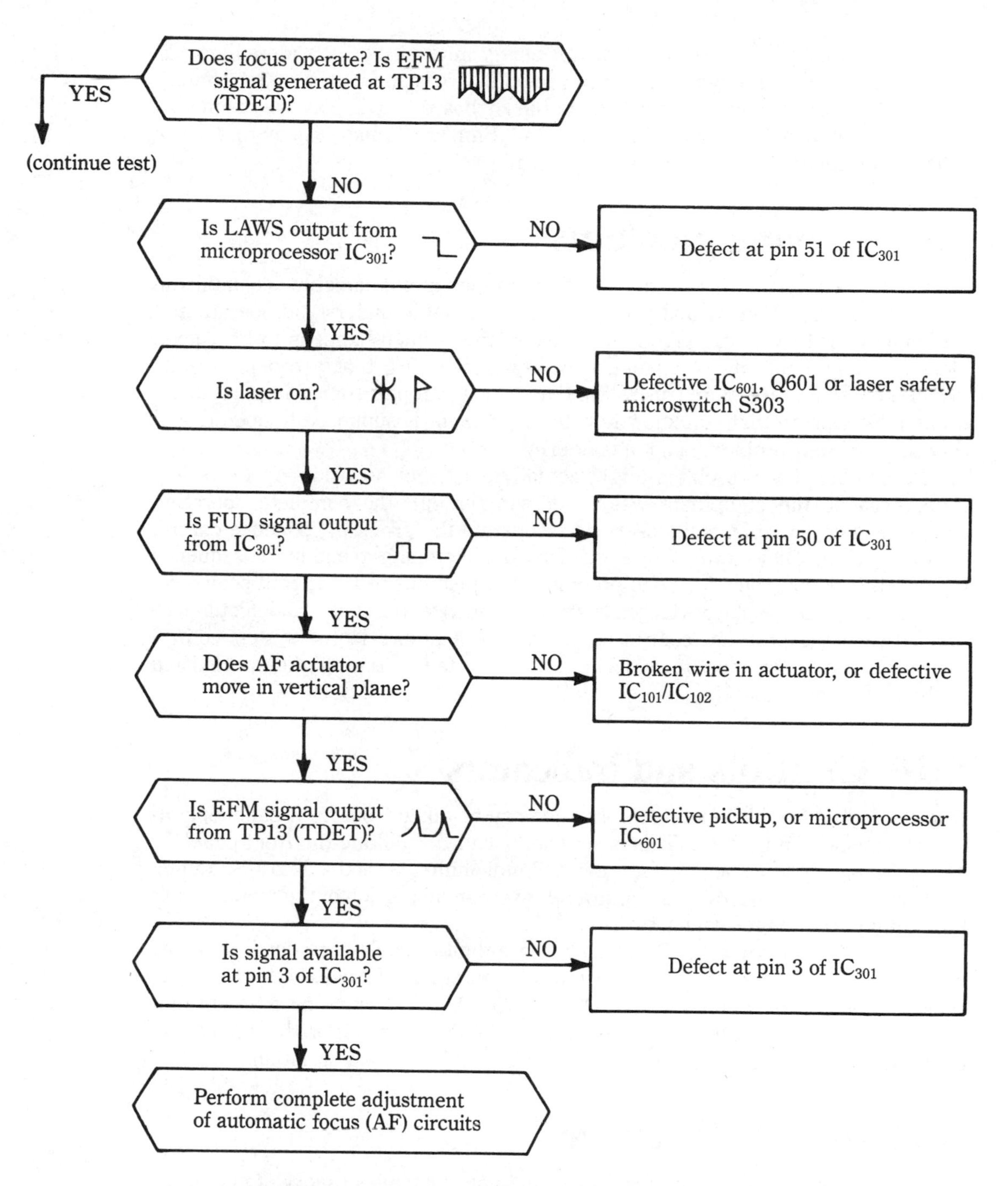

Fig. 6.1 Portion of a typical troubleshooting tree.

The interaction between *common scan circuits* must also be considered in making a troubleshooting diagnosis, when I/O circuits are scanned in sequence (multiplexed). For example, in a terminal where the keyboard and display are scanned, a stuck key can appear to make the display fail. Similarly, a bad display driver can cause what appears to be a keyboard problem.

6.9 Verifying problems

Most digital equipment is not only complex in function but sometimes complex in operation as well. This means that you must thoroughly understand operation of the equipment before starting any troubleshooting (which happens to be a good idea with any electronic equipment). Always make sure that a problem really exists. Few things are more frustrating than trying to repair something that is not broken. So be sure that you know how the equipment is supposed to operate and that the apparent problem is not a user error.

Remember that a problem might actually exist, but not show up for a long time, because the equipment was never run through the complete operating sequence. For example, some users never operate the program sequence control on a CD player. They simply play each disc from beginning to end in the sequence recorded. One day, they decide to program the selections in a different sequence, and discover that the player refuses to change sequence (because of a defect in wiring between the program control or because of a failure in the system-control microprocessor). You must recognize this as a true failure that might have existed in the player for a long time.

6.10 Controls and indicators

A great deal of troubleshooting information can be taken from front-panel operating controls and indicators. (This is commonly called "milking the front panel.")

As an example, if the only failure in a digital ohmmeter is the 0- to 10-Ω range, the problem can be narrowed down to a relatively small portion of the meter circuit (probably in the range-scale network).

As another example, if all indicators on a digital display are dead, with the power turned on, suspect a bad power switch, power cord, fuse, battery connections, or the power supply circuits. On the other hand, if only one segment of a display is dead, the problem is likely to be in the display itself or the circuit that drives the display. Chapter 7 describes such a troubleshooting situation.

6.11 Service literature

Well-written service literature for digital equipment includes theory of operation, schematics, block diagrams, troubleshooting trees, and test/adjustment procedures. The schematics of complex digital equipment often provide too much detailed information, making it difficult to see the big picture. For many troubleshooters, the block diagram can supply the right amount of information to understand how the various parts of the circuit work together.

Troubleshooting diagrams, such as those described in Chapter 9, are generally the best source of troubleshooting information for any digital device. Unfortunately, troubleshooting diagrams are not found in most microprocessor-based service literature.

Always take advantage of any designed-in performance verification or power-up test modes and diagnostic messages you find in the service literature. You might find many service aids and procedures in the manual just waiting for you to try (particularly in programmed equipment). In some cases, special service switches, jumpers, test fixtures, indicators, and test techniques can make the job much easier.

Try to understand the circuits and figure out where the major components are located. Check the theory or circuit descriptions, comparing what you read against the block diagrams, schematics, and troubleshooting diagrams (if any). Usually, you need not read every detail in the manual, but you must have a good idea of circuit operations.

As a minimum for any computer (or other programmed equipment), you must be able to identify the microprocessor, ROM, RAM, I/O address decoder, clock, bus, control or enable, and interrupt portions of the system. With control-type microprocessor equipment, look for all microprocessors (particularly the system-control microprocessor) and all interconnections to and from the microprocessors.

6.12 Product history

Product history can sometimes lead you to the best troubleshooting approach (but don't count on it). Of course, it's reasonable to assume that products in the field have all worked at one time (whereas anything can go wrong when a new product is first turned on at the factory).

Field failures are usually caused by components or connections that have failed. Always keep on the alert for any service bulletins or notes on digital equipment, particularly on equipment that you service regularly.

6.13 Typical problems

Always start looking for the easy things to test and repair. (Simple things are as likely to fail as the complicated ones.) For example, the power supply is one of the more failure-prone portions of many digital devices, and is one of the easiest to test (and is usually simple to troubleshoot). An abnormal voltage can cause erratic circuit performance. If you do not check all voltages first, it can take considerable time to find the problem.

Although solder and gold (copper) shorts on PC boards are not as common on field equipment as they are on production-line units, such faults do occur. Shorts can sometimes be removed with a sharp knife, if you know exactly where the shorts are located.

There is a technique (not always recommended) for removing such shorts whether you know the exact location of the short or not. The technique is sometimes useful where the short location is not accessible (such as inner layer shorts on multi-layer PC boards).

Charge a 100,000-μF (or larger) capacitor to 5 V (a safe voltage for most digital circuits). Then, with all cables solidly connected and proper polarity observed, discharge the capacitor and listen for a snapping sound on the board. Check continuity to see if the short is removed and, if not, try again.

Note that the capacitor-discharge method of removing shorts should be used with caution, because the technique opens the weakest link in the current path. (The weakest link might not be the fault source but might be a fine trace on the board.) The current tracer described in Sec. 5.6 provides a much safer means of finding shorts, if the shorts are accessible to the tracer.

Poor PC board and cable connections, broken wires, and loose parts can usually be found either visually or by touch. This is one reason for making a thorough mechanical or visual inspection early in the troubleshooting sequence. Look for improperly set switches, improperly connected jumpers, misloaded components (wrong components or components installed backwards) and cold-solder joints.

Edge connectors can cause problems (mostly in production-line equipment) when their borders are cut off-center, or when the connectors are accidentally covered with boardscaling spray or solder resist. Such problems usually show up during visual inspection.

Misregistration and contamination of the inner layers on multi-layer PC boards (which can cause high-frequency or leakage problems) can often be observed by holding the board up to the light. Since repair of the inner layers is often impossible, the entire board may have to be replaced.

6.14 Mechanical problems

Gently twisting and flexing the board and connectors can often help locate mechanical problems. Before you resort to such drastic treatment, make a visual inspection. Look for loose wires, broken traces, cracked ceramic ICs and resistor packs, bent wire-wrap posts, and dirty connectors. If the visual inspection does not reveal any problems, try reseating all PC-board edge connectors. If the connectors appear to be dirty, clean the connector contacts with a pencil eraser.

6.15 Interchanging components

If PC boards that are known to be good are available, and are easy to replace, some manufacturers recommend board swapping as a first troubleshooting step (or somewhere near the first step). Board swapping (one board at a time) has some merit. If a new board cures the problem, you have localized the trouble. If a complete set of new boards does not cure the problem, it is fair to assume that the problem is not on the boards, but is in the interconnecting wiring, control, and indicators, etc.

If a component is socket-mounted, try tapping (not pounding) the component before you install a new component. Socket-mounted components often become loose in the socket (especially after rough handling).

Do not always start by interchanging the microprocessor in any suspect circuit. Microprocessor failure rates are generally low compared to other components. This is also true of the LSI and VLSI chips used with microprocessors. Of course, if the microprocessors do not produce the correct outputs with correct inputs applied as described in Chapter 9, you must interchange the microprocessor.

If you are fortunate enough to have the identical equipment available (say an identical keyboard or terminal) start by making functional comparisons before you interchange any components. That is, measure voltages, signals, and so on at corresponding points of both the known-good equipment and the suspect unit. Such comparisons are especially helpful when you do not know the equipment well. You might suspect a hardware problem, when it might be an equipment idiosyncrasy or design limitation. A voltage or signal comparison answers the question immediately.

6.16 Stress testing

Stress testing can be very effective in dealing with marginal or intermittent failures, and can often cause these types of failures to temporarily improve or deteriorate. In addition to physical stressing, PC boards can be stressed *thermally* (alternate heating with an air gun or hair dryer and cooling with an aerosol freeze can) and *electrically* (by varying the supply voltage). These procedures (especially thermal stress) can show up intermittents (resulting from marginal ICs, lead bonds, solder joints, connections, and drive or timing circuits).

6.17 Shorts in power supplies

If you suspect a short in the power supply (say a particular voltage is abnormally low), remove one board at a time until the power supply is no longer shorted. The last board to be removed is the shorted one.

A technique for locating power-bus shorts is to supply a relatively high current (about 3 to 5 A) into the short. Be sure to maintain the same voltage polarity and not to exceed the supply voltage normally present. The current path to the short can often be determined by using a digital voltmeter (or DVM) with high resolution (0.01 μV or better) to look at voltage drops on the power buses. Voltages are developed along the traces in the path going to the short, but not in other paths, as shown in Fig. 6.2.

Another technique for finding the short on a faulty board is to inject current through two shorted lines with the logic pulser, and then follow the current to the short with a current tracer. Similar procedures are described in Chapters 7 and 8.

Before current tracers were available, the most common method to find power-supply shorts was to freeze the entire PC board, allow moisture to condense on the board, and then power the board with a 3- to 5-A supply (but at the normal supply voltage). As the board warms up and defrosts, the current path becomes visible

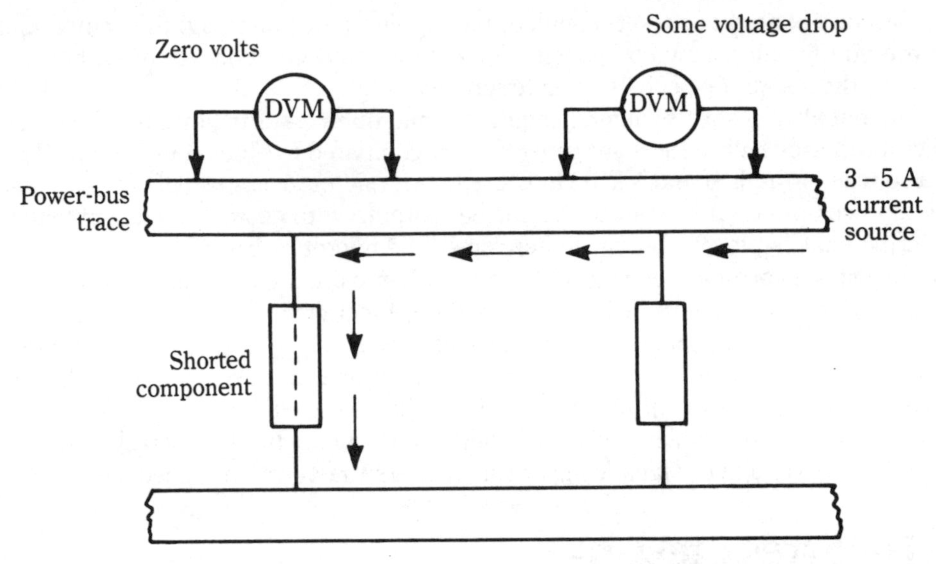

Fig. 6.2 Locating power-bus shorts with a digital voltmeter.

and in many cases pinpoints the short. Although this method is still recommended by the troubleshooting masters at Hewlett-Packard, others use it only where the power bus is inaccessible.

Remember that capacitors (especially electrolytics) have some current going into them when pulses are involved (as they are in most digital equipment). Shorted capacitors can be found using the current tracer to compare current levels going into identical capacitors on the same board. The capacitor that shows a much higher level of current going in than the others is likely to be shorted. This technique is particularly useful for finding shorted ceramic bypass capacitors.

6.18 Built-in troubleshooting aids

Be sure to take full advantage of any built-in and documented *circuit isolation features*, such as selective board removal, service jumpers, and special test modes or procedures. Also, it is very useful to separate the microprocessor system from any peripheral circuits. This permits you to troubleshoot each section separately.

If the microprocessor system is both digital and analog, use the *half-split* technique. The half-split technique involves choosing a point at the approximate center of the circuit and checking if the fault exists before or after the selected point. This is difficult in equipment with feedback paths (such as microprocessor-based computers). That is why Chapter 8 describes alternate procedures for breaking the feedback loop in microprocessor-based equipment.

When equipment is both digital and analog, you can use the half-split technique to isolate the digital portion by monitoring the clock, reset, control, and chip-enable lines with a logic probe or scope. If there is no activity on any of these lines, with the equipment supposedly running, you definitely have a problem in the digital portion of the circuits. (Of course, you might also have a problem in the analog circuits, but you now have a good starting point for troubleshooting.)

6.19 Failure patterns

One very common microprocessor failure is a shorted input pin to ground. This fault is often caused by a bad input-protection diode (in the IC) and appears as the pin being stuck low. A scope connected to the pin shows a voltage level near ground being pulled up, perhaps by a few hundred millivolts, when a logic 1 is produced on the pin. As discussed in Chapters 7 and 8, the current tracer is the best instrument for finding such problems.

Open inputs and outputs are also common problems in microprocessors. In many cases, there are thin wires connecting the package pins to the IC chip. If an output lead opens, the output pin floats, and the logic probe probably indicates a constant floating logic level (because of other device inputs connected to the pin). If an input lead opens, one or more corresponding outputs appear abnormal (typically stuck high, low or executing a logic function incorrectly).

If any output goes to a three-state bus, this can cause bus conflicts (*bus fights*). As discussed in Chapters 7 and 8, the current tracer is the most effective tool for finding the causes of a bus fight.

If a current tracer is not available, a sensitive, high-resolution DVM can be used to locate stuck inputs and outputs. Connect the DVM to the stuck pin and select the most sensitive dc voltage range. While monitoring the voltage, spray each IC connected to the stuck pin with a recommended freeze spray, one at a time, to change the IC temperature. A noticeable change in voltage (more than about 10 μV) on the point indicates that the IC being sprayed is drawing current. (If a freeze can is not available, use a heat source instead.) This technique relies on the properties of the semiconductor material used in the IC. Typically, when silicon is increased in temperature, current passing through junctions in the silicon also increases.

6.20 Troubleshooting tips

The *touch test* can be applied effectively to some digital equipment. That is, briefly touching each device on a PC board can pinpoint a component that is running hot (much hotter than others of the same type). When a particular device runs significantly hotter than others of the same type, a problem may exist.

Use care when making a touch test. A faulty component can run hot enough to burn your finger. Also, be aware that some good devices might run hotter than you expect (under normal operating conditions) and that temperatures may vary widely from one device to another.

A simple resistance test can sometimes be used to detect bus problems. Measure the resistance to ground (with the power off) of each line in the bus (data bus, address bus, etc.). The resistance of each line is usually the same. If any one line differs substantially, you might suspect a problem on this line.

If two lines show the same resistance (higher or lower from other lines), the two lines might be shorted together. In any event, before going further, check the schematic to see if the circuits connected to the line or lines could explain the differences.

If is often useful to isolate a pin from the rest of the circuit once a particular input or output pin is suspected. A quick, nondestructive way to do this is to suck the solder away from the area between the pin and PC board pad, using a vacuum desoldering tool (*solder gobbler*) or solder wicking braid. Then bend the pin so that it is centered in the pad hole, not touching the pad at any point. Use a continuity tester or ohmmeter to verify that the pin is no longer in electrical contact with the board.

Override interrupt lines and *chip-enable lines* on suspected devices can be used to verify that the IC is functioning correctly. This can be done by momentarily driving the appropriate pin high or low, as required. The logic pulser is the best tool for driving an interrupt or chip-enable line. If you do not have a pulser, it is possible to momentarily short the line to ground or to +5 V, but do so with caution.

Allowing a computer (or other programmed device) to *free-run* can sometimes pinpoint problems. For example, if some of the boards can be removed and still allow the basic microprocessor ROM/RAM (or system *kernel*) to function, a problem can be isolated to the kernel or to circuits outside the kernel. This subject is discussed further in Chapter 8.

Feedback paths and stuck buses should be removed from the main system when troubleshooting computers and other programmed devices. For example, if the kernel can be allowed to run open-loop (no feedback from the data bus), a free-run mode can be used to check the kernel and address-bus activity.

If available, use an extender board with switches on the bus and signal lines to break selected signals between the board and the rest of the system. An even simpler way to open selected signals going through a board edge connector is to place a tape or stiff paper on the edge connector fingers that you wish to isolate. (Be sure to remove the paper when the test is complete.)

6.21 Feedback problems

A feedback loop with a faulty output signal sends signals back to the inputs and thus produces more bad outputs. This makes feedback loops (in any equipment, analog or digital) difficult to troubleshoot. Opening the feedback path prevents the faulty output signals from going back to the input.

If controlled inputs to the loop can be generated, the signal flow from the input to the output can be observed. However, it is not always easy to provide such inputs (many lines might need to be controlled), and it is not easy to predict correct circuit operation.

If you have known-good equipment available (or at least a good board with the same circuits), try using the output of the good circuit to control the inputs of both circuits (suspected and known-good circuits). You then know that the circuit under test is getting the correct input signal; it is a matter of comparing the two circuits. A signature analyzer can be useful in making such comparisons.

6.22 Piggybacking

The *piggybacking technique* involves looking at suspected IC inputs with a scope (or signature analyzer) and then placing an identical IC package directly on top of the IC under test. The pins of the known-good or piggyback IC must be bent slightly to ensure that all pins of both ICs are making good electrical contact.

If no change is observed, and the outputs are not stuck, you can generally assume that the IC in question is not the problem. Be cautious of ICs with sequential circuits (such as shift registers and counters) that might cause output differences because of startup condition.

A better way of performing the piggyback test is to use an IC comparator as described in Sec. 5.8. The comparator provides an instant indication of differences between the two ICs.

7
Classic examples of digital troubleshooting

This chapter describes some classic examples of basic digital and microprocessor-based equipment troubleshooting. Because the circuits of a typical digital system are mostly IC rather than discrete, you do not have access to the circuits. You must determine if the IC is performing its function or failing without actually getting into the circuits. From a practical standpoint, it is very helpful if you can make this decision before you pull the IC from the board. The following paragraphs concentrate on basic test/troubleshooting techniques to do just that.

7.1 Timing

The classic method for troubleshooting discrete digital circuits is to introduce a pulse train at the circuit input while monitoring various points throughout the circuit with a scope. If any of the pulses are absent or abnormal (such as very low amplitude pulses or pulses that occur at the wrong time), this means a failure or degrading of the components between the pulse source (pulse generator or pulser) and monitor (scope).

As an example, a degraded diode or transistor can introduce an abnormally long delay and cause an output pulse to occur too late to open a buffer. A leaking capacitor can appear as *pulse jitter* on the scope display. However, when a microprocessor fails, there is usually total failure of the system. Microprocessor-timing parameters rarely degrade or become marginal after prolonged use. Once a program is debugged, and the microprocessor performs properly, timing remains good (or fails completely).

There are exceptions to this general rule. Microprocessors tend to speed up when power-supply voltages are high and slow down with low power-supply voltages. If the program timing is on the borderline, and there is a drastic change in power-supply voltage (particularly if the voltage is low), improper timing can result. Microprocessors are also affected by temperature extremes. Such extremes can affect microprocessor systems with marginal timing.

7.2 Tracing pulses

Basic point-to-point pulse tracing can be used in digital troubleshooting. Pulses are introduced on a line at some point in the circuit (with a pulse generator or logic pulser) and then monitored at another point on the line (with a scope, logic probe, or clip). Unfortunately, there are certain cases where this does not work too well with microprocessor-based equipment.

Consider the circuit of Fig. 7.1. This circuit is part of the output buffer (connected to one line of the data bus) in a RAM. (The circuit is sometimes referred to as a *transistor totem-pole,* as discussed in Sec. 3.1.4.) In either the high or low (1 or 0) state, the circuit output is the same as a saturated transistor connected to ground.

The totem-pole circuit presents a problem in pulse tracing. A signal source used to inject a pulse at a point driven by this output (such as trying to inject a pulse on that line of the data bus) must have sufficient power to override the low-impedance output state. A typical pulse generator (even a logic pulser) might not have this capability.

In some cases, it might be necessary to either cut the PC-board traces or pull IC leads to inject pulses. Both of these practices are time consuming and can pro-

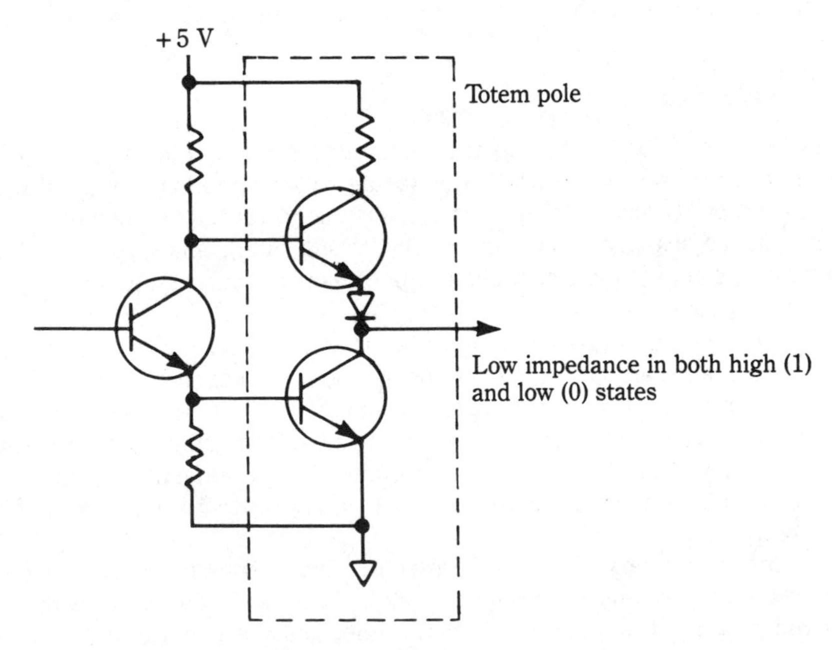

Fig. 7.1 RAM output-buffer (totem-pole) circuit.

duce unreliable repairs. However, there are basic digital troubleshooting procedures using logic probes and pulsers as well as current tracers that resolve such problems. These techniques are discussed in the following paragraphs. But first, the next section considers the how and why of digital IC failure.

7.3 IC failures

There are two classes of digital IC failure: (1) those caused by internal failure, and (2) those caused by a failure in the circuit external to the IC.

Four types of failure can occur internally to an IC: (1) an open bond on either the input or output, (2) a short between an input or output and supply or ground, (3) a short between two pins (neither of which is supply or ground), and (4) a failure in the internal circuits of the IC.

Four types of failure can occur in the circuit external to the IC: (1) a short between a point and supply or ground, (2) a short between two pins (neither of which is supply or ground), (3) an open signal path, and (4) a failure of an external component.

7.4 Digital/microprocessor troubleshooting sequence

Microprocessors are essentially digital devices, so all of the techniques used in digital troubleshooting can be carried over into microprocessor-based equipment. Thus far in this book, you have read about test equipment that is available for both microprocessor and other digital work and how to use the equipment effectively in testing circuits and components. Now consider how to combine these practical techniques with logical thinking to solve some problems in microprocessor-based equipment. (Practical digital/microprocessor troubleshooting is a combination of detective work or logical thinking, step-by-step measurements, and a tremendous amount of good luck.)

7.4.1 The basic troubleshooting sequence

The first step in any troubleshooting process is to narrow the malfunctioning area as much as possible by examining the observable characteristics of failure. For example, in a computer, this often involves inputting data at the keyboard and noting the response on the terminal screen. In the case of a nonprogrammed device such as a VCR, the initial sequence is to operate front-panel controls and note the corresponding response on the front-panel indicators.

The designers of some computers and other programmed devices have developed diagnostic routines or short programs that isolate troubles down to specific ICs or groups of ICs. (Chapter 8 describes some examples of simple diagnostic routines for a microprocessor-based system.) The routines found in the service literature should be followed religiously.

The logic analyzer is the most effective tool for observing characteristics of failure in programmed equipment, when display responses are meaningless. The logic analyzer also can be used in the absence of diagnostic routines (or in conjunction with the routines).

A logic analyzer makes it possible to observe the full transfer of data in a programmed system (such as a computer) for each step of the program. That is, you can see and compare the data word (on the data bus) at each address (on the address bus) for all program steps.

This checking of the data versus address can be done one step at a time, in groups of steps, or you can move quickly through the program steps to an area of the program where trouble is suspected. Based on observations of the logic analyzer (or display/keyboard and diagnostic routines), you can localize the failure to as few ICs or other circuits as possible (hopefully to one circuit or IC).

At this point, it is necessary to narrow the failure further to one suspected circuit by looking for improper key signals or pulses between circuits. The logic probe is a most effective tool for tracing key signals in digital circuits. In many cases, a signal completely disappears (no clock signal, no read/write signal, etc.). By rapidly probing the interconnecting signal paths (clock line, read/write line, etc.), a missing signal can be found.

Another important troubleshooting feature is the occurrence of a signal on a line that should not have a signal. The *pulse-memory option* found on some probes allows such signal lines to be monitored for single-shot pulses or pulse activity over extended periods of time. The occurrence of a signal is stored and indicated on the pulse-memory LED.

Although you have heard this before, it is essential that you understand all equipment circuits and operating characteristics. In this regard, well-written service literature is invaluable. Properly written manuals show where key signals are to be observed. The logic probe then provides a rapid means of checking these signals.

The logic probe, logic pulser, and current tracer can be used to observe the effects of failure on circuit operation, once a failure has been isolated to a single circuit or IC. This localizes the failure to the cause (either an IC or a fault in the circuit external to the IC). The logic clip and comparator can also be used if they can accommodate the number of IC pins.

The classic approach at this point is to test the suspected IC (or ICs) by substitution (or with a logic comparator if practical). However, the following steps are based on the assumption that neither of these approaches is practical (the comparator cannot accommodate the ICs and the ICs are very difficult to remove).

7.4.2 Checking for pulse activity

The logic probe (or scope) can be used to observe signal activity on inputs and to view the resulting output signals. This applies to both programmed and nonprogrammed microprocessor/digital equipment. From this input/output information, you can make a decision as to the proper operation of the IC.

For example, if a clock signal is occurring on the clock line to a RAM or ROM, and the enabling inputs (such as read/write signals, chip-enable signals, etc.) are in the enabled state, there should be signal activity on the data-bus lines. Each line should be shifting between high and low (1 and 0) as the program goes through each step.

The logic probe or scope allows the clock and enabling input to be observed and, if pulse activity is indicated on the outputs (each of the data-bus lines in this case), the ROM and RAM can be considered as operating. As discussed, it is usually not necessary to measure actual timing of the signals because ICs generally fail catastrophically. A possible exception is when an output buffer does not open and close at the proper time.

With few exceptions, the occurrence of pulse activity is usually a sufficient indication of operation. Of course, the ROM and RAM can have incorrect data bytes stored at various addresses, or the data bytes can be read out incorrectly to the data bus (because of open bonds, shorts, etc.).

The logic pulser can be used to inject input signals, and the probe can be used to monitor the response when more detailed study is desired or when input signal activity is missing. The logic pulser is also valuable for replacing the clock, thus allowing the circuit to be single stepped, while the probe or scope is used to observe changes in the output state (such as changes on the address and data-bus lines).

It is wise to check all of the output lines before you try to check the first fault you find. A premature study of a single fault can lead you to overlook faults that cause multiple failures, such as shorts between two lines on the data or address bus. This can lead to the needless replacement of a good IC and much wasted time (and money). The extra work can be minimized by systematically eliminating the possible failures.

7.4.3 Checking for open bonds

When there is an open output bond, as shown in Fig. 7.2*a*, the inputs driven by that output are left to float. In a typical TTL IC, a floating input rises to about 1.4 to 1.5 V and usually has the same effect on the circuit operation as a high or logic-1 level.

As a result of the high, an open output bond causes all inputs driven by that output to float to a bad level, because 1.5 V is less than the typical TTL high threshold level of about 2 V and greater than the low threshold level of 0.4 V (Fig. 3.11*b*). In TTL ICs, a floating input is usually interpreted as being a high level. The effect is that these inputs respond to the bad level as though it is constant high level.

When there is an open input bond, as shown in Fig. 7.2*b*, the open circuit blocks the signal driving the input from entering the circuit (a microprocessor data-bus buffer, in this case). The input line is thus allowed to float and responds as though the input is a constant high signal.

It is important to realize that because the open occurs on the input side of the IC, the signal driving this input is unaffected by the open. That is, the signal is blocked to the microprocessor buffer inside the IC, but there is no effect on other ICs connected to the same line. Of course, the microprocessor responds as if there is a static 1 on that line of the data bus and does not respond properly to those commands or words where a 0 is to appear on that line of the bus.

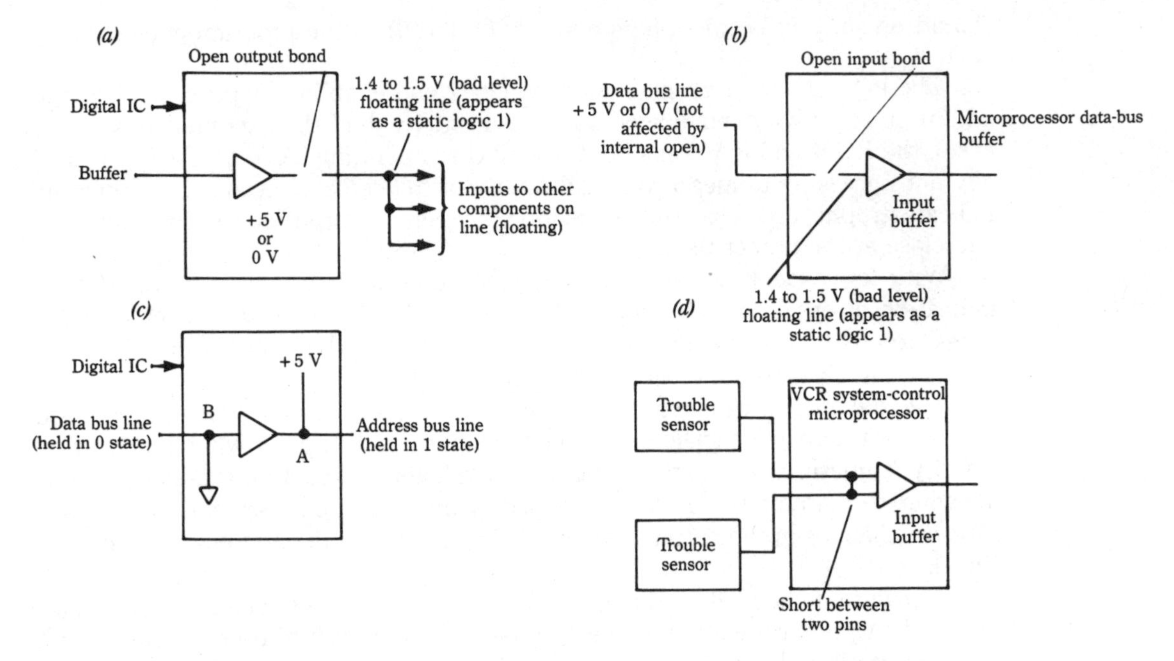

Fig. 7.2 Checking for open bonds and shorts in digital ICs.

Start the open-bond troubleshooting by testing for an open bond in the IC driving the failed point. The logic probe provides a quick and accurate test for such a failure. If the output is open, the point floats to a bad level. By probing the point, the logic probe quickly indicates a bad level. If a bad level is indicated, the IC driving the point is suspected of being the problem.

7.4.4 Checking for shorts to power or ground

A short, as shown in Fig. 7.2*c*, has the effect of holding all signal lines connected to that input or output either high (in the case of a short to power or supply) or low (if shorted to ground). In the case of Fig. 7.2*c*, the address line connected to point A is held in the high (1) state, and the data line connected to point B is held low (0). This results in a total disruption of the program and is thus one of the easiest types of digital failures to locate.

In troubleshooting power or ground shorts, test for a short to power or ground using the logic pulser and probe. (The current tracer can also be used effectively to find shorts, as discussed in Sec. 7.5.)

Although the logic pulser is powerful enough to override even a low-impedance output, a pulser is not powerful enough to produce a change in state on a power or ground bus. So, if the pulser is used to inject a pulse while the probe is used simultaneously on the same point to observe the pulse, a short to power or ground can be detected. (If you put both the pulser and probe on the same line, and see no response on the probe, that point is shorted to power or ground.) If you suspect a short, check the point with a voltmeter (power on) and/or ohmmeter (power off).

If you do get a pulse when the pulser and probe are connected to the same point, that point is not shorted. However, it is still possible that the point has a low-impedance condition (rather than a direct short).

If the point is shorted to power or ground, there are two possible causes. The first is a short in the circuit external to the ICs, and the other is a short internal to one (or more) of the ICs attached to the point. The external short should be found by an examination of the circuit.

If no external short is found, the cause is likely to be any one of the ICs attached to the point. The ICs can be eliminated one at a time. Also check for shorted capacitors and resistors on the line. (Note that resistors do not usually short, but their terminals can be shorted by mechanical vibration.)

7.4.5 Checking for shorts between points

Shorts, such as shown in Fig. 7.2*d*, are not as straightforward to analyze as the short to power or ground. When two pins are shorted, the outputs driving those pins (the two trouble sensors in this case) oppose each other when one attempts to pull the pins high, while the other attempts to pull the pins low.

In the situation of Fig. 7.2*d*, the output attempting to go high supplies current, while the output attempting to go low dissipates or sinks this current. Whenever both outputs attempt to go high (or low) simultaneously, the shorted pins respond properly. Whenever one output attempts to go low, the short is held low.

If you suspect a point or IC pin, and you have checked for shorts to power or ground, as well as for open bonds, the next step is to check for shorts between adjacent pins (especially if the adjacent pins show the same voltage or ground). This can be done as follows.

Use the logic pulser to pulse the suspected point or IC pin, and the logic probe to observe each of the adjacent points or pins. If a short exists between the point being studied and one of the other adjacent points, the pulser causes the point being probed to change state (the probe detects a pulse on one of the adjacent pins). To confirm that a short exists, reverse the probe and pulser and repeat the test. If a pulse is again detected, a short is definitely indicated.

7.4.6 Defective-internal circuits

The effects of internal failure in any digital IC (but particularly microprocessors) are difficult to predict. For example, a failure in an internal circuit such as shown in Fig. 7.1 has the effect of permanently turning on either the upper transistor of the buffer output (thus locking the output in the high state), or turning on the lower transistor (to lock the output in the low state). Of course, this failure blocks output signal flows and has a disastrous effect on microprocessor operation.

If you suspect a problem with the internal circuits of an IC, substitution is the only practical test (because you cannot get at the internal circuits). If substitution cures the problem, you have located the trouble. If substitution does not cure the problem, you have confirmed a good IC. (Unfortunately substitution of ICs is much easier on paper than on some inaccessible PC board.)

Also, it is often very difficult to tell an internal problem from an external problem. For example, a short between a point and power or ground, external to the IC, cannot be distinguished from a short internal to the IC. Both cause signal lines connected to the point to be either always high (shorted to power) or always low (shorted to ground).

When this type of failure is found, only a very close physical examination of the circuit can show if the failure is external to the IC. The current tracer is an effective tool for locating such shorts.

7.4.7 Checking for open signal paths

An open signal path in the external circuit has an effect similar to that of an open output bond driving the point or pin. As shown in Fig. 7.3*a*, the microprocessor input to the right of the open floats to a bad level, and appears as a static or constant high level (in typical TTL operation). Those inputs to the left of the open (display, etc) are unaffected by the open and thus respond as expected.

In troubleshooting open signal paths, when no points are indicated as failing (no shorts to power, ground, or other pins), but there is a definite failure of the circuit, suspect an open signal path (or possibly an internal open bond). From a practical standpoint, it is generally easier to check for open signal paths than to replace ICs.

The logic probe provides a rapid means of not only detecting, but also physically locating, an open circuit. For example, as shown in Fig. 7.3*b*, an open signal path causes the input to the right of the open (point B) to float to a bad level. Once an input floating to a bad level is detected, the probe can be used to follow the circuit back from the input while looking for an open.

The circuit to the left of the open is at a good level (either high, low, or pulsing), while the circuit to the right is at a bad level. By probing back along the signal path (from right to left) you can locate the open at the point where the probe indication changes from bad to good. Of course, if other inputs show a bad level, or if a good level cannot be found any place on the line back to the source (a trouble sensor in this case), the problem is likely an open output bond (Sec. 7.4.3) at the source.

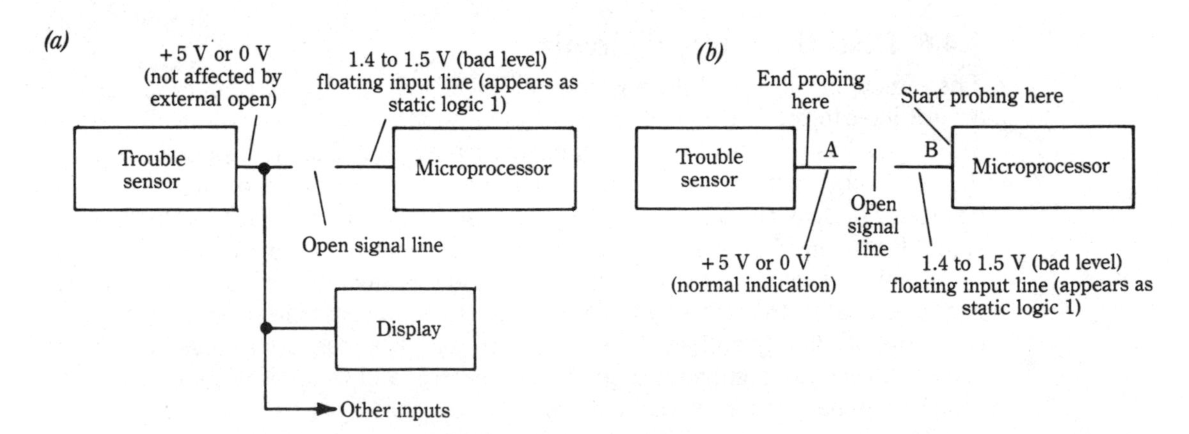

Fig. 7.3 Checking for open signal paths in digital ICs.

7.5 Using current tracers

The operating procedures for a current tracer are presented in Sec. 5.6. This section concentrates on the basic current-tracer troubleshooting approach for some typical digital-circuit problems. This approach is based on the idea of stimulating a shorted point or IC pin with a logic pulser and then following the current with a tracer.

7.5.1 Getting to know the current tracer

If you have never used a current tracer before, it might take time to familiarize yourself with the tracer capabilities and limitations. This is primarily because most electronic technicians are not used to thinking in terms of current and the information current provides simply because this information is not available in convenient form.

Most troubleshooting techniques involved measurement of voltage and resistance. Without a current tracer, it is necessary to open a line and insert an ammeter. This is not practical in most situations, so current is determined by calculations (based on observed voltage and resistance).

Use of the current tracer also requires some special skills to avoid the *crosstalk* problem. If a small current is being traced in a conductor or line that is very close to another line carrying a large current, the sensor at the tracer tip might respond to the current in the nearby line. The current-tracer sensor is designed to minimize this effect, but crosstalk can never be entirely eliminated. However, by observing the variation of the current tracer display as the tracer is moved along the line, you can learn to recognize interference or crosstalk from the nearby line.

7.5.2 Ground-plane currents

Defective ground planes or wires (Fig. 3.2) can cause problems in digital circuits. It is possible to determine the effectiveness of a ground plane by tracing current distribution through the plane. This is done by injecting pulse current into the plane from a logic pulser (or any pulse generator) and tracing the current flow over the plane. Generally, the current flow should be substantially even over the entire plane. However, it is possible that the current flows only in a few paths, particularly along the edges of the ground plane.

7.5.3 Power-to-ground shorts

Locating the exact point of a power-to-ground short can be quite difficult, even though the existence of such a short is readily apparent. Usually, there is a drop in power-supply voltage or a complete failure of the power supply.

To use the current tracer in this application, disconnect the power supply from the power line and pulse the power-supply terminal using the logic pulser, with the supply return connected to the ground lead of the pulser. Even if capacitors are connected between power and ground, the current tracer usually shows the path carrying the greatest current.

7.5.4 Troubleshooting stuck-line problems

The current tracer is very effective in troubleshooting stuck-line problems, whether caused by a dead driver or a shorted input. The current tracer can also be most useful in locating problems in three-state buses.

A stuck three-state bus, such as a microprocessor data or address bus, can present a very difficult troubleshooting problem, especially with voltage-sensing measurement tools (such as a logic probe). Because of the many bus terminals (typically 8 or 16), and the fact that several ROMs and RAMs might be connected, it is very difficult to isolate the one element (ROM or RAM) holding the bus in a stuck condition.

If the current tracer indicates high current at several outputs of a ROM or RAM, it is likely that one (and most likely only one) element is stuck in a low-impedance state. The defective element is located by placing the ROM or RAM control-input line to the appropriate level for a high-impedance output state (the off condition) and noting whether high current flow persists at the ROM/RAM buffer output. If so, the ROM/RAM is stuck. Repeat this for each ROM/RAM until the bad one is located.

If the current tracer indicates the absence of abnormally high current activity on all elements, yet the bus signals are known to be incorrect, this problem is likely an element (RAM or ROM) stuck in the high-impedance state. This can be found by placing a low impedance on the bus, such as short to ground, and using the current tracer to check for the element that fails to show high-current activity.

If the current tracer indicates high currents at only two elements (say two RAMs), the problem is likely a bus fight. That is, both RAMs are trying to drive the bus at the same time. This is probably caused by improper timing of control signals to the RAMs (one RAM buffer is opened before the other RAM buffer is closed).

When a line (particularly a data or address bus line) appears to be stuck high or low (no pulse activity on that line), several questions can arise. For example, is something (such as a short) clamping the line to a fixed value (such as to ground or power)? If not, is the driver (or other signal source) dead?

These questions are answered by tracing current from the driver to other points on that line, as shown in Fig. 7.4*a*. If the driver (source) is dead, the only

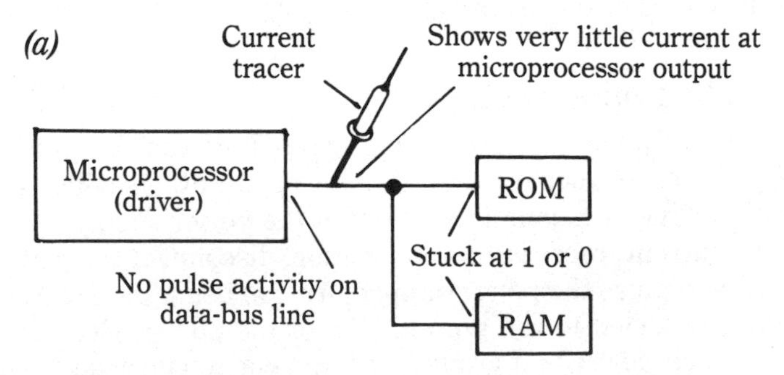

Fig. 7.4 Troubleshooting stuck-line problems.

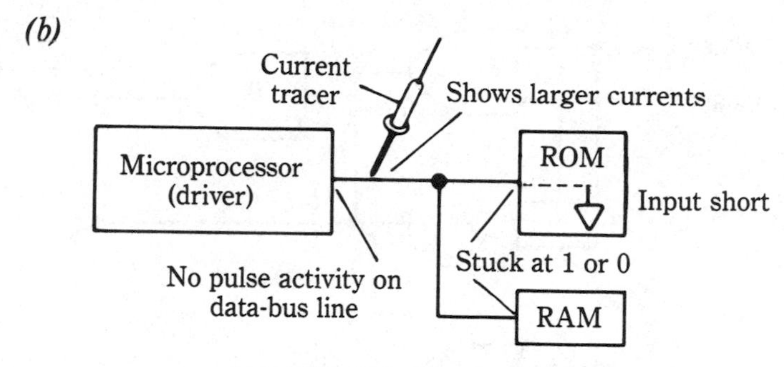

Fig. 7.4 Continued

current indicated by the tracer is that caused by stray coupling from any nearby currents, and is much smaller than the normal current capability of the driver. On the other hand, if the driver is good, normal short-circuit current is present and can be traced to the short or element (resistor, capacitor, etc.) clamping the line.

These same symptoms can be produced by an input short, as shown in Fig. 7.4*b*. However, when the line is stuck by an input short, the current tracer indicates a large current flowing from the driver, making it possible to follow the current to the cause (a shorter input). The same procedure also finds the fault when the short is on the interconnecting line (for example, a solder bridge to another line).

7.6 Basic logic-circuit troubleshooting

Figure 7.5 shows a very simple logic circuit. (This is the half-adder discussed in Chapter 3. Note that Fig. 3.3 shows the half-adder in block form, whereas Fig. 7.5 shows the gates that go to make up the half-adder function.)

Circuits such as shown in Fig. 7.5 are rarely found in discrete form in present-day digital equipment. However, the circuit is included here to show the basics of logical thinking required for troubleshooting digital equipment.

To test the circuit of Fig. 7.5, inject a pulse (true) to the addend input (digit A) and check for a pulse (true) condition at the sum output as well as a false (no pulse) condition at the carry output. If the response is proper, both the OR gate and the B AND gate are functioning normally. To confirm this, inject a true (pulse) at the augend input (input B) and check for a true at the sum output, as well as a false condition at the carry output.

If the response is not proper, the A AND gate is probably not at fault. Instead, either the OR gate or the B AND gate are the logical suspects.

To localize the problem further, inject a pulse at either addend (input A) or augend inputs, and check for an output from the OR gate. If there is no output, or the output is abnormal, the problem is in the OR gate. If the output is normal, the problem is probably in the B AND gate.

Now inject simultaneous pulses at the addend and addend inputs and check for a true (pulse) at the carry output and a false (no pulse) at the sum output. If the response is proper, all gates can be considered as functioning normally. If the

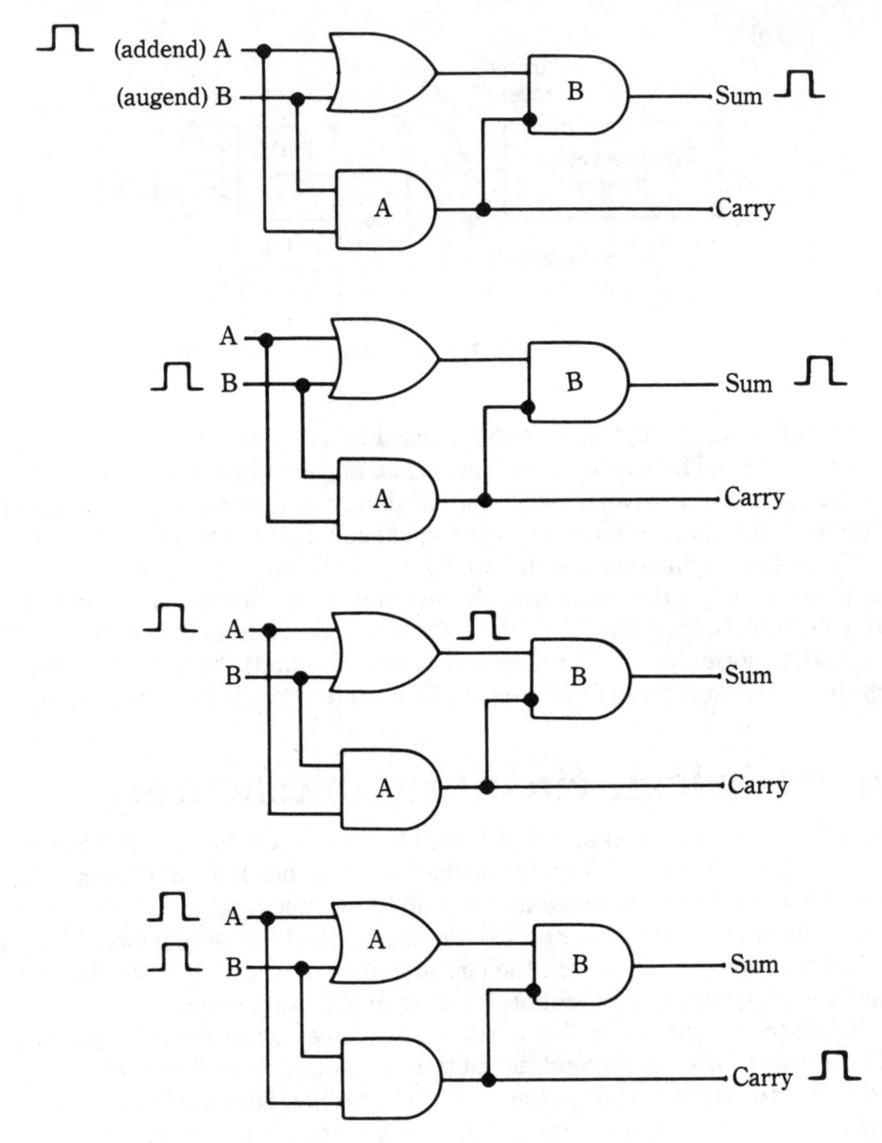

Fig. 7.5 Basic logic-circuit troubleshooting.

response is not correct, the nature of the response can be analyzed to localize the fault.

For example, if the carry output is normal (true) but the sum output is also true, the B AND gate is the most likely suspect. The A AND gate produces a true output only when both inputs are true. The B AND gate requires one true input from the OR gate and a false input from the AND gate (because there is one inverted input at the B AND gate). This produces a true condition at the sum output (because of the inversion). If the input to the B AND gate (from the A AND gate) is true, but the sum output shows a true condition, a defective B AND gate is indicated.

Now assume that the carry output is false no matter what the condition of the sum output. This points to a defective A AND gate. (The A AND gate should produce a true output, with two true inputs, no matter what the state of the OR gate or the B AND gate.)

A possible exception is where there is a short in the carry line (possibly in the PC wiring) that holds the carry line low. This can be checked with the logic probe, pulser, or current tracer as previously described.

7.7 Example of digital-circuit troubleshooting

Figure 7.6 shows a typical circuit that can be found in many types of digital or microprocessor-based equipment. The circuit is a *decade counter and display*, used in VCRs and CD players to display timing or programming functions.

Each digit is displayed by a separate seven-segment LED display. In turn, each LED display is driven by a separate BCD/seven-segment decoder and storage IC. (Figure 3.7 shows the truth table for a seven-segment display and the corresponding BCD input.)

The three decoder/storage ICs are turned on by a clock or enable pulse at regular intervals or on demand. Each decoder/storage IC receives BCD information

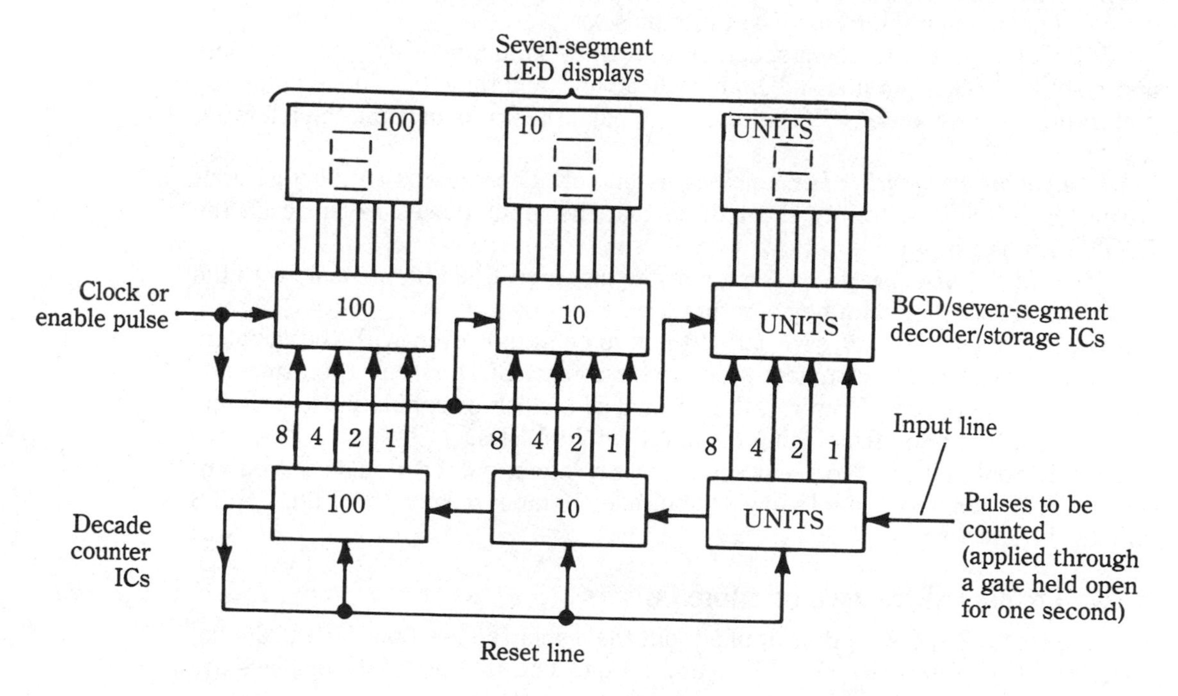

Fig. 7.6 Example of digital-circuit troubleshooting.

from a separate decade counter IC. Each counter IC contains four flip-flops and produces a BCD output that corresponds to the number of input pulses occurring between reset pulses.

The maximum readout possible is 999. At a 1000 count, the output pulse of the 100-decade IC is applied to all three counter ICs simultaneously as a reset pulse.

Now that you know how the circuit works, try solving some basic trouble-shooting problems based on evaluating symptoms and monitoring pulses.

Start by assuming that an input pulse is applied through a gate, and that the gate is held open for one second. The count shown on the LED display then indicates the frequency in hertz (Hz). For example, if the count is 798, this shows that there are 798 pulses passing in one second, and the frequency is 798 Hz. (This is the basic function of a digital frequency counter, in case you are interested.)

7.7.1 Display remains at zero

Assume that a 300-Hz input is applied (at the UNITS decade counter) but the display is 000,000,000, and so on. The LEDs turn on with full brilliance, but remain at 000.

This indicates that the LEDs are receiving power and are operative. If not, the LEDs could not produce a 000 indication (or any indications). The most likely causes for such a symptom are: (1) no input pulses arriving at the UNITS counter, (2) a simultaneous reset pulse with the first input pulse (or a short on the reset line), (3) no enable pulse to the decoder/storage ICs, or (4) a defective UNITS counter (not responding to the first input pulse).

With this symptom, use a scope and monitor (simultaneously) the input line and reset line, the input line and enable (clock) line, and the input line and the output of the UNITs counter IC. This should pinpoint the problem using the following logic.

If there are no input pulses, there is no output. The same is true if the input wiring is defective (possibly shorted) so that the input pulses never reach the UNITS counter input.

If the input pulses are present, but a reset pulse arrives simultaneously with the first input pulse, there can be no output.

If there is no enable or clock pulse, there is no output even with the counters operating properly. The counters produce the correct BCD output, then store the output in the decoders. However, the absence of a clock or enable pulse prevents the stored information from being displayed on the LEDs.

If the input pulses are present and there is no abnormal reset pulse, the output of the UNITS counter should show one pulse for 10 input pulses. If not the UNITS counter is defective.

7.7.2 Display skips two or more counts

Assume that a 300-Hz input is applied, but the display shows 000, 001, 002, 003, 004, 005, 006, 007, 010, 011, 012, and so on. The 008 and 009 displays are skipped.

This is probably a readout function (rather than a counter function) failure. Here's why.

If there is a failure in the UNITS counter, the 10 counter does not receive an output from the UNITS counter. Because there is some readout from the 10 counter (the readout goes above 011), you can assume that the UNITS counter is functioning.

There are several possibilities if the readout function is at fault. For example, if the UNITS counter output lines are shorted or broken, the UNITS decoder receives no input (or an abnormal input). Likewise, the UNITS decoder output lines can be shorted or broken, so that the UNITS LED receives an abnormal input. Or it is possible that the UNITS LED is defective.

The first practical step in troubleshooting this symptom depends on what test equipment is available. If you have a logic probe or scope, monitor the UNITS decoder output to the LED segments first. If pulses are present on all segments, but there is no 8 or 9 display, the LED is at fault. (Figure 3.7 shows the segments involved to make up an 8 or 9 display. All segments must be on for an 8.)

If you have a logic pulser or pulse generator, you can inject a pulse on all segments of the LED simultaneously, and check for an 8 display. If this is not practical, you can inject a pulse on each segment in turn and make sure that particular segment is capable of producing a display.

If you find that one or more pulses required to make an 8 or 9 display are missing to the LED, check for an 8 or 9 input to the UNITS decoder. As shown in Fig. 7.6, an 8 input is produced when there is a pulse at the BCD 8 line (between the counter and the decoder). A 9 input to the decoder requires simultaneous pulses on the 8 and 1 lines.

The UNITS decoder is probably at fault if the 8 and 9 inputs are available to the decoder, but there are no 8 or 9 pulses to the LED segments. (As shown in 3.7, an 8 display requires that all segments receive pulses. A 9 display requires all segments to receive pulses, except the d and e segments.)

If you have a logic clip (that can accommodate the IC pins) and a logic pulser, you can check operation of the UNITS decoder on a single-step basis. Inject input pulses at the BCD 8 line (from the counter), while simultaneously enabling the decoder, and check that the outputs on all LED segments go true.

7.7.3 Display is double the correct value

Assume that a 400-Hz input is applied, but the display shows 800 Hz (double the correct value). This is probably a counter function (rather than a readout function) failure. Here's why.

When flip-flops are involved (as they are in most counters) one flip-flop might be following the input pulses directly. The normal flip-flop function is to go through a complete change of states for two input pulses. Instead, the faulty flip-flop is changing states completely for each input pulse. This is similar to operation of a one-shot multivibrator.

If the decade counter (containing such a faulty flip-flop) produces two output pulses to the next counter for every 10 input pulses (the counter divides by 5 instead of 10), the count is double. All decades following the defective stage (the decade counter in this case) receive two input pulses instead of one.

The first practical step in troubleshooting this symptom again depends on what test equipment is available. Use either a logic probe or scope to monitor each decade input and output in turn. Start with the UNITS decade. The decade that shows two outputs for 10 inputs, or one output for five inputs, is at fault.

7.7.4 Display is stuck at an incorrect count

Assume that a 300-Hz input is applied, but the display count is 000, 001, 002, 003, 004, 005, 000, 001, 002, 003, and so on. The count is stuck at 005, and never reaches 006 (much less 300).

This is probably a counter function (rather than a readout function) failure. Here's why.

If the 10 or 100 readout is defective, a good UNITS readout still produces a display beyond 005. If the UNITS readout is defective (in some way so that the count is stuck at 005), the 10 and 100 readouts can still produce counts starting at 010. Under these conditions, the count would be 000, 001, 002, 003, 004, 005, 010, 011, and so on. Of course, it is possible that all three readouts are defective simultaneously, but this is not likely.

There are three possibilities for this symptom (the count stuck at 005): (1) the input pulses never reach more than five (not likely), (2) there is a reset pulse occurring at the same time as the sixth input or any time after the fifth input pulse (possible), and (3) the UNITS counter simply does not count beyond five (most likely).

Start by eliminating the reset-pulse possibility. Using a dual-trace scope, monitor the input-pulse line and reset the line of the UNITS counter as shown in Fig. 7.7*a*. Adjust the scope sweep frequency so that about 10 input pulses are displayed. If a reset pulse does occur after the fifth pulse, the problem is pinpointed. Trace the reset line to the source of the unwanted (improperly timed) reset pulse.

If there is no reset pulse before the sixth pulse (which is probably the case), monitor the input line and the 4 line of the UNITS counter, as shown in Fig. 7.7*b*. The 4 line should go true at the sixth input pulse (as should the 2 line). Then monitor the 8 line.

The connections of Fig. 7.7*b* should pinpoint the problem. Typically, with this symptom, the 2 and 4 lines will show no output to the UNITS decoder/storage IC, indicating that the UNITS counter is defective. Of course, you might get outputs at the 2 and 4 lines of the UNITS counter, but the outputs are abnormal. For example, outputs from the counter can be below the threshold of the decoder/storage IC.

If you have a logic probe (or a logic clip that can accommodate the IC pins) and a logic pulser, you can check operation of the UNITS counter on a single-step basis. Inject input pulses and check that the 2 and 4 lines go true when the sixth input pulse is injected, as shown in Fig. 7.7*c*.

If circuit conditions make it possible, disconnect the 2 and 4 lines and recheck operation of the UNITS counter. It is possible that the 2 or 4 line is shorted to power or ground, or that the lines are shorted together (Sec. 7.4). As discussed,

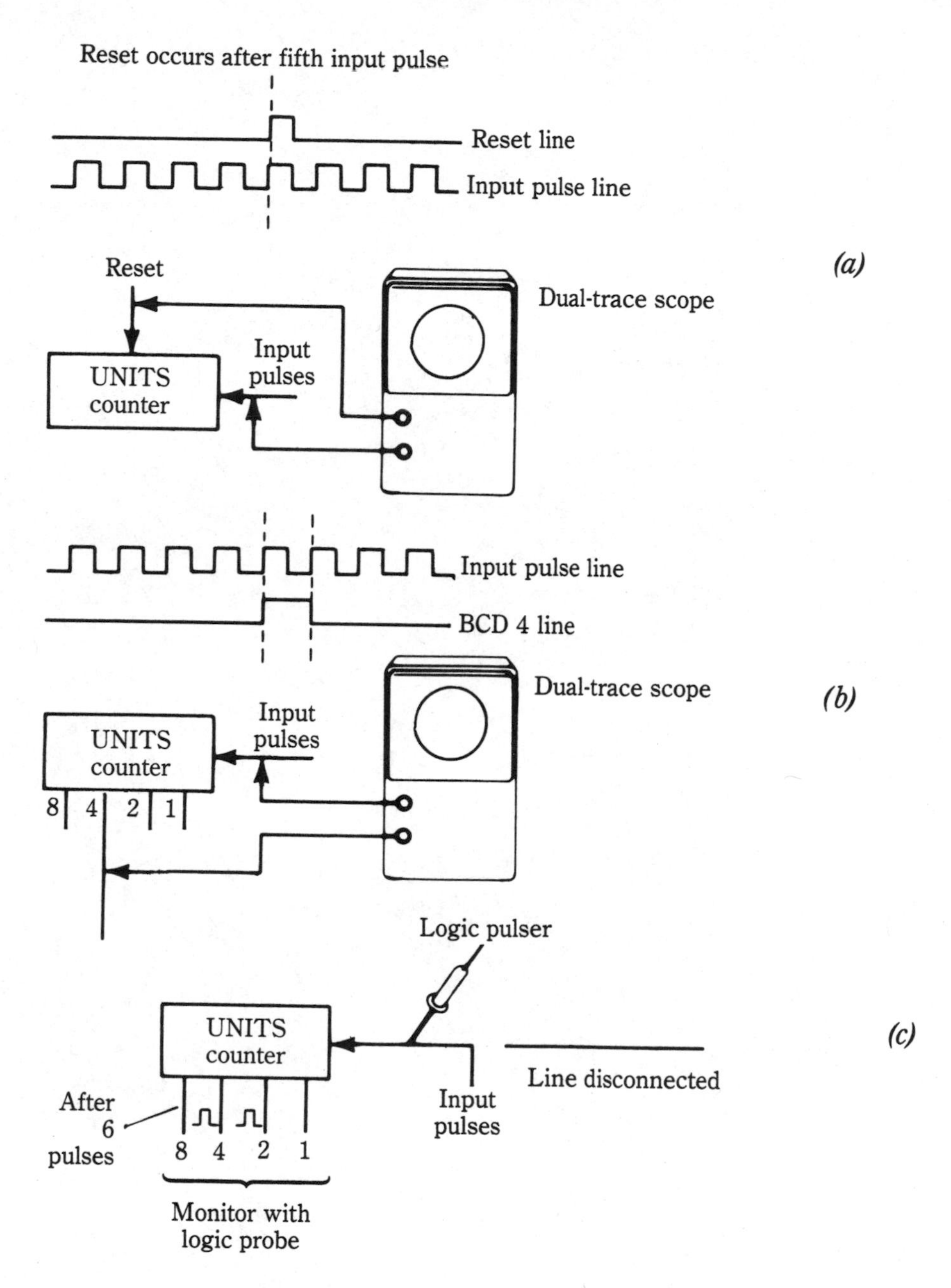

Fig. 7.7 Monitoring counter-circuit functions.

with most present-day digital circuits, the lines or wiring to the ICs are in PC form. This makes it impractical (at least difficult) to disconnect individual lines or leads. The IC package must be checked by substitution (or by comparison if a logic comparator is available for that IC).

8 Advanced digital troubleshooting techniques

This chapter describes some additional examples of digital troubleshooting. Many of these examples are more advanced than those described in Chapter 7 and involve some subtle troubleshooting decision-making using combinations of test equipment.

This chapter covers such subjects as stimulus-response testing, power and ground shorts from unexpected causes, more on gate and IC troubleshooting, clock-line problems, distinguishing between internal and external shorts, current flow analysis, flip-flop troubleshooting, register troubleshooting, wired-AND/OR problems, and three-state bus problems.

The chapter also describes a ''when-all-else-fails'' procedure for resolving the situation of an experimental microprocessor-based system that does not respond to routine troubleshooting.

8.1 Stimulus-response testing

Stimulus-response testing involves the use of a logic pulser in combination with other troubleshooting tools to check microprocessor/digital circuits. In the simplest of terms, the logic pulser is used to stimulate a circuit, and the response is monitored with a current tracer, logic probe, or logic clip. As an example, if the problem is a shorted line, pin, or other point (node), the pulser is used to stimulate the circuit, and the current tracer is used to follow current pulses to the physical location of the short.

Table 8.1 summarizes stimulus-response testing procedures described throughout this book. Note that the logic pulser is used as the stimulus for all faults shown in Table 8.1. However, the normal circuit signals or pulses can also be used (in some cases) if more convenient. Also note that the current tracer can be

Table 8-1 Summary of digital stimulus-response testing.

Fault	Stimulus	Response	Test method
Shorted node*	Pulser†	Current tracer	Pulse node and follow current pulses to short.
Stuck data bus	Pulser†	Current tracer	Pulse bus line and trace current to device, holding the bus in a stuck condition.
Signal-line short to power or ground	Pulser†	Probe, current tracer	Pulse and probe-test point simultaneously. (A short to power or ground cannot be overridden by pulsing). Follow current pulses to short with tracer.
Power-to-ground short	Pulser†	Current tracer	Remove power from test circuit. Disconnect electrolytics. Pulse across power and ground. Trace current to fault.
Suspected internal open	Pulser†	Probe	Pulse device input and probe output for response.
Solder bridge	Pulser†	Current tracer	Pulse suspected line and trace current pulses to fault. Tracer lamp goes out when solder bridge is passed.
Sequential logic fault in counter or register	Pulser†	Probe, clip	Disable clock circuit and use pulser to enter desired number of pulses. Use clip (or probe) to verify truth table.

*A node is any interconnection between two or more ICs (or other logic element).

†Use the logic pulser to provide stimulus. Normal circuit pulses will not override shorts (or they are too fast to be counted).

used to test circuit response in all cases except when an open IC is suspected or when monitoring the response of counters and registers.

The use of pulses for stimulus-response testing, rather than a continuous current, prevents burnout of components on the line. Pulses also provide a more realistic test, because many lines in digital equipment operate with pulses. Of course, many of the microprocessors used with noncomputer type equipment described in Chapter 9 operate on fixed-level signals (typically 5 V or ground) and must be so tested. In such cases, the current tracer cannot be used, because the tracer responds only to changing currents (ac or pulses).

8.2 Power-to-ground shorts

There are several ways to locate power-to-ground shorts on a PC board. For example, you can (1) supply a high current to the circuit and see which traces change color, delaminate, or burn up (not generally recommended); (2) measure microvolt drops across the active supply traces and see where the current is flowing (as described in Sec. 6.17, Fig. 6.2); (3) replace all capacitors on the board (one at a time); (4) replace all ICs on the board (one at a time); (5) set the board aside and troubleshoot it later; or (6) put the board in the nearest trash container.

This last suggestion might not be as unrealistic as it sounds, because the time locating the short might cost more than the value of the board. Generally, power-to-ground shorts are caused by shorted capacitors (decoupling or bypass), and it might be quite costly to find the defective capacitor (if you must replace all capacitors one at a time).

The following procedure, using a logic pulser and current tracer, can be used to find power-to-ground shorts caused by shorted decoupling capacitors. Remember that this procedure can be used for any type of board, even if the board is part of control-type equipment (such as described in Chapter 9). Of course, it is assumed that you have removed (or at least disconnected) the board from the equipment.

1. Remove power from the circuit. Power the pulser and tracer from a 5-V supply.
2. Lift one side of all electrolytic capacitors on the power-supply bus.
3. Pulse the power line as shown in Fig. 8.1, starting with pins or components in the corners of the PC board. Use any cables supplied with the pulser, so your hands are free to move from point to point with the tracer. Moving the pulsing point around from corner to corner while tracing current from the pulsing point helps speed fault location.

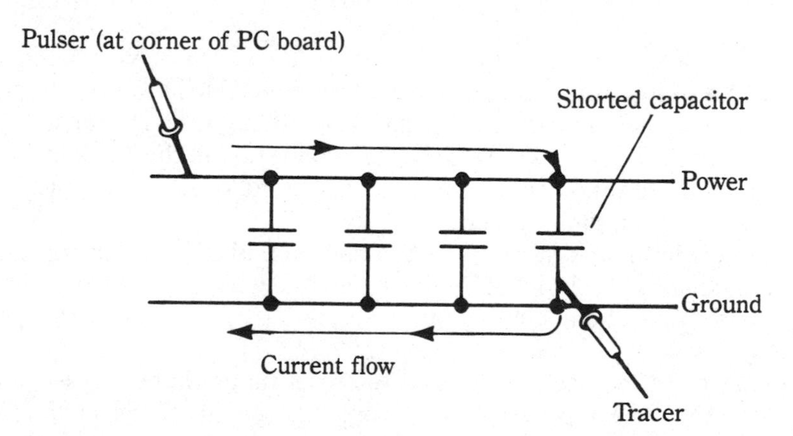

Fig. 8.1 Power-to-ground short troubleshooting.

4. Because you are pulsing a short, current is high, so set the tracer sensitivity to 1 A (or whatever is the highest current sensitivity of your particular tracer).
5. Remember that parallel current paths can exist on PC boards where power enters from more than one connector. This is another reason for moving the pulsing point around the board.
6. Sometimes a current path seems to disappear. There can be several reasons for this condition. The PC-board trace can become wider, causing current to fan out and lessen the intensity under the tracer tip (as shown in Fig. 5.5*c*). The current might pass through a plated-through hole in the PC board. Likewise, the current might branch and go several different places via several different paths, thus lessening the current density in the path you are tracing.
7. When you think the fault is located, move the pulsing point to the suspected short (or shorted capacitor) and check for current paths anywhere else on the board. If you have set the tracer for a high current, no current paths should be detected elsewhere on the board when you pulse directly at a short (unless there is more than one short).
8. Once you are convinced that you have found the short (say a bad capacitor), remove the capacitor and check that the power-to-ground short is cleared.
9. Replace the defective capacitor. Make certain to reconnect all electrolytic capacitors removed during step 2. As discussed, this test is not practical for electrolytic capacitors, because such capacitors will pass leakage current and appear to be shorted (even when operating normally).

8.3 Troubleshooting specific digital functions

The following paragraphs describe some additional examples of troubleshooting for various types of digital ICC and gates.

8.3.1 Flip-flops

In the example of Fig. 8.2*a*, there are two D-type FFs. One FF works properly, but the other FF does not change output states (even though the input conditions are supposedly identical). Note that both FFs are connected to the clock line and both reset inputs are open (no connection or NC). The following is a typical troubleshooting sequence for such a situation.

Pulse the D input and monitor the Q output of both FFs to confirm the symptom. Assume that the Q output of FF_1 changes state when pulses are applied to the D input (correct operation), but that the Q output of FF_2 does not change (incorrect operation).

The next point to check could be at the reset input or the clock input. However, because the clock inputs of both FFs are connected to the clock line, and FF_1 is working properly, you can skip the clock inputs for now. Instead, use a logic probe to check the reset input of FF_2, which should be floating (not high or low).

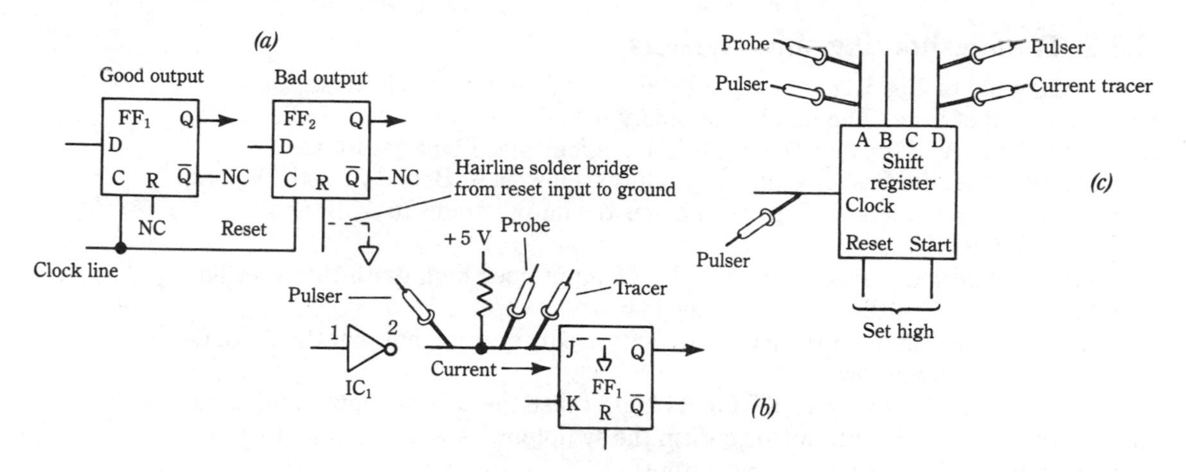

Fig. 8.2 Flip-flop and register troubleshooting.

If you are uncertain as to what the logic-probe indication should be at the reset input of FF_2, check the indication of FF_1. Both reset inputs should be identical.

In the example of Fig. 8.2*a*, the reset input of FF_2 is stuck low because of the short to ground (say a hairline solder bridge from the reset input to ground).

The final step is to pulse and trace the reset input of FF_2 to see if current is flowing towards FF_2 (indicating an internal short) or away from FF_2 (indicating an external short).

In the example of Fig. 8.2*b*, a JK-type FF is driven (at the J input) by an inverter. Although there is supposed to be pulse activity at the inverter input, there is no activity at the FF output (the FF output does not change states). The following is a typical troubleshooting sequence for such a situation.

Pulse the IC_1 input and monitor the Q output of FF_1 to confirm the symptom. Assume that there is pulse activity at $IC_{1\text{-}1}$, but not at the Q output of FF_1.

The next point to check is at the line between IC_1 and FF_1. Use a logic probe (or scope) to check for pulse activity. In the example of Fig. 8.2*b*, there is no activity, and the line is stuck low (because of a short within FF_1).

At this point, you are not certain why there is no activity on the line. The fault could be a defective IC_1, preventing inverted pulses from passing to the line. To remove doubt, try pulsing and probing the line simultaneously to see if the state can be changed from low to high. (If the state can be changed, the most likely cause is a defect in IC_1.)

In the example of Fig. 8.2*b*, the state cannot be changed, indicating that the line is shorted to something. Because the line is stuck in the low state, the line is shorted to ground. (If the line is stuck high, the short is to power.)

The final step is to pulse and trace the line to see if current is flowing toward IC_1 or FF_1. With a short in FF_1 as shown, there should be little or no current flow between the pulser and IC_1, but the considerable current flow between the pulser and FF_1. Also, the current indication should remain as the tracer is moved along the line to the FF_1 input pin (indicating an internal short).

8.3.2 Troubleshooting shift registers

In the example of Fig 8.2*c*, a shift register is used to hold clock pulses during a certain period of time. The number of pulses held during this time is indicated by the state of the A, B, C, and D outputs using a four-place binary format.

For example, the first clock pulse produces a low at A, B, and C, with D going high (0001). The second clock pulse changes the output states to A, B, and D low, with C high (0010) and so on.

The time period starts when the START input goes high (with the reset line high) and stops when the reset line goes low.

Now assume that all of the inputs are supposedly normal, but that the A, B, C, and D outputs remain low.

Set both RESET and START inputs high. Pulse the CLOCK input and monitor the A, B, C, and D outputs to confirm the symptom. Assume that all four outputs remain low (no activity on any output).

At this point, you are not certain why all outputs are low. The fault could be a defective register (not producing the correct output) or a short on the output lines (external to the register).

To remove doubt, try pulsing and probing the output lines simultaneously. Pulse and probe each line in turn. Now assume that none of the lines can be changed from low to high with simultaneous pulsing and probing. This indicates a short on the lines rather than a defective register IC, but are you certain?

To remove this doubt, try pulsing each output line and monitoring for current with the current tracer, as shown in Fig. 8.2*c*.

Finally, assume that there is a current indication between the pulser and the IC on all four output lines, but little or no current on the other side of the pulser. This indicates a defect in the register IC rather than in the lines. In any case, the register IC must be replaced, because you do not have access to internal circuits.

8.3.3 Troubleshooting combination circuits

In the example of Fig. 8.3*a*, the line between $IC_{1\text{-}13}$, $IC_{2\text{-}13}$, and $FF_{1\text{-}4}$ (which is the clock line) appears to be stuck low ($IC_{1\text{-}13}$ low and no activity on $IC_{2\text{-}11}$ or $FF_{1\text{-}Q}$).

To confirm this symptom, monitor for clock pulses on the line and on the input to IC_1 at pin 11. Assume that there are clock pulses at $IC_{1\text{-}11}$, but not at any point on the line.

Note that IC_1 is located on PC_1 (PC board 1), and IC_2 and FF_1 are located on PC_2. Assume that both boards are easily replaced (plug-in cards), so you decide to localize the problem by substitution (or using a logic comparator). You find that the problem is removed when PC_2 is replaced.

As a next step, you could probe and pulse the line to see if the line can be driven from low to high. However, because you have already localized the problem to PC_2 (probably a short), you can go directly to current tracing on the PC_2 clock line.

Start by pulsing at some point along the line and see which way current flows using the current tracer. In the example of Fig. 8.3*a*, current flows from the pulser toward IC_2 (which has an internal short).

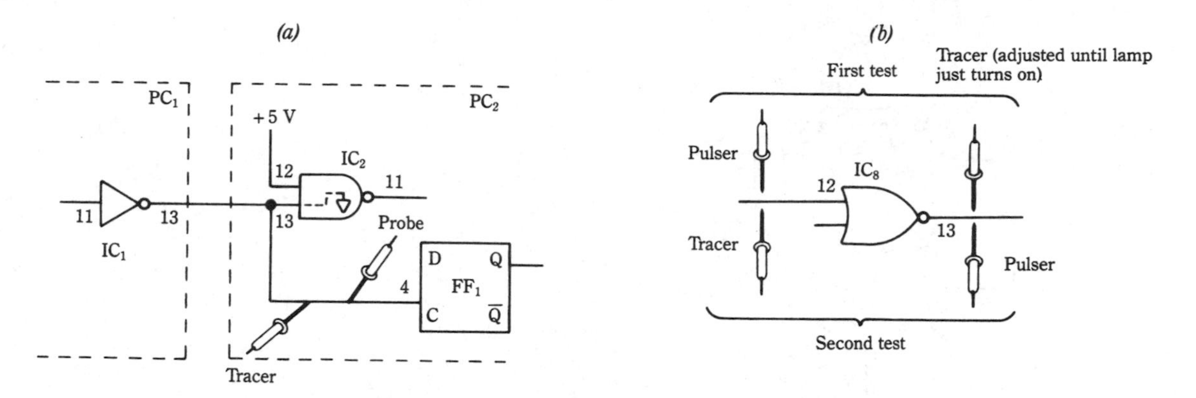

Fig. 8.3 Combination-circuit troubleshooting.

8.3.4 Distinguishing between internal and external shorts

It is very helpful to know if the line is shorted or the IC is shorted, before you pull an IC from the board. For example, assume that the NOR-gate circuit shown in Fig. 8.3*b* is found to be bad using a probe and pulser. That is, positive pulses at pin 12 appear as positive pulses at output pin 13 (instead of reversing polarity at the output).

Such a symptom can be caused by a short within IC_8 or by an external short. One way to help tell the difference is to pulse IC_{8-12} and monitor current at IC_{8-13}. Set the current-tracer sensitivity so that the tracer just turns on. Then reverse the test instruments (pulse IC_{8-13} and monitor IC_{8-12} for current).

If the current is essentially the same in both directions, the short is likely to be external (say a solder bridge between the two pins, 12 and 13). If the current is present, but substantially different in each direction, the short is likely internal.

8.4 Current flow analysis

Knowing the amount of current flow between logic elements during normal high and low states can help when troubleshooting digital equipment.

For example, Fig. 8.4*a* shows the current flow between two NAND gates in a typical TTL circuit during both the high and low states. If current flow exceeds these values by a substantial amount, there is probably a fault current being added to the normal current.

As discussed in Sec. 5.6, current-tracer sensitivity can be adjusted to see transitions as low as about 300 μA. (Remember that current tracers have an inductive pickup and see only transitions, not steady currents.) Generally, a current tracer cannot see the high-level current transition of a TTL circuit (about 40 μA) but can see low-level transition (1.6 mA) if tracer sensitivity is properly adjusted.

High-level output current is the current flowing from the output while the output voltage is at logic 1 (and is generally listed as IOH). In TTL, IOH is about 40

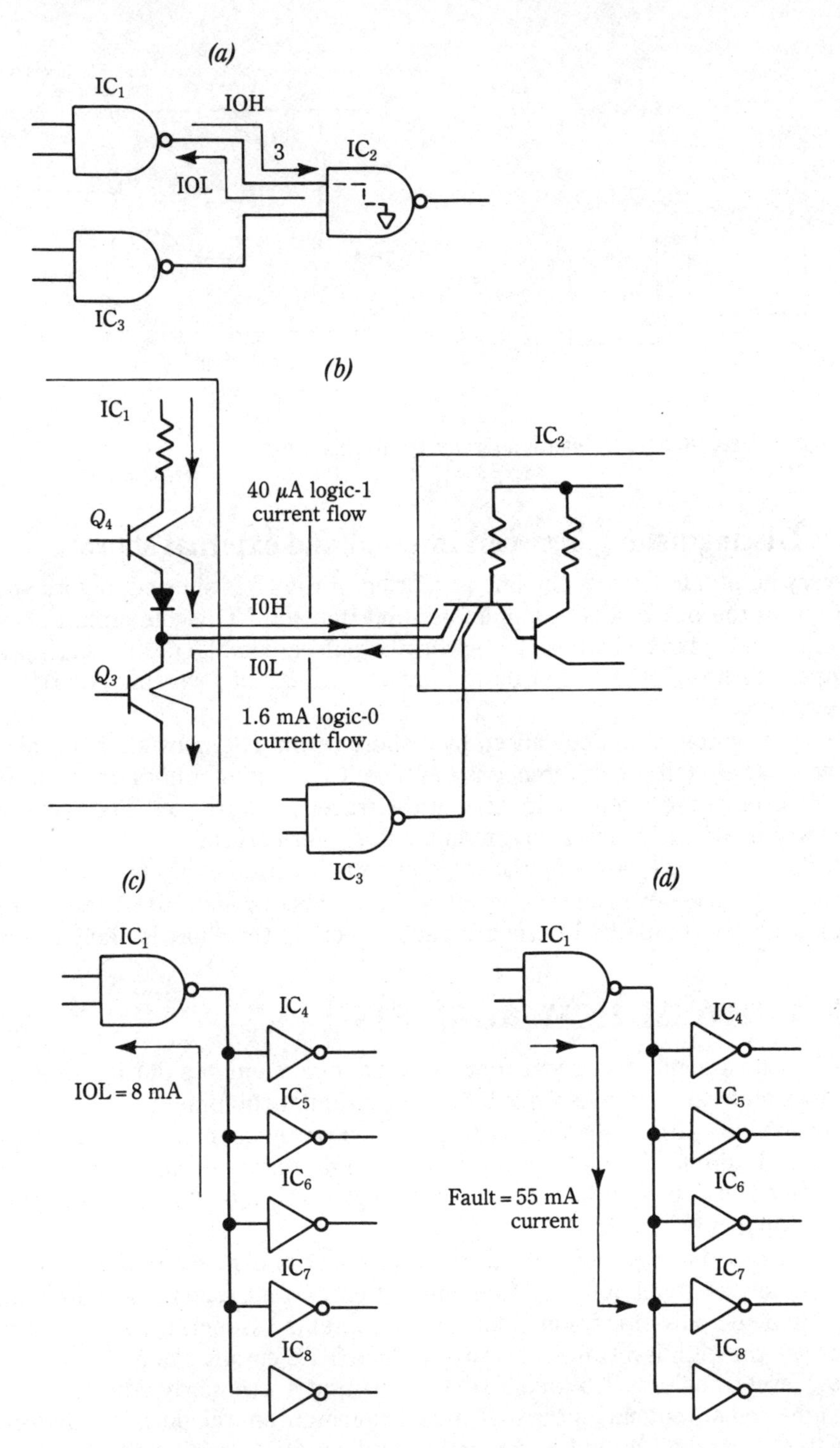

Fig. 8.4 Current-flow analysis in troubleshooting.

μA. Low-level output current is the current flowing into an output while the output voltage is at logic 0. In TTL, IOL is about 1.6 mA (with a fanout of 1).

Fortunately for troubleshooters, fault currents are generally much higher than normal pulse-transition currents. So if you adjust the current tracer sensitivity to barely detect normal pulse activity, a fault current is easy to spot.

As an example, assume that input 3 of IC_2 in Fig. 8.4*a* is shorted internally. This causes an excessive current (about 55 mA for a typical TTL IC) when IC_1 tries to go from low to high. Such an excessive current produces a brilliant glow of the current-tracer lamp. The fact that current is flowing from IC_1 to IC_2 indicates that the short is in IC_2 (or in the direction of IC_2).

Figure 8.4*b* shows the internal structure of the gates illustrated in Fig. 8.4*a*. Figure 8.4*b* also shows the current flow for normal operation. When the output of IC_1 is low, Q_3 is a saturated transistor to ground, drawing about 1.6 mA. For a typical TTL, Q_3 has the capability to drive up to 10 gates and thus sinks up to 1.6 mA of current during normal operation.

When IC_1 goes low, while driving five gates as shown in Fig 8.4*c*, the output transistor in IC_1 sinks 1.6 mA from each inverter ($IC_4 - IC_8$). Normal current is then about 8 mA with five inverters. A short anywhere on the line between IC_1 and the inverters drastically alters the direction and magnitude of current, as shown in Fig. 8.4*d* (where there is a short in IC_7). In this example, the fault current flows toward the inverters (instead of the normal current flow to IC_1) and increases from about 8 mA to at least 55 mA.

Remember that the examples shown here are generally less complex than the problems in real circuits. However, these examples demonstrate two important rules: (1) fault currents generally exceed all other currents on the board by a wide margin; (2) knowing the average or typical source and sink currents of a circuit (operating normally) can quickly pinpoint a fault.

8.5 Open-collector, wired-AND, and wired-OR problems

Before the wide use of buses such as those found in present-day computers, the wired-AND circuit using open-collector gates was quite proper in TTL design. This same configuration is still used in equipment where a microprocessor functions as a controller (such as the applications described in Chapter 9).

As shown in Fig. 8.5, the wired-AND uses an open-collector circuit to connect a number of gates, any one of which can go low and control the entire output line. Open-collector gates differ from other gates in that the open collector does not have an active high current source. Instead, the collector of Q_3 is left untouched (internally).

From a troubleshooting standpoint, it is important to remember that wired-AND can sink current (in the low condition) but cannot supply current (in the high state). To provide current in the high state, the output line is connected to power through load resistor R_1. So if you monitor current from wired-AND gates to the

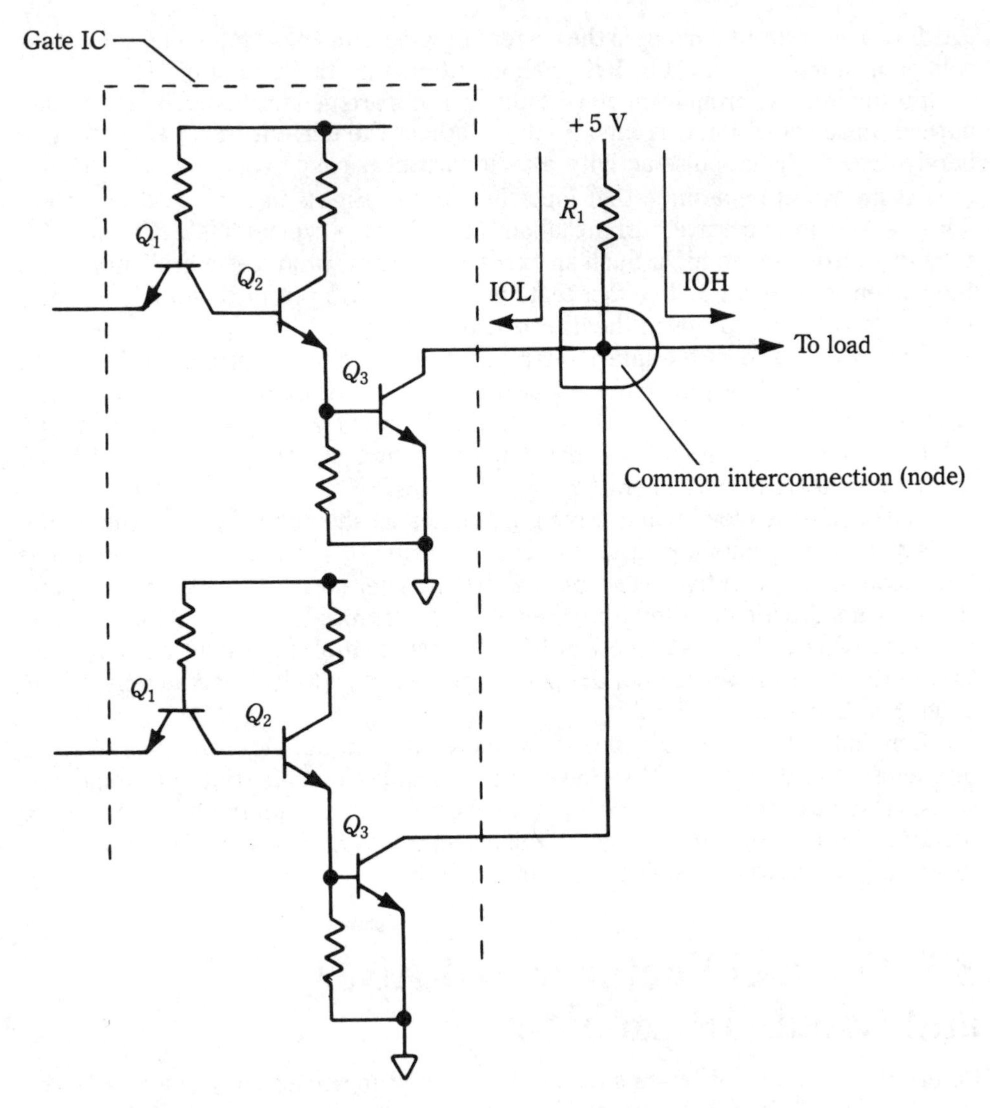

Fig. 8.5 Open-collector, wired-AND, and wired-OR troubleshooting.

load during troubleshooting, the current is coming from the power line (through the load resistor) and not from the gates.

There are other points to consider when troubleshooting wired-AND gates. As long as transistor Q_3 in *every gate* is turned off, the voltage at the common interconnection is near the supply (or high). When Q_3 in any one gate is on, the common interconnection voltage drops to near zero (low). The result is that the common interconnection acts as an AND gate in itself (the interconnection is high only when all of the inputs are high). This is usually referred to as implied, dot, or wired-AND, depending on the literature you read. (The word dot is used since schematics usually show wired-AND with a dot at the junction center, within the AND symbol.)

Also note that the circuit of Fig. 8.5 can be a wired-OR circuit if used in a system where the true (logic 1) state is represented by a low (near zero volts), and the logic 0 state is represented by a high (near the supply, 5 V in this case). Obviously, any gates with active pull-up resistors (not the open collector of Fig. 8.5) should never be connected for wired-OR. When one gate has a 0 input and the others are at 1, the resulting logic output is unpredictable.

Although the wired-AND gate is not used extensively in present-day computers, the wired-AND principle is used frequently in the three-state bus circuit, which is discussed next.

8.6 Bus troubleshooting

Figure 8.6 shows the truth table, internal wiring, and typical gate interconnections for an IC with three-state bus drivers. Such drivers are built into computers' ICs and some microprocessors to allow common bus usage by all digital devices.

For example, a ROM has a three-state driver on each of the data-output lines. When the ROM is enabled, data bits are put on the data line by turning on the internal three-state driver outputs. When several drivers are fused together, a form of wired-AND is created.

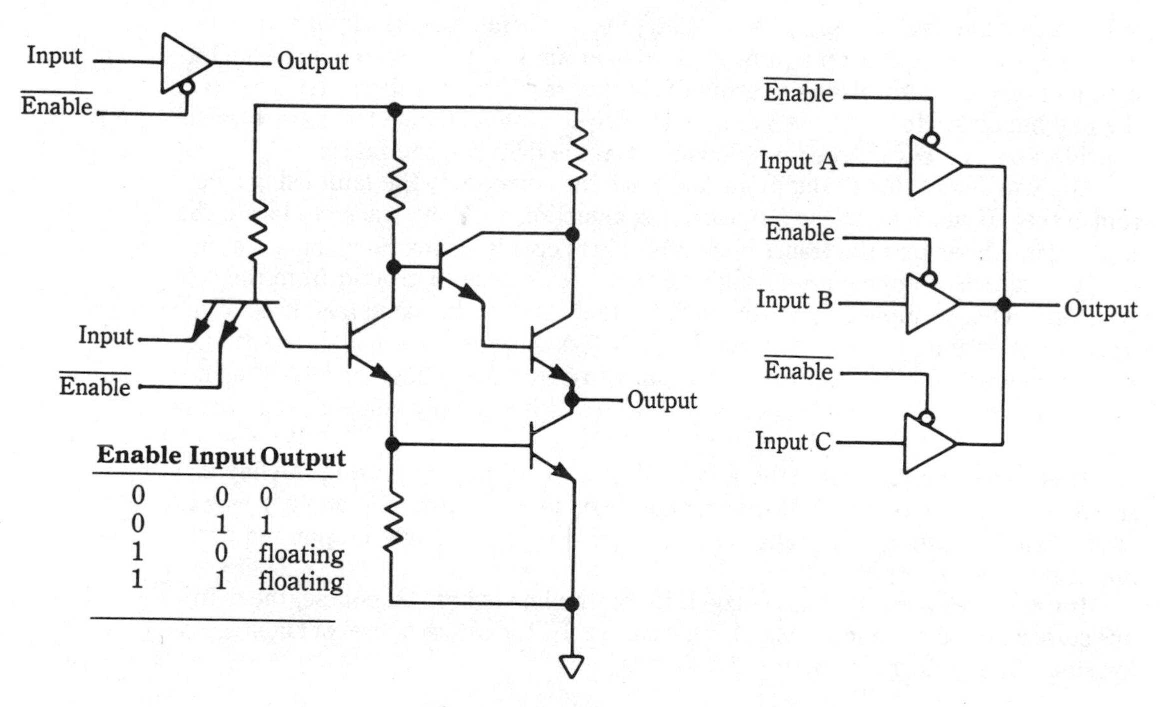

Enable	Input	Output
0	0	0
0	1	1
1	0	floating
1	1	floating

Fig. 8.6 Bus troubleshooting.

Three-state bus drivers (and wired-AND configurations) are often difficult to troubleshoot. This is because the output can be controlled by one faulty device, which causes the entire output line to be faulty. Because all points on the bus are at the same voltage, voltmeters, scope, and other voltage-measuring instruments cannot indicate which component is causing the bus to be stuck.

The first decision to make when troubleshooting bus structures is to decide if the fault exists in the bus drivers (the talkers) or other components receiving data bits (the listeners). This is not an easy task, but the following tips should help you make the decision.

Start by answering the following questions. Do you find pulse activity on the bus-driver inputs? Is the driver enabled? Does the driver respond to a stimulus? These questions can be answered using a probe and pulser to see the initial state of the circuit and then see if the state changes when the circuit is pulsed.

Be sure you can enable or disable the driver without stressing the circuits physically or electrically. Also, check drivers with multiple inputs, to make sure you have control when trying to stimulate the circuit. (Some drivers require more than one data input and/or enable input, Fig. 4.7*a* for example.) If you cannot check the driver first, check each listener in turn.

If you suspect a particular bus – say through the use of logic analyzers (Sec. 5.9) or by means of incorrect signatures found with a signature analyzer (Sec. 5.10) – here are some procedures to help you locate the fault.

8.6.1 Bus line stuck high or low

When a bus line is stuck in one logic state, the line is not necessarily inactive. For example, a line stuck low might be shorted to ground at the input to one of the bus listeners (say a solder bridge to ground). In this case, the bus driver tries to drive the line but fails. However, even though the driver cannot change the state, there is considerable current on the line (flowing from the driver to the fault).

You can detect these conditions and trace the currents to the fault using a current tracer. If necessary, use a pulser in conjunction with the tracer to locate the fault. Remember that the tracer responds only to current transitions (pulses or ac).

You can also remove power and check the resistance to ground from the suspect line. Then compare this resistance to other lines in the same bus. If all of the lines are the same, the bus is probably good. If one or more lines is different from the remainder, that line is suspect. You can also check the voltage on all of the lines of the bus. If any one line is always at +5 V (or other supply voltage), that line is suspect.

If you are convinced that there is no short on the line, try pulsing the line and see if the listeners respond. If this cannot be done easily (there is no easy way to check listener response in many cases), try pulsing at the driver output and probing at the listener input.

If the pulses pass, the line is good. If not, the line is bad. Of course, the driver or listener can be open internally. In that case, a TTL *floats* to a level of about 1.5 V (or static high), as shown in Fig. 7.2.

8.6.2 Buses with open collector (wired-AND/OR)

When troubleshooting buses that have ICs with open collectors (no matter how they are wired) remember that such ICs can sink current, but cannot source current. Any current you find is from supply through the external load resistor.

Start by disabling the driver inputs, then pulse the driver outputs while monitoring current on the corresponding line. The line drawing the most current has the faulty driver (assuming an internal short in the driver IC).

8.6.3 Drivers with source and sink capability

There are two ways to check single-state drivers with source and sink capability (not open-collector drivers).

First, pulse the inputs and probe the outputs for corresponding state changes. For example, if you inject a pulse train at the driver input, there should be a corresponding pulse train at the output. The outputs of all drivers on a bus should be the same when pulsed.

An alternate method is to pulse the inputs and monitor the outputs with a current tracer. The amplitude and direction of the current flow should be the same for all lines on the bus. If one line is different, you can then pulse and trace current on that line to find the fault.

8.6.4 Three-state drivers and buffers

There are two ways to check three-state buffers with source and sink capability (not open-collector drivers).

First, disable the driver inputs, pulse the driver output lines, and monitor the lines for current. Typically, none of the lines should show current. (If current is indicated, the current should be the same for all lines in a bus without fault.) If one line shows current (or an abnormal current), trace the fault with the pulser and current tracer.

An alternate method is to enable all of the drivers (with the appropriate enable signal) and pulse driver inputs individually. If one driver output fails to indicate activity, it is probably defective (possibly open). If open, the output should float to a static high level (of about 1.5 V for a typical TTL circuit).

Note that the circuit of Fig. 8.6 is enabled by a low at each enable input. This low can be provided by simply connecting the enable inputs to ground.

8.7 Experimental system troubleshooting

This section describes a general procedure for resolving problems when an experimental microprocessor-based system does not respond to the troubleshooting procedures described thus far. This is a "when all else fails" technique based on breaking the feed back loop into sections. Each section is turned on independently of all other sections.

The technique described here permits you to prove each section, good or bad, on an individual basis. Once a section is proved good, you go on to another section until all sections in the feedback loop are operating properly. If each section in the loop is normal, but the system fails to work, the problem is one of the interconnections (unless the program is simply not practical).

The basic problem in troubleshooting any microprocessor-based system, where the cause of trouble is not readily apparent, is one of feedback. For example, in a computer, the microprocessor operates in response to data words appearing on the data bus. In turn, the data word depends on the ROM/RAM address selected by the word on the address bus. Going further, the address word depends on the microprocessor address register (which depends on the data bus).

In a typical system, the microprocessor increments to a new address each time the data word is changed. If there are any words or parts of words appearing on the I/O, the words can be applied to the address and data buses.

From a troubleshooting standpoint, assume that one data or address line (only one) opens so that the line is always at 0 (say one pin of an IC is broken internally). Or assume that two data lines are interchanged (in a development system).

With either of these conditions, the microprocessor places an incorrect address in memory (possibly an unused address). In turn, memory plays back an incorrect data byte (possibly a meaningless instruction). Even if everything else is in perfect order, including the program, the result is (at best) unpredictable and usually causes total failure of the system.

The following procedures isolate the problem areas in a microprocessor-based system. Although the procedures are particularly effective in the development stage, they can be applied to any microprocessor-based system.

Figure 8.7 shows all essential elements of a typical computer. The following procedures are based on breaking the feedback loop of such a system to isolate each section in turn (starting with the microprocessor).

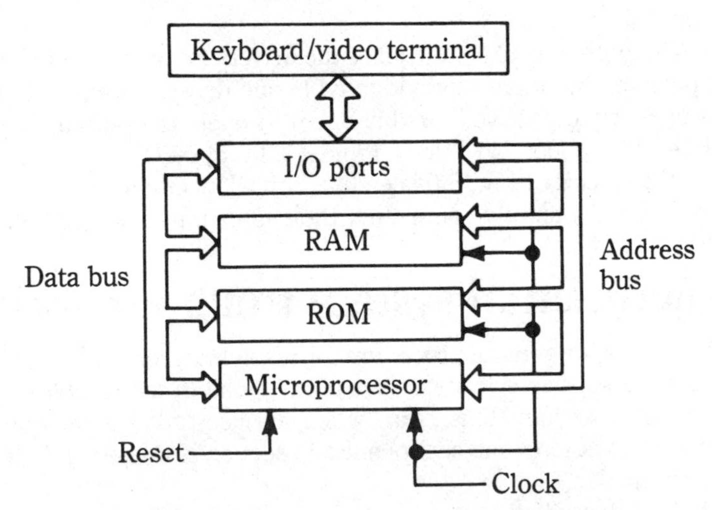

Fig. 8.7 Essential elements of a typical computer.

8.7.1 Breaking the feedback loop

Because the microprocessor is the heart of the system, begin troubleshooting by breaking the feedback loop to isolate the microprocessor. The simplest way is to connect only the microprocessor, and leave the RAM, ROM, and I/O out of the system. This is generally no problem for experimental equipment because the ICs are often mounted in sockets. However, there are problems in connecting just the microprocessor when the system is assembled in final form.

Remember that you should use the procedure described here only when all else fails. Any inconvenience of disconnecting the ROM/RAM and I/O is usually worth the final results.

Before you break the loop, make sure that you have power applied to the circuits (that is always a good idea) and that the clock and reset lines are functioning properly.

Reset The reset line is usually connected only to the microprocessor. Generally, the reset line is held at 0 V until the reset function occurs. This action applies 5 V to the microprocessor reset pin, resetting the entire system. Verify the reset operation by monitoring the address bus with the reset signal applied. If there is no reset button or control in the system, pulse the reset line or apply 5 V to the reset line (momentarily), whichever is more practical. In most systems, the address bus goes to 0000 when the reset line is activated.

Clock In most systems, the clock line is connected to all elements. Verify this by checking for a stream of good clock pulses at each element using a logic probe or scope. The clock pulses must be of correct amplitude (typically about 5 V), width, and frequency. If there are two or more clocks, the clock pulses must be shifted in phase from each other (typically about 90°).

8.7.2 Actuating the system with a no-operation instruction

Once you are certain that the clock and reset functions are normal, actuate the system with a repetitive command. Generally, this is done by applying a *no-operation (no-op)* word or instruction on the data bus. The instruction can be called by many names (NOP, IDLE, WAIT, etc.) in various microprocessors. However, most microprocessors have some form of no-op instruction.

Apply the instruction by connecting each line in the data bus to 5 V or 0 V (ground) as needed to form the correct binary word (machine language), as shown in Fig. 8.8.

Note that resistors are used between the supply or ground and the data bus. The resistors are included to prevent damage to the microprocessor. Many microprocessors try to put data onto the bus during an operating cycle. If the data bus is wired directly to the supply or ground, the data output buffer in the microprocessor can be damaged.

With a fixed no-op instruction word on the data line, the program counter in the microprocessor increments repeatedly. That is, the microprocessor executes a no-op, increments the program counter, executes the next no-op, and so on.

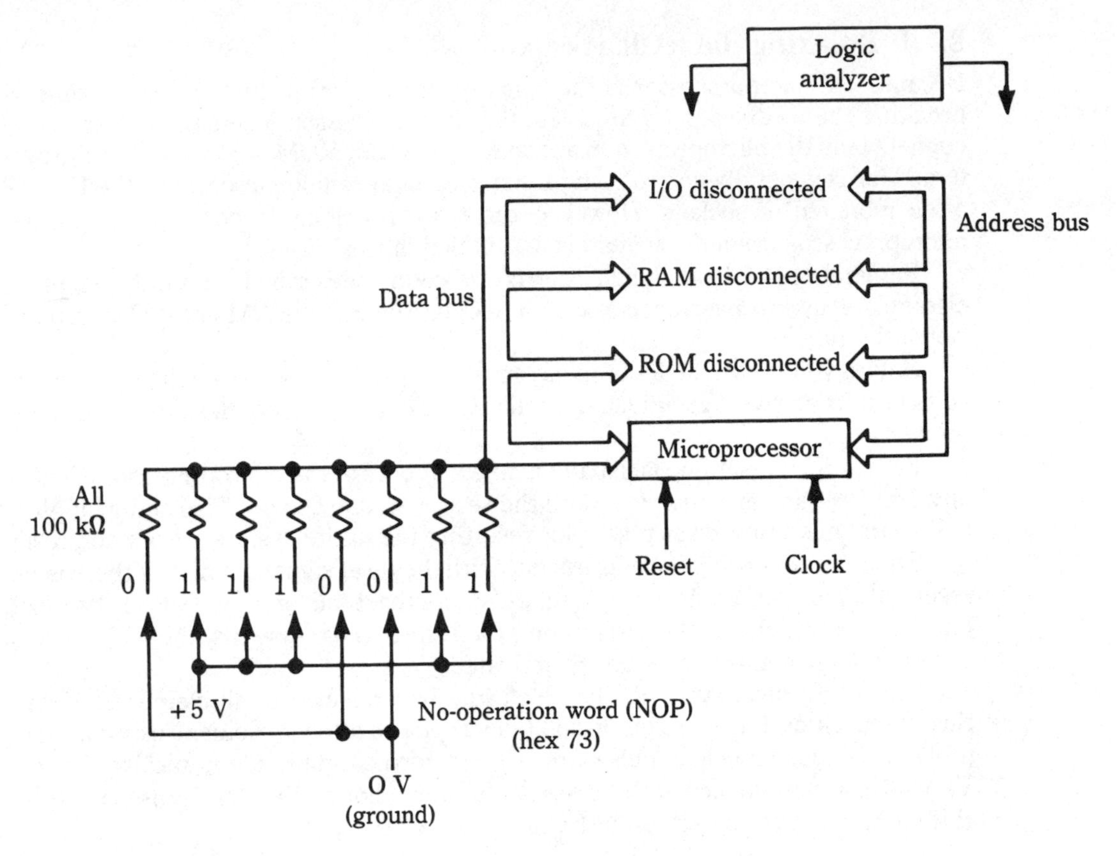

Fig. 8.8 Applying a no-op instruction.

8.7.3 Verifying the count function

Microprocessors used in computer applications should increment continuously through all the addresses. This can be verified by connecting the input probes of an analyzer to all lines of the address bus and setting the analyzer controls for a *count function*. (Most logic analyzers have some form of count function.) Make the connection at the ROM end of the address bus to ensure that correct addresses can be transmitted to and from the ROM.

Start by actuating the reset line and checking for an address of 0000. Then return the reset line to 0 V and check that the count goes through FFFF in sequence. In most systems, the microprocessor continues to execute no-ops indefinitely with the Fig. 8.8 setup. When FFFF is reached, the address register continues on to 0000 again and repeats all addresses.

This procedure verified that the microprocessor is executing no-ops and that the addresses are correctly transmitted to the ROM. If the addresses do not form a counting sequence, examine the address pattern to determine if address lines are interchanged, are inactive (fixed at 0, 1, or high impedance), are shorted, or if the microprocessor is executing an unexpected branch instruction.

8.7.4 Verifying in the timing functions

Timing should be checked as each new block or section is added, and immediately after verifying the count function. Although such testing might seem repetitive, it takes very little time if there is no problem. If there is a problem, a check at each stage can save a great deal of time.

Timing can be checked by examining the waveforms on all buses and control lines (read, write, chip enable, etc.) using a scope or analyzer with a *timing display* (Fig. 5.6). All timing functions should be compared against the timing diagrams shown in the service literature. Any incorrect timing, marginal voltage levels, noise, or crosstalk should be eliminated before proceeding.

Marginal voltage levels can be a problem, particularly on the data and address buses. For example, assume that 5 V normally represents a 1, but that a data-bus input buffer (say at the microprocessor) recognizes anything above about 3 V as a 1. If the pulses appearing on the data bus drop to something on the order of 2.75 V, the pulses might be recognized as a 0.

The timing of control signals is particularly important. It is obvious that a control signal (such as read, write, etc.) must be applied at exactly the right time. For example, if a data-bus buffer is held open too long, all or part of another data word might appear on the data bus. Likewise, if the data-bus buffer is closed too soon, all data-word bits might not appear in sequence.

Noise and crosstalk can also be a problem on data and address buses and non-control lines. For example, if a line has noise pulses (or crosstalk, which is a noise picked up from adjacent lines), and these pulses are near the 2.75-V level, the pulses might appear as 1s. The results are obvious if that particular line is supposed to be at a 0 when the noise pulse appears.

8.7.5 Verifying the ROM data

The ROM can be connected into the system once you are certain that the microprocessor is operating properly (executing no-op commands, issuing all control signals such as read/write and chip enable, incrementing through all addresses in proper sequence, etc.). As shown in Fig. 8.9, make all system connections to the ROM except for the data bus.

Monitor the data output of the ROM with an analyzer, but do not connect the analyzer or ROM to the data bus. Instead, leave the fixed no-op instruction word on the data bus (via the resistors). With these connections, the microprocessor continues to see a no-op instruction and continues to increment or count through all the addresses as discussed.

With the connections of Fig. 8.9, the ROM cycles through all possible addresses so that you can measure the ROM data outputs with an analyzer. It is assumed that the ROM has some known stored information that you can check on an address-by-address basis. If not, make up a test PROM by programming predetermined words at each address and then inserting the PROM into the ROM socket.

If you are troubleshooting an existing system, the ROM has its own program. (The PROM is used primarily for systems under development.) It is also assumed

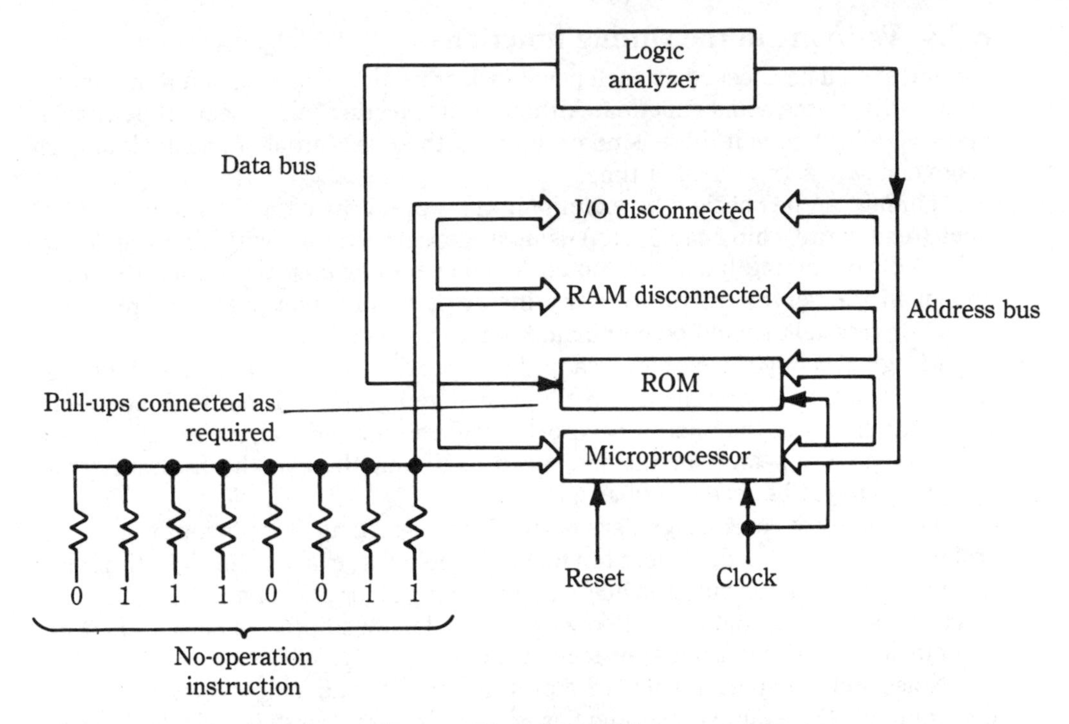

Fig. 8.9 Verifying ROM data.

that if there is more than one ROM, each ROM is checked in turn. If the analyzer has the capacity, continue to monitor all the address lines as well as the data lines.

It might be necessary to connect some temporary pull-ups to the ROM data outputs, because not all the addresses are ordinarily allocated to the ROM. This is done by connecting the lines to 5 V through 100-Ω resistors. With such pull-ups, an address outside the allocated ROM address generates a known data word (all 1s).

You might not have to monitor all possible ROM outputs, although a check can be helpful. However, if there is more than one ROM, you should check sufficiently to verify that the correct ROM is selected and that every ROM is addressed correctly (in sequence).

You should also check some addresses outside those of the ROM to verify that the ROMs are off when not being addressed (this is often overlooked). Remember to check the timing (Sec. 8.7.4) on the address bus and control lines, particularly on the ROM control lines. It is possible for two ROMs (or a ROM and a RAM) to be on at the same time.

8.7.6 Verifying ROM/microprocessor compatibility

The ROM/microprocessor link can be verified once you are certain that the ROM is being addressed in sequence and is delivering the correct outputs. Remove the no-op instruction word (by removing the resistors on the data bus) and connect the ROM to the data bus, as shown in Fig. 8.10. Note that the RAM and I/O are still not connected at this stage.

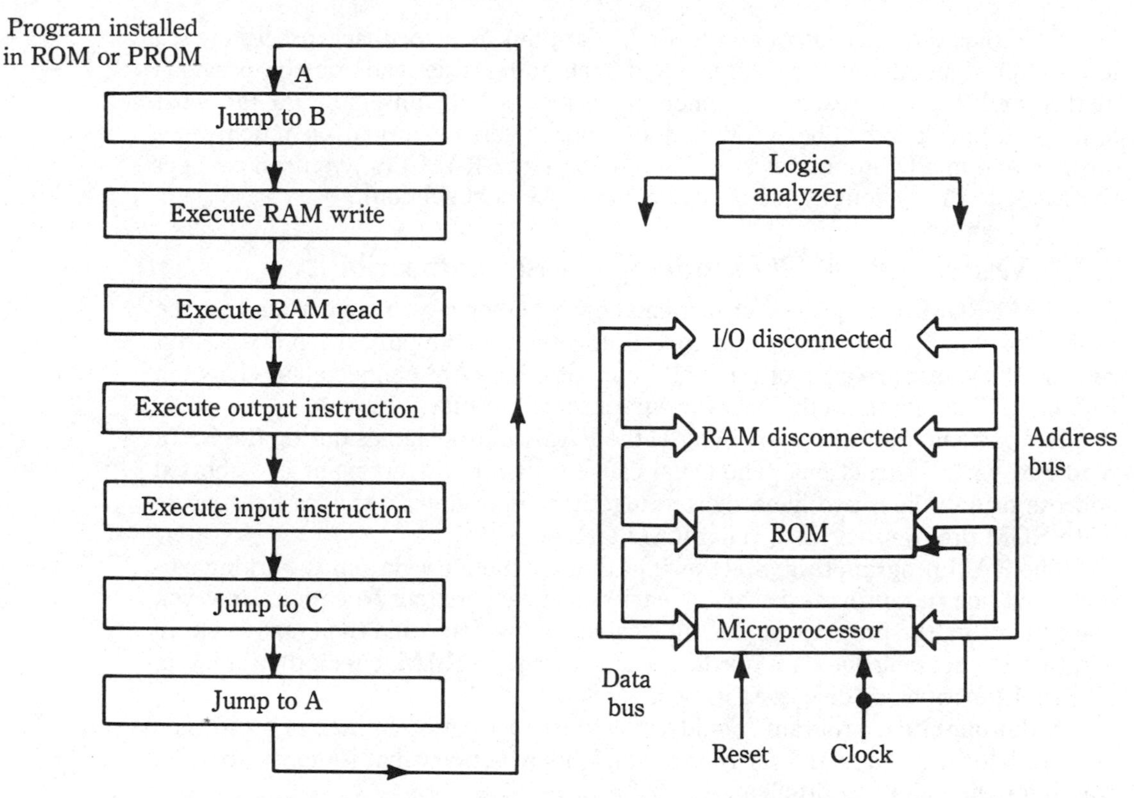

Fig. 8.10 Verifying ROM/microprocessor compatibility.

Install a ROM (or test PROM) with a simple diagnostic routine or program. The test program should contain several unconditional jumps. (A flowchart of such a program is included in Fig. 8.10.) Verify operation of the test program by monitoring the address bus with the analyzer.

The program of Fig. 8.10 includes a RAM access and I/O instructions so that the RAM and I/O control cycles can be checked before the RAM and I/O devices are installed. At this point, RAM and I/O timing should also be checked.

Generally, it is not necessary to monitor the data bus at this point unless there is a problem. This is because the sequence of program addresses is ample to verify proper execution of the program. If you have put each of the steps in Fig. 8.10 in addresses that are in sequence and each address appears in that sequence on the address bus, the correct word must be appearing on the data bus.

The program of Fig. 8.10 is generally adequate to test the microprocessor-to-ROM data link as well as the RAM and I/O control cycles of a basic computer system. Of course, if another more elaborate program is available in the service literature, use the manufacturer's program. As in the case of other electronic troubleshooting, when all else fails, follow instructions!

Do not use any branches on the RAM or I/O instructions at this point. (The RAM and I/O blocks have not yet been turned on and checked out.) Such checks are made in the next two steps of this procedure.

Note that the microprocessor-to-ROM data link is a feedback process (each instruction depends on the address, and each address depends on the previous instruction). For this reason, the microprocessor-to-ROM link is by far the most tedious to check out. The RAM and I/O blocks can be turned on much more directly and in any order. However, if you choose the RAM first, you can connect the RAM to the system in one operation (as in the next section).

8.7.7 Verifying RAM/ROM/microprocessor compatibility

The RAM/ROM/microprocessor link can be verified once you are certain that the ROM and microprocessor are functioning together. To simplify the process, first run the ROM test program of Fig. 8.10, but with the RAM connected as shown in Fig. 8.11. Then go on to the RAM program shown in Fig. 8.11.

Pay particular attention to timing of the RAM control signals during the RAM read and write instructions. The usual cause of failure at this point is a shorted address or data line, two lines shorted together, or an unwanted RAM response. With ROM program verified, run a RAM test.

The RAM program (Fig. 8.11) verifies that the memory system is working correctly but does not provide a check of each cell at each memory location. However, you can design the program so that all locations are written and then read back. In addition to checking the data words at each address in RAM, check the timing of all RAM functions as discussed in Sec. 8.7.4.

A thorough test program should write to every location in memory and then read each location back and verify the data. With a memory that is eight bits wide, watch for the following problem.

The eight bits of memory represent only 256 states. Conventional memories are usually much longer. This means that each possible data pattern must be written several times to fill the memory. If the same data wires are written into each block of 256 words, an error in any of the higher-order addresses can be masked.

An extreme example of such masking is the case where all address lines are disconnected. Any slight change in the 256-word pattern, such as shifting the pattern one word location in each block, reveals the problem. For example, if you count from 0 to 255 in the first block, you should count from 1 to 255, then go back to 0, in the next block. Next count 2 to 255 and go back to 0 and 1 in the next block, and so on.

8.7.8 Verifying I/O compatability

The I/O operation can be verified once you are certain that the RAM,ROM, and microprocessor are functioning together. (The I/O should be checked before any peripherals are connected to the system.)

Start by running the ROM program of Fig. 8.10 but with the I/O connected (to the system but not to the peripherals) as shown in Fig. 8.12. Then go on to the I/O program shown in Fig. 8.12.

The ROM program of Fig. 8.10 verifies that the control timing is correct with the I/Os in the system. Once you are satisfied that the timing is correct and that

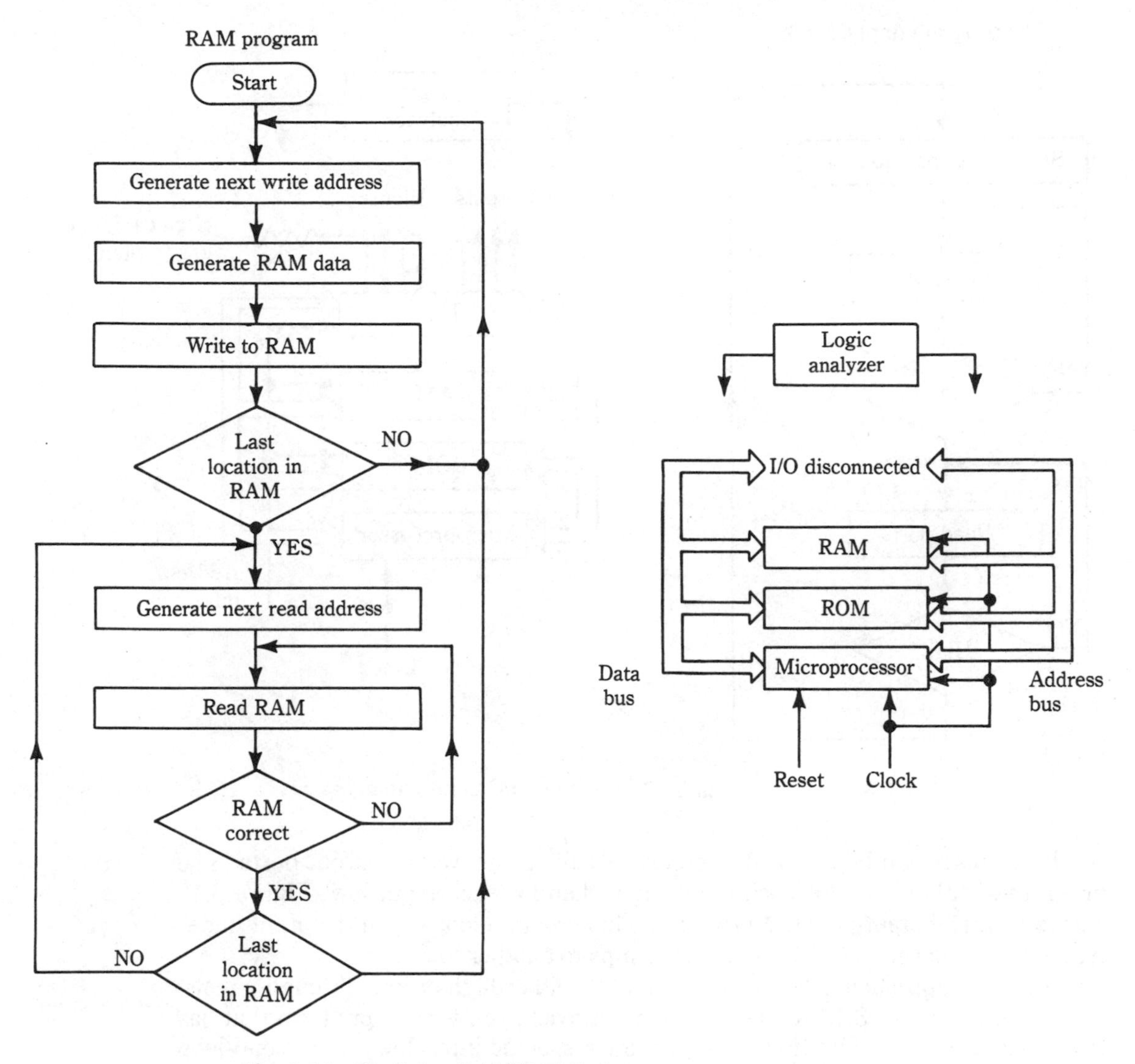

Fig. 8.11 Verifying RAM/ROM/microprocessor compatibility.

input/output instructions can be executed, the next step is to test the I/O functions with diagnostic programs.

Output ports can be checked with a simple program that first sets all the ports to 0, then sets each port in turn to 1, and finally sets each port back to 0, one at a time. When testing the output ports, connect the analyzer to one block at a time. If the analyzer has sufficient channels, connect the analyzer to the address bus as well as to the input/output ports, as shown in Fig 8.12. The object of this exercise is to see if the output ports are connected in proper order and can be set both high and low.

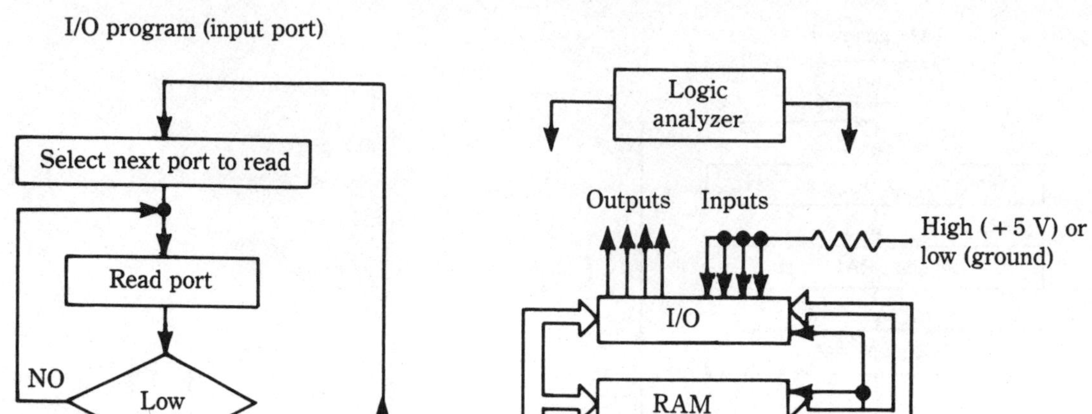

Fig. 8.12 Verifying I/O compatibility.

Input ports can be checked in essentially the same way as output ports. The program should check for each input high, then for each input low. Figure 8.12 shows a typical input-port test program. This program loops until the input under test is forced to the desired state, then jumps to another loop.

A simple approach is to pull all the inputs, either high or low, through a resistor as shown in Fig. 8.12. Using the high approach, you write a program that has two loops as shown. The first loop tests for a specific input low; the second loop tests for that same input high.

While the analyzer monitors the address bus and at least one input under test (preferably all the inputs if the analyzer has the capability), force the monitored input low with a ground wire. Under these conditions, the analyzer shows which loop the microprocessor is in exactly when the input goes low and (in a second pass) when the input goes high.

Although this process might seem tedious, the time required to write the test programs must be spent only once. The programs are then available at every phase of troubleshooting and system development.

9

Troubleshooting microprocessor-based devices

This chapter describes step-by-step procedures for troubleshooting the microprocessor-based circuits of a typical noncomputer electronic device. A videocassette recorder (VCR) is an example. The *system control* functions and many of the other circuits in a VCR are under the direct control of a microprocessor.

This chapter concentrates on such circuits and describes (1) how the microprocessor controls the circuit and (2) how to troubleshoot the circuit with a minimum of test equipment (preferably with only a scope or logic probe).

This chapter concentrates on such circuits and describes (1) how the microprocessor controls the circuit and (2) how to troubleshoot the circuit with a minimum of test equipment (preferably with only a scope or logic probe).

Although the techniques described here are for a VCR, they also can be applied to any microprocessor-based device (for example, to the frequency-synthesis system of a TV tuner or to the system-control functions of a CD player, CD-ROM drive, or CD-video (Laserdisc) player.

9.1 Communications buses

Operation of the system-control circuit for any modern VCR (or CD, CD-ROM, CD-V, etc.) is determined primarily by microprocessors. Often, VCRs use several microprocessors in system control. The microprocessors accept logic-control signals from the VCR operating controls (typically feather-touch pushbuttons) and from various sensors (such as tape sensors). In turn, the microprocessors send control signals to video, audio, servo, and power-supply circuits as well as drive signals to solenoids and motors.

Because more than one microprocessor is used in many cases, some form of communications is required among the microprocessors. Typically, a communications bus is used. Note that the communications buses used in controller applications are quite similar to buses used in computers.

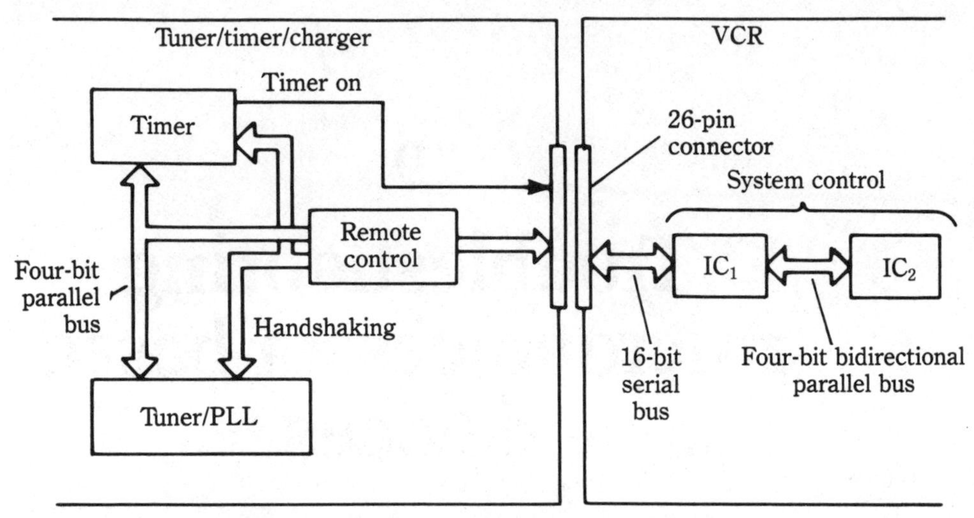

Fig. 9.1 Microprocessor communications bus.

Figure 9.1 shows the microprocessor communications bus system for a typical VCR. As shown, the VCR has a total of five microprocessors that communicate to each other through a variety of parallel and serial communications buses.

There are three microprocessors within the tuner/timer/charger: *remote-control*, *tuning/PLL* (phase-locked loop), and *timer*. Note that the remote-control is the master microprocessor in this group. The remote-control and timer pass channel-selection data to the tuning/PLL microprocessor on a four-bit parallel data bus. Handshaking control lines are used to maintain communications control.

The system-control functions are performed by two microprocessors IC_1 and IC_2. (IC_1 is considered the master of IC_2.) Microprocessors IC_1 and IC_2 communicate to each other on a four-bit bidirectional parallel bus. In turn, IC_1 communicates with the tuner/timer/charger remote-control on a 16-bit serial data bus just routed through a 26-pin connector (located at the rear of the VCR). Data bits are passed between the remote-control and IC_1 during the time that either the infrared (IR) remote, or front panel, is turned on. The 16 serial data bits, along with handshaking signals, are transferred on the 16-bit bus.

The conventional troubleshooting procedures used in servicing radio, TV, stereo, and the nonmicroprocessor circuits of VCRs cannot be applied to servicing digital-data communications systems using communications buses. A conventional scope (even a storage scope) is not of much value when trying to monitor the various complex signal patterns passed between microprocessors.

The logic analyzer (Sec. 5.9) is the best instrument to monitor buses. Unfortunately, most TV/stereo shops do not have a logic analyzer. Alternate methods are described here so that you can get around the problem of data-communications troubleshooting using less exotic (and less expensive) test equipment. But first, see how the microprocessor-based system-control functions are performed.

9.2 System control

Figure 9.2 shows overall operation for the system-control circuits of a typical VHS VCR. Microprocessor A IC_1 receives the various input commands from the tuner/remote-control IC (through the 16-bit serial data bus), the camera and remote input, and/or the front-panel operating buttons. IC_1 also controls the power on-off circuits, which latch the power relay. Microprocessor IC_1 communicates to microprocessor B IC_2 (through the four-bit data bus).

Microprocessor B IC_2 drives the mechanical systems (Sec 9.11) of the VCR deck and determines the condition of the deck (tape loading, cassette loading, etc.) in all modes of operation. Microprocessor IC_2 also drives the LCD counter-display module, (Sec. 9.6) and monitors the various trouble sensors (Sec. 9.5) which protect the system should a malfunction occur.

Microprocessors IC_1 and IC_2 function together as one microprocessor (and are physically mounted on the same system-control board in this VCR). As discussed, it is very difficult (nearly impossible) to monitor the data communications between IC_1 and IC_2. So the manufacturer recommends that both microprocessors be replaced as a package, along with the system-control board. This is a practical approach because both ICs are *flat-pack* types and very difficult to replace on an individual basis.

Even though the microprocessors are replaced as a package, it is still helpful in

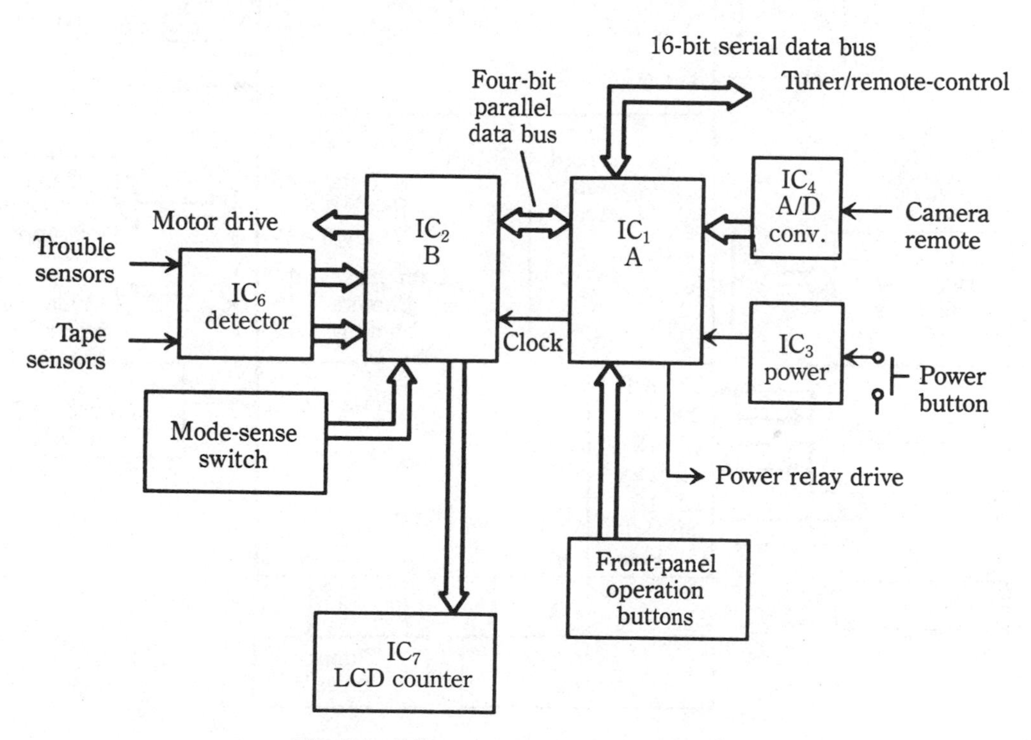

Fig. 9.2 Microprocessor system control.

troubleshooting to know what inputs are applied to which microprocessor and how both microprocessors produce output in response to these inputs. The following paragraphs summarize some typical functions that affect the circuits of this VCR.

9.2.1 Mode-detector microprocessor

Figure 9.3 shows the eight functional sections of IC_1. Pins 25, 34, and 35 are used during the on/off function of the VCR (Sec. 9.3). The power-hold IC_{33} is connected to the on/off switch. When the switch is pressed, a low is applied to pin 25. This signals IC_1 that the on/off command is requested, and that if the VCR is in the off mode, IC_1 is to apply power and turn on the VCR.

Note that if the VCR is already on, the information is interpreted to turn the VCR off. The same function is found on the system-control ICs of many microprocessor-based devices. In this case, the function is done by IC_1 generating output phase-0 (0 0) and phase-1 (0 1) signals at pins 34 and 35, respectively.

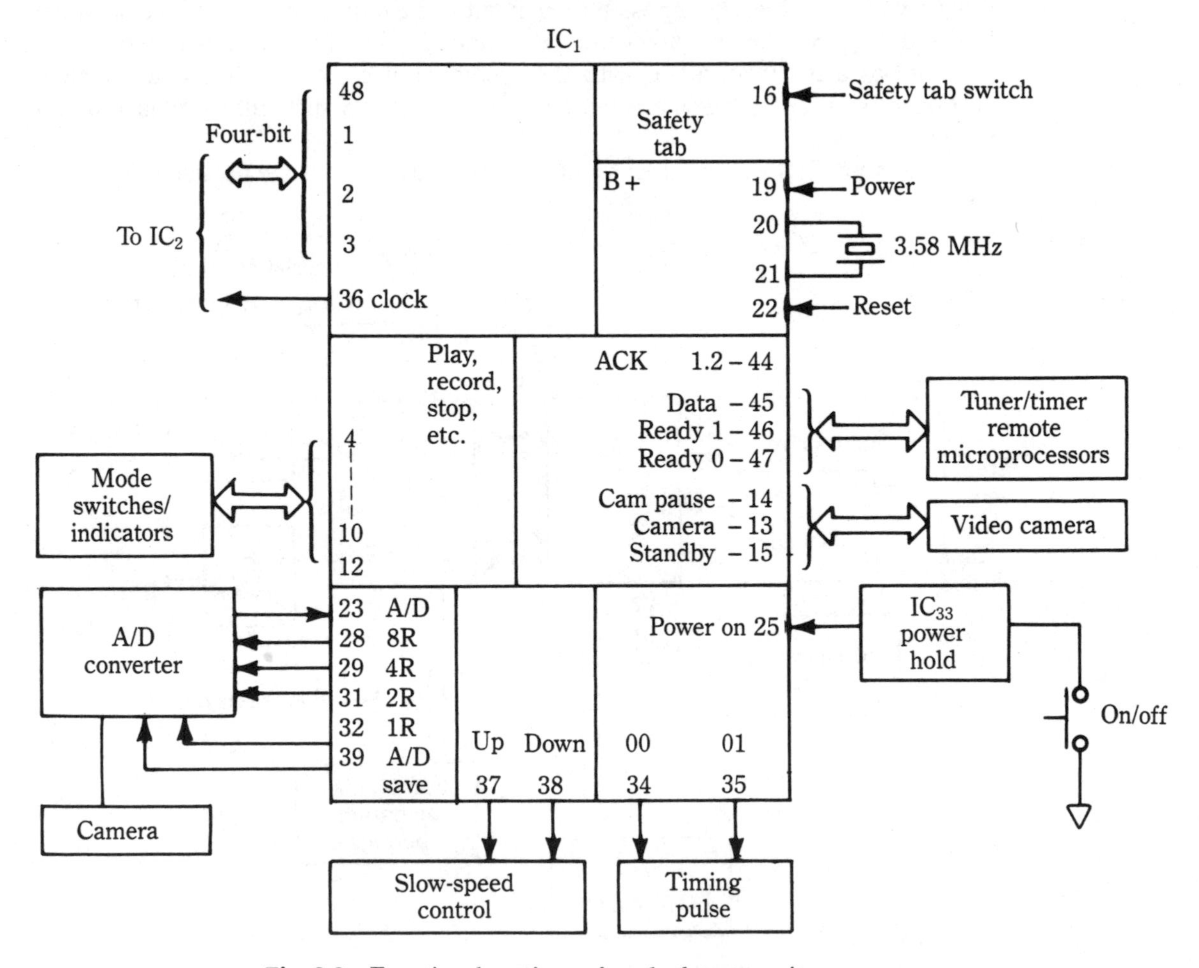

Fig. 9.3 Functional sections of mode-detector microprocessor.

When IC_1 goes into the on condition, pins 19, 20, and 21 become active with the application of power. When power is applied, reset occurs and the 3.58-MHz oscillator begins operation.

The cassette safety tab is connected to pin 16 of IC_1. If the tab is removed, IC_1 recognizes this condition and prevents the VCR from going into the record mode.

If you are not familiar with VCRs, both Beta and VHS cassettes have provisions for preventing accidental erasure of recorded material. Typically, safety tabs engage a plunger-rod or switch when the cassette is inserted and the compartment lid is closed. Actuation of the switch disables record operation, and thus prevents accidental erasure.

Pins 4 through 12 are multiplexed input/output pins used to accept signals from the mode switches (front-panel operating buttons) and to drive the front-panel indicators. Pins 44, 45, 46, and 47 provide communications to the tuner/remote-control. Pins 13, 14, and 15 of IC_1 monitor the external video camera (Sec. 9.4) and determine when the camera is connected. The camera can request pause and camera-standby through these pins.

9.2.2 Mechanical-drive and condition detector microprocessor

Figure 9.4 shows the many functional sections of IC_2. Microprocessor IC_2 monitors the VCR-deck mechanical-mode sense switches and troubleshooting sensors, informing IC_1 when a malfunction occurs. Responding to mode commands from IC_1, IC_2 passes the information to the various mechanisms within the VCR deck. The functional sections of IC_2 are as follows.

When the tuner/timer/charger is programmed for a timed recording, and the appropriate time has occurred, the information is transmitted from the timer to pin 30 of IC_2. This instructs IC_2 to place the VCR in record mode.

The mechanical position or operational mode of the VCR is monitored at pins 33, 34, and 35 of IC_2. Various trouble sensors are input at pins 3 through 6 and pins 28/29.

The battery voltage is monitored at pin 31 of IC_2. When battery voltage falls below 10.5 V, the VCR is placed in stop mode. The power-save function (which is required during portable operation) is contained within IC_2. During camera operation, if the VCR is left in stop or pause modes for five minutes, the power-save line goes high. This turns off a switching regulator, placing the VCR in power-save mode to conserve battery power.

Control of both loading and reel motors is done at pin 2 and pins 50 through 53. The servo circuit is controlled by a variety of outputs from IC_2 at pins 7, 10, 12, and 16 through 20.

The LCD display is controlled by signals at pins 42 through 47. The LCD display is a self-contained digital counter and display-driver on one modular circuit board (in this VCR). The assembly cannot be serviced and must be replaced as a package. (This is typical of many VCRs.)

The audio circuits are controlled by pins 8 and 9 during the audio-dub mode. The various power signals found within the VCR are developed from the outputs at pins 13, 14, and 15 through mode-switching circuits.

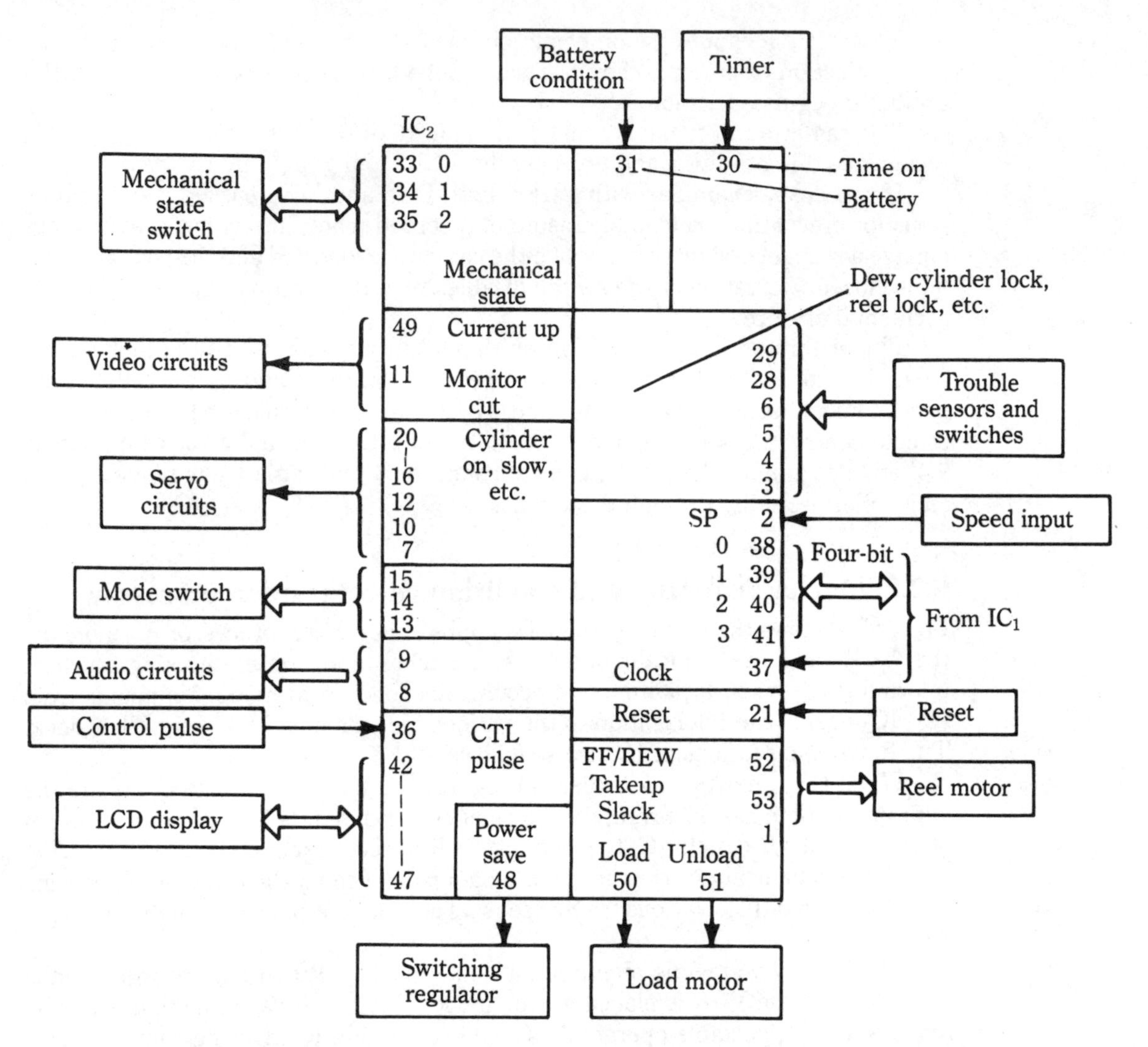

Fig. 9.4 Functional sections of mechanical-drive and condition-detector microprocessor.

9.3 On/off operation

Figure 9.5 shows the on/off circuits for our VCR. On/off operation is controlled through relay RL_1.

When the front-panel power on/off switch S_9 is pressed, or a command is sent from remote control, a low is applied through diode D_1. This pulls pin 16 of the power-hold IC and pin 2 of IC_{74} to ground. When pin 2 of IC_{74} is grounded, IC_{74} energizes relay RL_1. This causes the +12 V from the power supply or battery to be applied to the input of the switching regulator, causing the regulator to output various dc voltages.

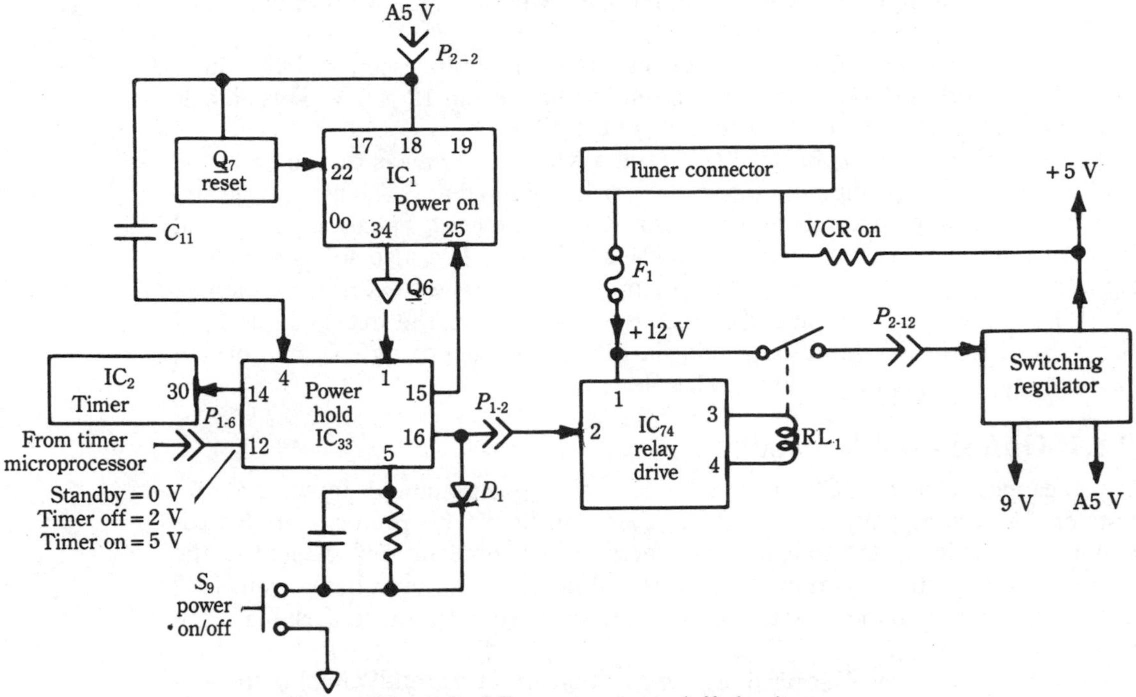

Fig. 9.5 Microprocessor on/off circuits.

One voltage output from the switching regulator is an A5 voltage. The letter *A* signifies that the voltage is always present when the VCR is turned on. The A5-V output supplies power to IC_1 and the reset circuits of Q_7.

Voltage A5-V also charges capacitor C_{11}, generating a momentary pulse at pin 4 of power-hold IC_{33}. The pulse set an internal latch, causing pin 15 of IC_{33} to go low. If the voltage at pin 12 of IC_{33} is not 2 V (timer off), the latch circuit does not generate a low at pin 15.

The low at pin 15 of IC_{33} is input to pin 25 of IC_1, the power-on input. With power at pin 13, a reset at pin 22, and a low at pin 25, IC_1 generates phase-0 pulses at pin 34. The phase-0 pulses are inverted by Q_6 and applied to pin 1 of power-hold IC_{33}.

If inverted phase-0 pulses are present at pin 1 of IC_{33}, indicating that power has been turned on, IC_{33} latches RL_1 (through IC_{74}) in the on mode. This is done by a latch circuit within IC_{33} that pulls pin 16 low. The low is maintained at pin 2 of IC_{74} for the duration of the VCR-on time.

When switch S_9 is pushed again (to turn off the VCR) or an off command is received from the remote-control IC, pin 5 of IC_{33} is pulled low. With pin 5 low, IC_{33} resets the internal latch, allowing pin 16 to change to a high. This high is applied to pin 2 of IC_{74}, which de-energizes RL_1 and disconnects the 12-V power from the switching regulator.

During the power-on mode, IC_{33} monitors the timer-on line at pin 12. One of three different voltages appears at pin 12, as shown in Fig. 9.5. If the timer has not been preset for a time recording, and the timer switch is in the off position, a 2-V level appears at pin 12.

If a programmed time is input to the timer microprocessor, and the timer switch is turned to the on position, the time-on line at pin 12 is 0 V. This signals IC_{33} not to allow manual on/off operation of the VCR.

The third voltage at pin 12 of IC_{33} occurs when the timer is turned on at the start of a timed recording, and the line at pin 12 is pulled to 5 V. This voltage signals IC_{33} to pull pin 15 low. Microprocessor IC_1 then generates phase-0 pulses, and IC_{33} goes low, turning the VCR on. At the same time, IC_{33} also sends a time-on signal to IC_2 at pin 30. This signal instructs IC_2 to place the VCR in record mode.

When the end or off time of the timed recording occurs, the voltage at pin 12 of IC_{33} returns to the 2-V level, IC_{33} pulls pin 16 high, de-energizing RL_1 and turning off the power-supply circuit to the VCR deck.

9.3.1 On/off troubleshooting

If you cannot turn the VCR on (or off), the first step is to monitor inputs and outputs on the system-control board. Remember that the two microprocessors IC_1/IC_2 cannot be serviced, and that the manufacturer recommends replacement of the entire system-control board in the event of malfunction. However, before you order a new system-control board (at mind-boggling expense to the customer), make the following checks.

Check for 2 V power at pin 6 of plug P_1. If incorrect, carefully unplug the system control board and check for 2 V at pin 6 of the socket that mates with P_1. If you do not get 2 V, the problem is likely in the timer circuits and not the system control. Fortunately, many of the timer circuits for most modern VCRs are replaceable. Of course, if 2 V power is available at pin 12 of IC_{33}, the system-control board is probably at fault.

If the voltage at pin 6 of P_1 is correct, then confirm that the voltage level at P_{1-2} is 11 V (when the power switch S_9 is not pushed), but goes to about 0.7 V (when S_9 is pushed and held in). If not, suspect S_9 and the system-control board.

Next, look for +12 V at P_{2-12}. If it is missing, suspect F_1, IC_{74}, or RL_1 (all of which are replaceable). Then look for +5 V at P_{2-2}. If it is missing, suspect a defective switching-regulator module.

If the voltage at pin 12 of P_2 is present when S_9 is pressed, but does not remain when S_9 is released, suspect the system-control board.

9.4 System-control inputs

Figure 9.6 shows the system-control input circuits for a typical VCR with microprocessor control. The VCR operating mode is selected from three different sources: the front-panel keyboard (push buttons), the tuner and remote IC, or the camera. Microprocessor A IC_1 monitors these three input sources.

Keyboard operation occurs when IC_1 generates phase-1 scanning pulses at pin 35. These pulses are amplified by Q_5 and applied to keyboard switches. IC_1 also generates phase-0 pulses, which are output at pin 34 and amplified by Q_6 to drive

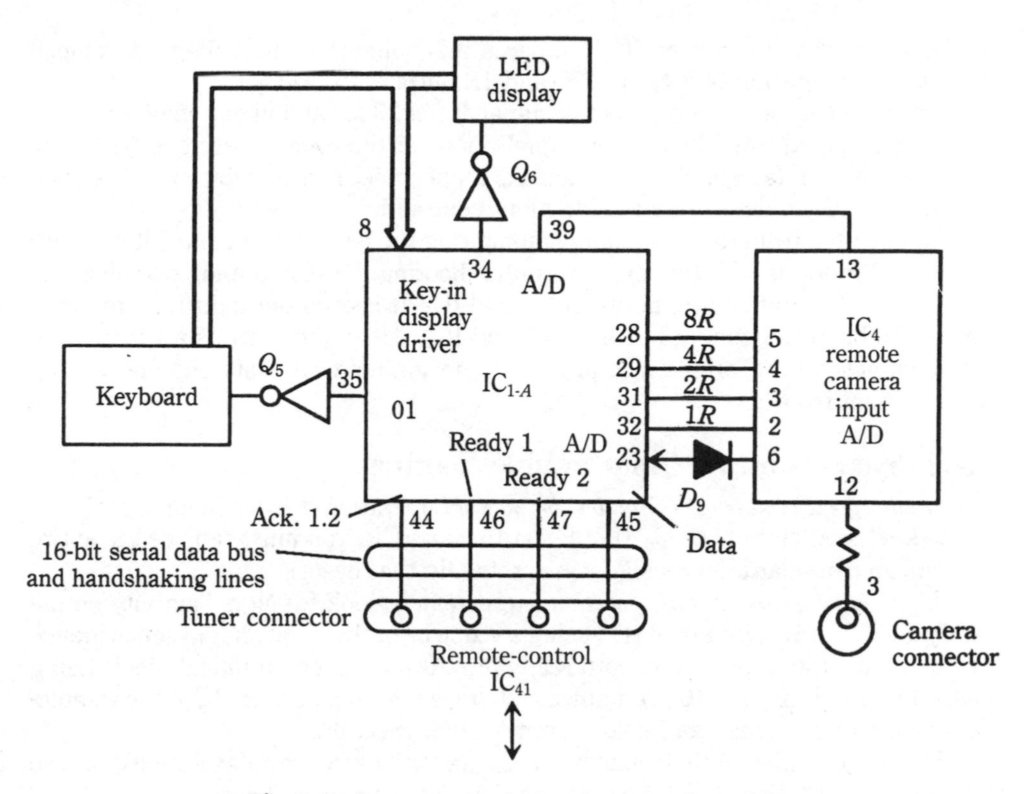

Fig. 9.6 System-control microprocessor inputs.

the LED displays on the front panel. The return signals from the keyboard and LED displays are routed to eight key-in and display-drive input pins on IC_1.

Microprocessor IC_1 multiplexes the function of keyboard input and display drive. During the time when phase 0 is active, LED-drive operation is being performed. When phase 1 is active, keyboard scanning occurs, and IC_1 monitors for a command or function input.

Microprocessor IC_1 also monitors for input from remote and tuner IC_{41} (over the 16-bit serial data bus and handshaking lines) at pins 44 through 47. The ready-1 (pin 46) and ready-2 (pin 47) lines inform the microprocessors that one or the other is ready to send data. After receipt of the ready signal from the transmitting microprocessor, the receiving microprocessor outputs an acknowledge signal on the acknowledge 1.2 line. The transmitting microprocessor then acknowledges this signal and output serial data on the data line (pin 45). This sequence occurs for each 16-bit word.

The third source of input is applied through pin 3 of the external camera connector to IC_4 (the camera input A/D converter). Microprocessor IC_1 passes four scanning pulses (pins 28, 29, 31, 32) to IC_4 (pins 2, 3, 4, 5) and monitors the A/D signal at pin 23 (through D_9). The camera connects a predetermined value of resistance from pin 3 to ground for the mode required. A constant-current source in IC_4 is passed through the camera resistance, developing a voltage at pin 3 proportional to the resistance. Converter IC_4 compares this voltage to the voltage developed from the combination of 8R, 4R, 2R, and 1R outputs from IC_1.

Microprocessor IC_1 monitors the signal at pin 23 for an indication of when the voltage developed from the four-bit signals matches the voltage developed from the external resistor (at pin 3 of the camera connector). The four-bit signal is then decoded by IC_1 to determine the function requested.

You can see from this that simply monitoring the levels on the four-bit bus with a scope or probe is of little value in troubleshooting. You need a logic analyzer to monitor the four-bit coding between IC_1 and IC_4. It is even more difficult to monitor the 16-bit serial coding between IC_1 and IC_{41}. However, there is a way around this troubleshooting problem in any VCR, no matter what data-bus and coding arrangement is used.

9.4.1 System-control input troubleshooting

If you have good remote operation but no manual operation, the problem is probably in the keyboard circuits or Q_5. (In this particular VCR, you must replace the entire system-control board, increasing your profits, in that event).

If you have good manual operation but no remote, look for signal activity on the ready-1 and ready-2 inputs of IC_1, using a scope or probe. (You should see evidence of rapidly changing pulses on both ready lines if normal communications is being passed between IC_1 and IC_{41}.) If pulse activity is missing, suspect IC_{41}, the remote-control circuits, or the hand-held IR remote unit (Sec. 9.7).

If you get pulse activity on the ready lines, look for similar activity on the acknowledge 1.2 line. If missing, replace the system-control board.

If you get pulse activity on the acknowledge 1.2 line, look for similar activity on the data line. If missing, suspect the remote IC_{41} (or IR remote unit).

If you get pulse activity on the data line, try substituting the system-control board. Of course, it is possible that either IC_{41} or the IR remote unit is producing some unintelligible code that cannot be interpreted by IC_1, but this is a long shot.

If you get good manual and remote but no camera function, try a different camera. If this is not practical, try substituting various resistors between pin 3 of the camera connector and ground. The resistor values are given in the service-literature schematic.

If you do not get the correct mode with a resistor mounted at pin 3, suspect IC_4 or IC_1 (in that order). Either way, you must replace the entire system-control board.

If you get the desired mode of operation with the correct resistance connected at pin 3 (for example, this VCR should go into slow motion when 15.5 kΩ is placed between pin 3 and ground), look for problems in the camera cable or in the camera circuits (if you feel courageous enough to tackle a video-camera repair job).

9.5 Trouble sensors

Figure 9.7 shows the system-control trouble-sensor circuits for a typical VHS VCR. Microprocessor IC_2 monitors the various trouble sensors and detectors: dew sensors and rewind or forward, tape-end reel-rotation, cassette-up, and cylinder-lock detectors.

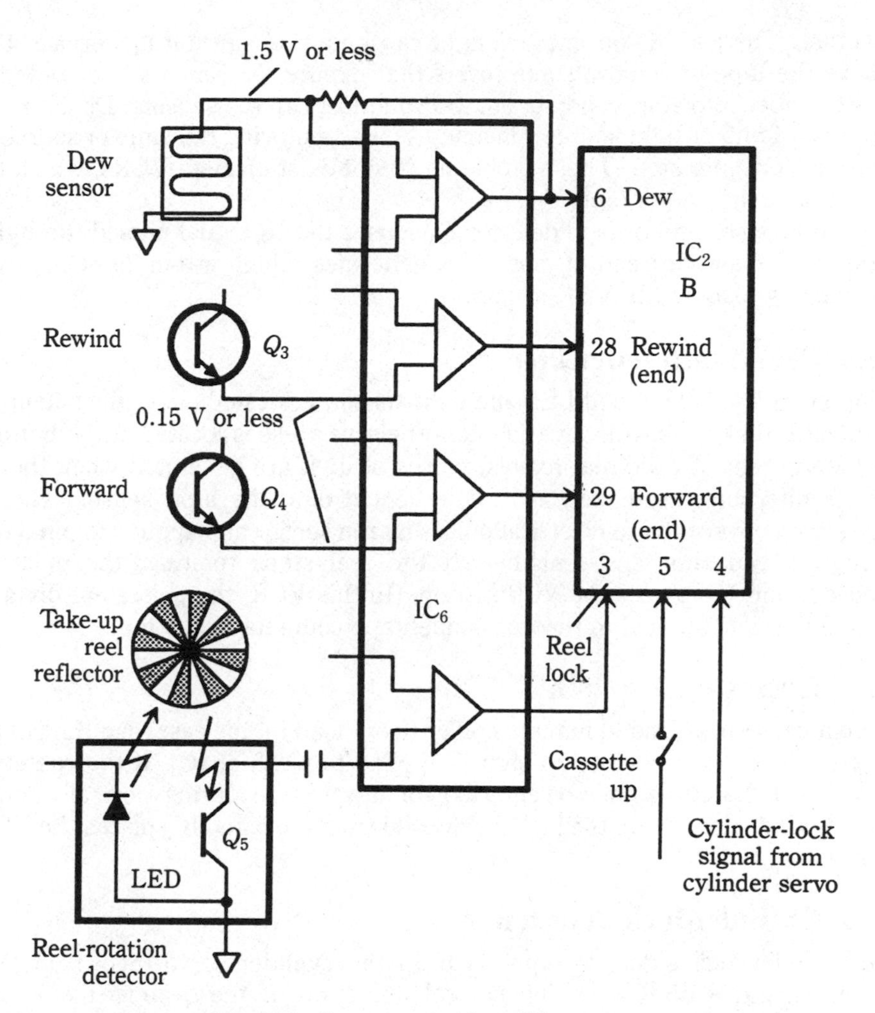

Fig. 9.7 System-control trouble sensors.

9.5.1 Dew sensor

If there is low moisture in the VCR, the resistance of the dew sensor is high. This produces a high at pin 6 of IC_2, which responds by doing nothing. However, if there is considerable moisture within the VCR, the resistance of the dew sensor drops, producing a low at pin 6 of IC_2, and the VCR is placed in the stop mode.

9.5.2 Rewind or forward tape-end detectors

The rewind tape-end detector Q_3 senses the infrared light that is passed through the clear leader of the VHS tape just before the end of the tape is reached. This generates a high at pin 28 of IC_2, which responds by placing the VCR in stop.

Note that most modern VHS VCRs use infrared end-of-tape sensors and light sources. This is not true of early-model VCRs, which use ordinary light and photodetectors. The use of conventional light can be a problem during service. If you remove the tape or remove some covers that expose the sensors to outside light, the VCR goes into stop, ready or not. So you must cap the sensors. Do not remove the sensor lamp or light source. On many VCRs, removing the lamp or source triggers the VCR into stop. These problems do not exist on Beta VCRs, which use a metal foil at the end of tape.

The forward end-of-tape detector Q_4 senses the IR signal passed through the leader at the forward end of tape. This generates a high at pin 29 of IC_2, which responds by placing the VCR in stop.

9.5.3 Reel-rotation detector

Component Q_5, a combined LED and light sensor, is located under the takeup reel. A reflector disk with reflective and nonreflective areas is located at the bottom of the takeup reel. As the reel rotates, pulses of light are generated when the LED light is alternately reflected and not reflected onto the light sensor. The light pulses are converted into electrical pulses by the sensor and applied to pin 3 of IC_2 through a capacitor and amplifier. If the reel stops rotating, the pulses are removed, and IC_2 places the VCR in stop. (In this VCR, the pulses are divided in IC_2 and are also applied to the front-panel tape counters Sec. 9.6.)

9.5.4 Cassette-up switch

When a cassette is loaded into the holder (front load in this case) and the holder is moved to the down position, a high is applied to pin 5 of IC_2. If the cassette-up switch is actuated (even momentarily) for any reason during normal operation (play, rewind, slow, etc.), the high is removed from pin 5 and IC_2 places the VCR in stop.

9.5.5 Cylinder-lock detector

The cylinder-lock signal is supplied from the cylinder servo (Sec. 9.11.1) and applied to pin 4 of IC_2. During normal operation, if the cylinder-motor speed decreases, the cylinder servo generates a high at pin 4 of IC_2, causing IC_2 to place the VCR in stop.

9.5.6 Trouble-sensor troubleshooting

If the VCR does not go into play, check the inputs from all trouble sensors. Look for a voltage of less than 1.5 V at the dew sensor and less than 0.15 V at end-of-tape sensors Q_3 and Q_4. Also look for a high at pin 5 of IC_2 from the cassette-up switch (make sure that there is a cassette in place).

If any of the trouble-sensor inputs are incorrect (voltages substantially higher than 1.5 V for the dew sensor and higher than about 0.15 V for the end-of-tape sensors), trace the signal back to the source. If all signals are normal, replace IC_2 (or the board).

If the VCR does not make any attempt to load tape from a cassette (after the cassette is pulled in and down) the reel-rotation detector and cylinder-lock detector are probably not at fault. These two trouble sensors place the VCR in stop only after the cassette is loaded, the tape is loaded from the cassette, and the tape starts to move.

If the VCR loads tape from the cassette and then immediately unloads the tape and stops, look for pulses from Q_5 (about 1 V) at the time when the tape just starts to move (takeup reel rotating). If the pulses are missing, suspect Q_5, reel-motor problems, or capstan-servo problems. If the pulses are present, check for a cylinder-lock signal at pin 4 of IC_2. The cylinder-lock line should be high while the tape is being loaded from the cassette, and then go low at completion of the tape load (when the tape starts to move). If not, suspect cylinder-servo problems.

9.6 Counter/record timing

Figure 9.8 shows the system-control counter/record-timing circuits for a typical VHS VCR. In this particular VCR, the counter/record-time indicator is considered part of system control rather than a separate circuit. Also note that the VCR in Fig. 9.8 uses an LCD display rather than a fluorescent display (found on many other modern VCRs).

The LCD display and associated circuits shown in Fig. 9.8 are contained on a display module located on the front of the system-control board (the module cannot be serviced). The display performs three modes of operation: record-time indication, counter indication, and battery-empty indication (battery E).

During record-time and counter operation, the control track pulse (CTL, Sec. 9.11) at pin 7 of P_3 is amplified by Q_2 and applied to pin 36 of IC_2, which divides the pulse by 30. (For those not familiar with VCRs, the vertical sync pulses of the TV broadcast signal are used to synchronize the rotating heads with tape movement. The TV sync pulses are converted to 30-Hz control signals. These CTL signals are recorded on the tape by a separate control track head, usually on the same stack as the audio head.) The divided-down CTL pulse (now at 1 Hz) is applied to pin 13 of the display module and used to count the time in record mode of operation.

The output from the reel-rotation detector (amplified by IC_6) is applied to pin 3 of IC_2, where the signal is divided by 8, producing a signal having one pulse per reel rotation at pin 47 of IC_2. This signal is applied to pin 14 of P_6, and used as the tape-counter signal to drive the four-digit tape counter.

The circuits within the display module must be able to count both up and down, and must be able to recognize when the VCR is in record, playback, fast forward, or reverse. This is done by mode-control signals applied to the display module from IC_2 (at pins 43, 44, and 45).

As an example, if all three mode lines (0, 1, and 2) are high, the VCR is in stop mode, and the counter stops.

In play or forward search, mode line 0 is low, and mode lines 1 and 2 are high. Under these conditions, the counter counts up, and the counter-memory function is effective.

During various modes of operation, the counter-memory circuit is turned on to tell the VCR when a reading of 0000 is reached and to place the VCR in stop mode. The memory function is activated when the display-module memory switch is pressed. This applies a 0000-signal to pin 42 of IC_2 when the count reaches 0000.

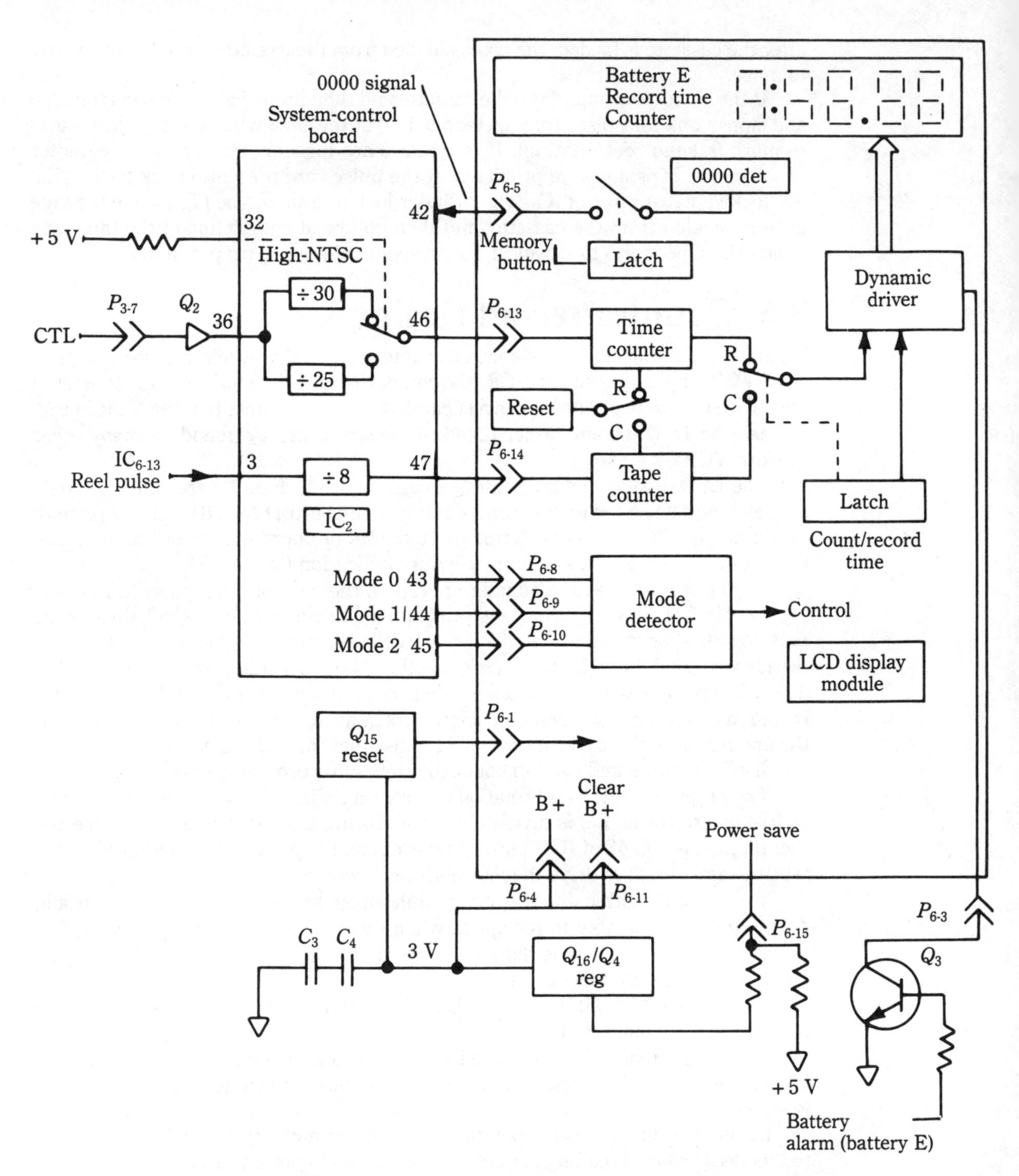

Fig. 9.8 System-control counter/record-timing microprocessor.

In reverse-play or reverse-search, mode lines 0 and 1 are low, and mode line 2 is high. Under these conditions, the counter counts down, and the counter-memory function is effective.

In fast forward, mode lines 0 and 1 can be either high or low, and mode line 2 is low. Under these conditions, the counter counts up, the time indication is switched over to tape counter indication, and the counter-reset and counter-memory functions are effective.

In rewind, mode lines 0 and 2 are high, and mode line 1 is low. Under these conditions, the counter counts down, the time indication is switched over to the tape-counter indication, and the counter-reset and counter-memory functions are effective.

The battery-E signal from Q_3 is applied to pin 3 of P_6. when the battery charge falls below the serviceable level (typically 10.5 V), an alarm is applied to Q_3. The battery-E display appears when pin 3 of P_6 goes low.

Power for the display module is taken from a two-transistor voltage regulator circuit on the system-control board. Transistors Q_4 and Q_{16} provide a 3-V regulated supply that is filtered by C_3 and C_4. These capacitors are the power-backup system in the event power is lost. Capacitors C_3 and C_4 can store enough charge to maintain LCD display modulation for about one hour. This is done by the power-save-sense line at pin 15 of P_6. If the 5 V supply is low, the counter is turned off, reducing power consumption.

9.6.1 Counter/record-timing troubleshooting

Because the LCD display cannot be serviced, the first step in troubleshooting is to check inputs to the display. If any input is absent or abnormal, trace the line back to the source. If all the inputs are present and appear to be normal, replace the display module. The following notes should be helpful in troubleshooting counter/record-timing.

If there is no display in any mode, check the power inputs at pins 4, 11, and 15, and the reset input at pin 1 of P_6. Also look for shorted C_3/C_4 or defective Q_4/Q_{16}.

If there is no battery empty (battery-E) display, check the input at pin 3 of P_6 and Q_3.

If there is no tape-counter mode, but record-time is good, check the input at pin 14 of P_6.

If there is no record-time mode, but tape-counter mode is good, check the input at pin 13 of P_6.

If the display appears to be operative in some modes, but not in one or more modes, check the mode-control signal inputs at pins 8, 9, and 10 of P_6. For example, if the counter counts up (during play or forward-search) but does not count down (during reverse play or reverse search), check that mode 0 and mode 1 (pins 8 and 9 of P_6, respectively) are low and that the mode 2 (pin 10 of P_6) is high. If the mode-control signals are not good, replace the system-control board.

If the display goes to 0000, but the VCR does not stop when the memory button is pressed, check for an output from pin 5 of P_6 to pin 42 of IC_2 (on the system-control board). This output, the 0000 signal, should be present when the count is

0000 and the memory button is pressed. If not, replace the display module. If the 0000 signal is present at IC_2, but the VCR does not stop, replace the system-control board.

9.7 Remote control

Many modern microprocessor-based products (VCR, TV, CD player, etc.) use some form of IR remote control. Similarly, most present-day remote-control systems use some form of *digital position modulation* or P-M. So both IR and P-M remote control are covered in this section.

9.7.1 Remote-control transmitter circuits

Figure 9.9 shows the circuits of a typical hand-held IR remote transmitter. The circuits shown are unique in one respect. The reference frequency of the transmitter is 255 kHz (rather than the more common 455 kHz). This variation in transmission frequency prevents the transmitter from interfering with remote operation of other devices.

The digital code representing a given function appears as a series of *bursts*

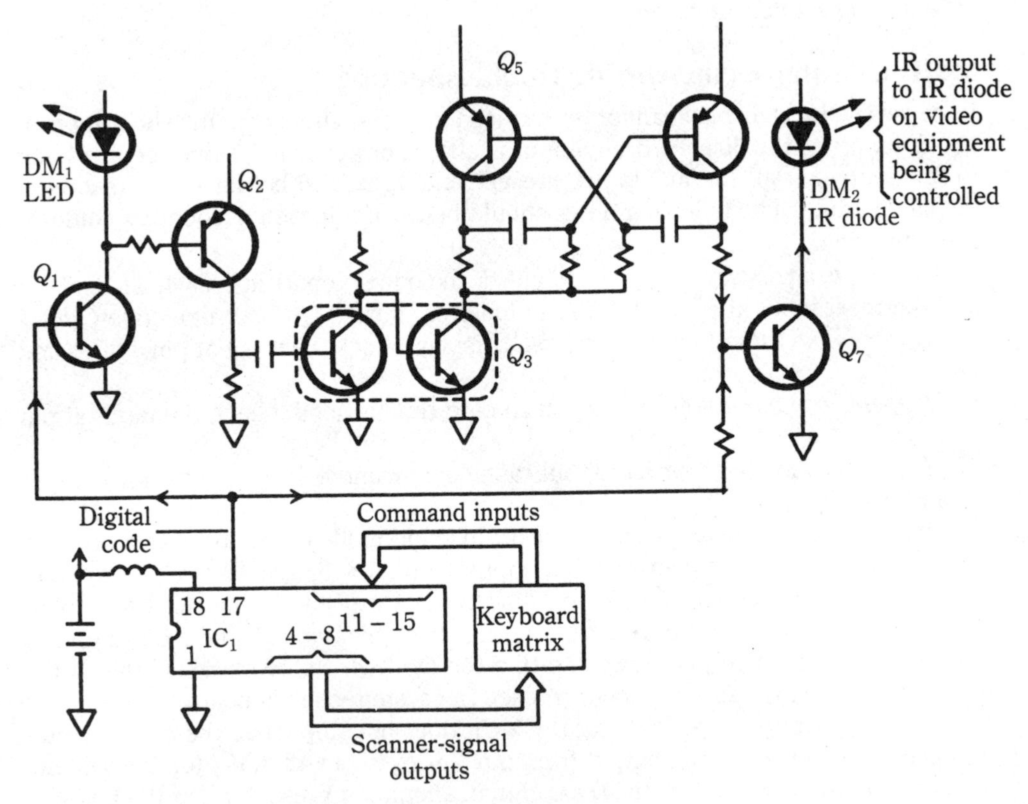

Fig. 9.9 IR remote transmitter.

from the transmitter. Each burst contains 10 pulses of 50-μs duration each, at a frequency of 20 kHz (rather than the more common 38 kHz). The duration of each burst is 500 μs.

The *position* or distance between recurring bursts identifies the digital 1s and 0s that make up the digital code. In this case, position or distance is determined as the amount of time between the rise of the first burst and the rise of the succeeding burst, as shown in Fig. 9.10. A digital or logic 0 is 2 ms, and a logic 1 is 4 ms.

The function of IC_1 is (1) to generate an oscillator signal for the creation of scanner-signal outputs (at pins 4 through 8), which are applied to specific inputs (at pins 11 through 15) of IC_1 (through the keyboard) to identify specific functions and (2) to decode the information received at the inputs and produce a digital code (at pin 17) representing the selected function.

No scanner output is available unless a key is pressed. When a given key is pressed, the oscillator in IC_1 is turned on, and generates the appropriate scanner signals. The scanner outputs at pins 4 through 8 of IC_1 are connected to one contact of specific keyboard switches (which close when the corresponding key is pressed). The remaining contact of each switch is connected to a specific input at pins 11 through 15 of IC_1.

When a given key is pressed (enabling the oscillator), a scanner signal is applied to the corresponding input through the closed switch contacts. A specific scanner signal is applied to a specific input for each switch closure. Similarly, each keyboard switch is represented by a unique combination of scanner output and signal input. The scanner signal applied to a particular input is decoded by IC_1 and produces a specific digital code representing the function of the corresponding keyboard switch.

The digital code (representing the selected function) at pin 17 of IC_1 is applied to the base of Q_1 and Q_7. The code signal is amplified by Q_1 and Q_2, delayed for 2.2 ms, differentiated, and applied to Q_3, which conducts for 550 μs, enabling a 40-kHz multivibrator Q_5. The output of Q_5 is applied to the base of Q_7, along with the output from IC_{1-17}.

The combination of both signals turns on Q_7, activating the IR diode DM_2. (Because two signals are required to activate DM_2, erroneous transmission is prevented.) Diode DM_2 is pulsed and the IR output is sent to the light-sensitive IR diode on the equipment being controlled (Sec. 9.7.3). The output from DM_2 con-

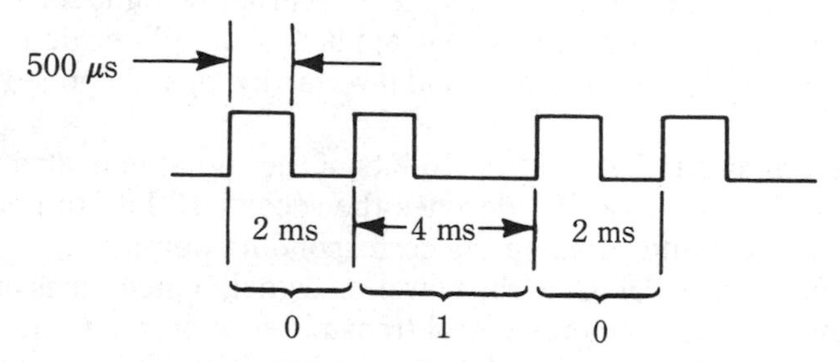

Fig. 9.10 Remote-control pulses with microprocessor control.

sists of a series of bursts, with the position or spacing between bursts determined by the digital code being transmitted.

9.7.2 IR transmitter troubleshooting

Once you have definitely pinpointed the problem to a remote-control transmitter, by trying a different transmitter with the same equipment being controlled, look for weak or defective batteries. (Also look for any switch on the equipment being controlled that disables the remote-control function, as is the case on the author's TV set, which he forgot about.)

Batteries are the most common causes of trouble in any remote-control transmitter. So start troubleshooting by putting in a known-good battery. The next most common problem is a defective keyboard switch contact. Defective switches usually show up when one or more functions are absent or abnormal but other functions are normal.

Note that when Q_1 is turned on by the output at pin 17 of IC_1, the remote-transmission LED DM_1 is turned on. This indicates that the transmitter is sending commands to the receiver diode. (Of course, it is possible that the transmitter is operating properly and only DM_1 is defective. However, this is not likely.) So, if DM_1 does not turn on when a key is pressed, look for a defective battery. If the battery is definitely good, suspect IC_1 or Q_1.

If the batteries and switches appear to be good, but you cannot transmit any commands with the remote transmitter, press the keys while monitoring pin 17 of IC_1. Although it is not practical to determine the actual digital code being transmitted, the presence of pulse bursts usually indicates that IC_1 is good.

Next, try monitoring the pulse bursts through from Q_1 to DM_2. Also, look for 40-kHz square waves at the collectors of Q_5. If pulse bursts appear at DM_2 but the transmitter cannot transmit commands, suspect DM_2.

9.7.3 Remote-control receiver circuits

Figure 9.11 shows the circuits of a typical IR remote receiver. IC_1 has the dual function of decoding digital commands from the remote transmitter and converting the commands into signals (that are applied to system-control, tuner-control, and power circuits). These signals replace or supplement the commands applied by push buttons on the equipment being controlled.

The command in IC_1 is taken from the signal transmitted by the remote-control transmitter. The transmitted IR signal is received by light-sensitive diode PD_1 and converted into an electrical signal applied to amplifier/detector IC_{91}. The detected signal from IC_{91} is amplified and inverted by Q_1 and applied to the input of IC_1 at pin 12.

The code transmitted at the first 16 bits of the signal indicates to IC_1 that the signal is for VCR operation. IC_1 decodes the second 16 bits to determine which function is requested and turns on the corresponding output(s).

The output at pin 36 of IC_1 is the power-on signal, which turns on Q_5 whenever the power button on the remote-control transmitter is pressed. Transistor Q_5 is in parallel with the manual on/off push button of the VCR. The output at pins 17, 29

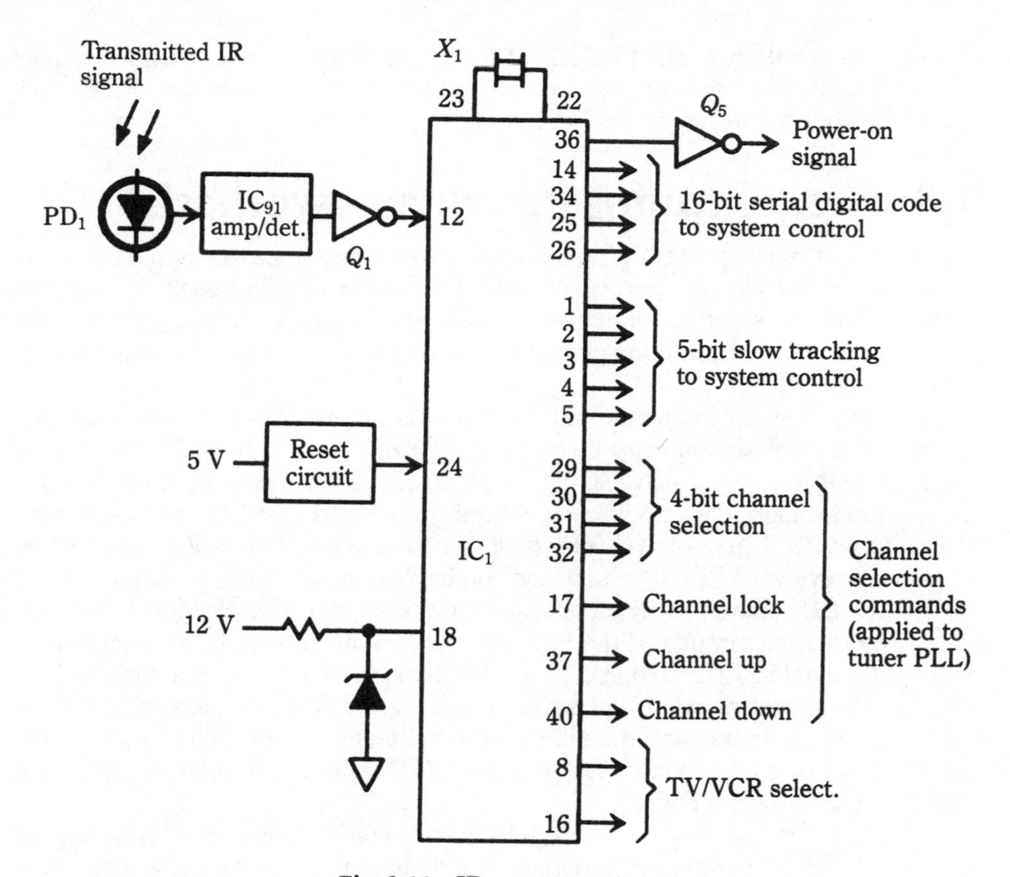

Fig. 9.11 IR remote receiver.

through 32, 37, and 40 of IC_1 are the channel-select commands. (These commands are applied to a tuner, such as described in Sec. 9.8.)

9.7.4 Remote-control receiver troubleshooting

If you do not get proper remote-control operation, substitute a remote transmitter that is known to be good. Next, confirm the presence of the signal at pin 12 of IC_1. This signal is a 32-bit serial data stream that cannot be intelligently monitored with a scope. That is, there is no easy way to determine which digital code is being transmitted by monitoring the signal. However, it is reasonable to assume that if a signal is present, the code is correct.

If the signal is present, check for 5 V at pin 18 of IC_1. If present, check for a 3.57-MHz clock signal at pin 23 of IC_1, and for a reset high at pin 24 of IC_1. If these inputs are correct, but there is no remote-control operation (but manual operation is good), suspect IC_1.

If you get remote-control operation for some but not all functions, check the corresponding output from IC_1. For example, if you get channel-up operation but no channel-down operation (with a known-good remote transmitter), check for a channel-down output at pin 40 of IC_1. If the output is absent or abnormal, suspect IC_1. If the output is present, the problem is likely to be in the tuning system and should be checked as described in Sec. 9.8.

9.8 Digital tuning (frequency synthesis)

The tuners of most modern VCRs and TVs use a microprocessor-based *frequency synthesizer* (FS). This is often called *digital tuning* or possibly *quartz tuning.* No matter which it is called, digital or FS tuning provides convenient pushbutton channel selection with automatic channel search and automatic fine-tune (AFT) capability.

The key element in any digital/FS system is the *phase-locked loop* (PLL) that controls the *variable-frequency oscillator* (VFO) of the TV or VCR tuner. It is assumed that you are already familiar with PLLs. (If not, and you must troubleshoot digital tuners, you should read several good books on PLLs, such as *Lenk's Video Handbook, Lenk's Audio Handbook*, and *Lenk's RF Handbook*.) This section has a brief review of PLLs, just in case you have forgotten their operation.

Figure 9.12 shows the digital tuning circuits of a typical VCR. Figure 9.12 also shows the internal circuits of the PLL microprocessor (in simplified block form). Note that a single PLL microprocessor IC_1 is used to control the entire digital system. IC_1 also performs other functions, such as supplying digit- and segment-drive signals for a two-digit channel display on the VCR front panel. Before you get into these functions, review some PLL basics as they apply to the internal circuits of the PLL microprocessor IC_1.

PLL is a term used to designate a frequency-comparison circuit in which the output of a VFO is compared in frequency and phase to the output of a very stable (usually quartz-crystal controlled) fixed-frequency oscillator. In the case of a TV or VCR tuner PLL, the VFO is the local oscillator used in the tuner. When a PLL is used, the local oscillator is actually a *voltage-controlled oscillator* or VCO, but functions as a VFO. (That is, the variable oscillator frequency is controlled by the variable voltage applied to the oscillator.)

Should a deviation occur between the two compared frequencies (local oscillator and fixed-frequency reference oscillator), or should there be any phase difference between the two oscillator signals, the PLL detects the degree of frequency error until both oscillators are locked to the same frequency and phase. The loop is said to be "locked" at this time. The accuracy and frequency stability of the reference oscillator (and the crystal that controls the reference oscillator) determine the accuracy and frequency stability of the PLL.

The system shown in Fig. 9.12 is generally called an *extended* PLL and holds the local-oscillator frequency to some harmonic (or subharmonic) of the reference oscillator. When the loop is locked, the PLL microprocessor maintains a fixed phase relationship between the local oscillator and reference oscillator (although the two are not necessarily exactly in phase). Under locked conditions, the local-oscillator frequency is, however, held exactly to that of the crystal-controlled reference.

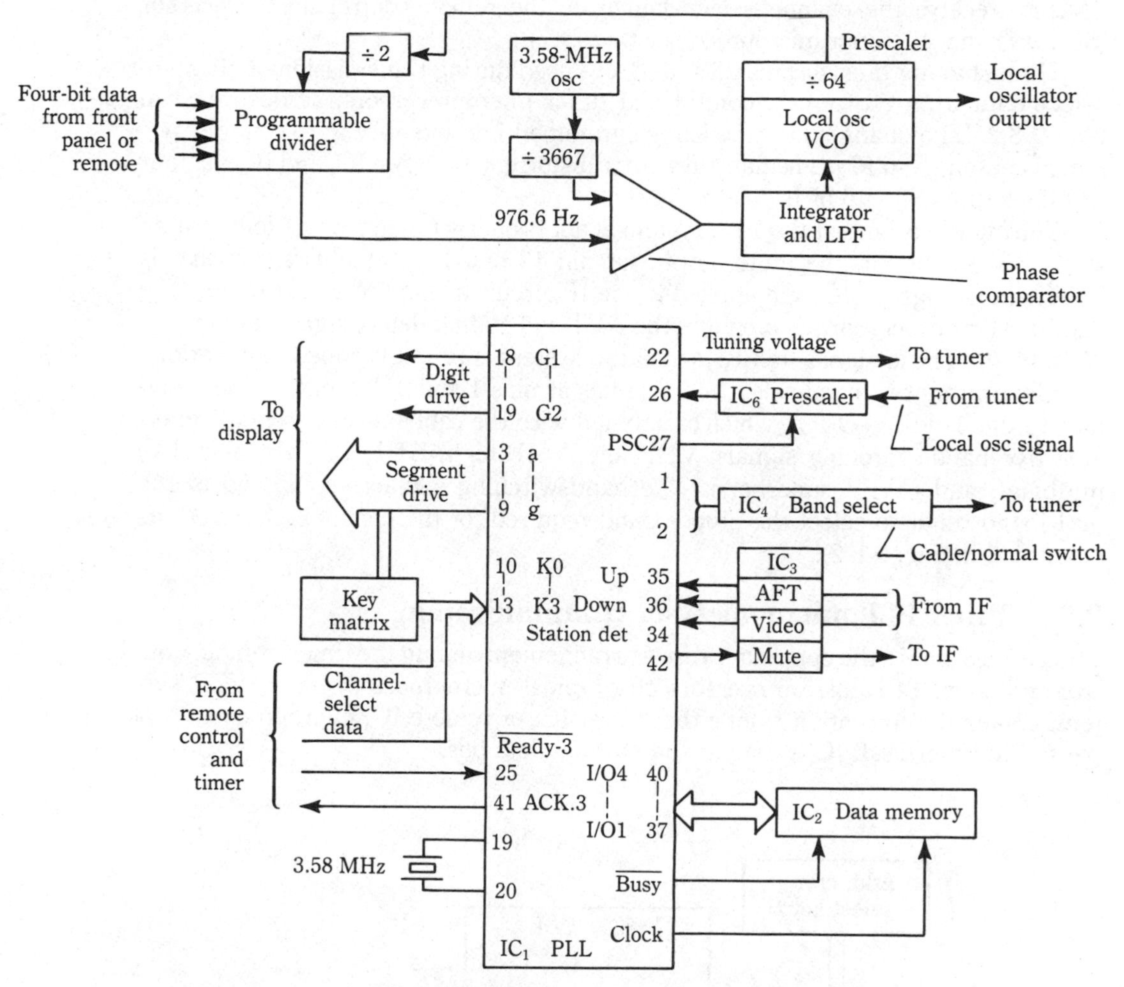

Fig. 9.12 Digital-tuning circuits.

A variable divider is used between the reference oscillator and phase comparator to reduce the reference-oscillator frequency to a more workable, easily compared frequency, while still maintaining the maximum frequency range.

A channel change is done by varying the division ratio of the programmable divider with four-bit data commands. The steps of the programmable divider are made small, and a low-frequency reference to the phase comparator (976.6 Hz) is used. Both of these factors provide for more accurate tuning.

Microprocessor IC_1 (PLL) uses the segment drive to scan the keyboard and then monitors the selected signals at inputs K0 through K3 (pins 10 – 13 of IC_1). These signals contain channel-data information (and are also applied to the LCD display, which shows the selected channel). The K0 through K3 ports are also used as input ports to receive the channel-select data from the remote-control microprocessor (Sec. 9.7) and the timer microprocessor (Sec. 9.7).

Handshaking lines at pins 25 and 41 are used during transmission of channel-select data from the remote-control and timer microprocessors, as described in Sec. 9.8.2. The channel information programmed into memory in the tuning system is contained in IC_2. The data bits are transformed between IC_1 and IC_2 via four I/O lines (pins 37 – 40 of IC_1).

During the time that the PLL system is not properly tuned to a station, or if no station is on the air, the mute output (at pin 42 of IC_1) generates a low that is inverted to a high by IC_3 and applied to the IF circuit within the VCR (to mute the audio). Microprocessor IC_1 monitors the AFT and station-detect signals at pins 34, 35, and 36. These signals tell IC_1 when the system is properly tuned to a station.

Generating two band-select code signals at pins 1 and 2, IC_1 passes these signals to the band-select IC_4, which (combined with the cable/normal switch) generates five band-switching signals: VHF low, VHF high, UHF, CATV (cable TV) midband, and CATV superband. The bandswitching signals are passed to the multiband tuner to select the proper band required for the channel requested (as discussed in Sec. 9.8.2).

9.8.1 Tuner PLL microprocessor communication

Figure 9.13 shows the communications arrangement among the timer, remote-control and tuner/PLL microprocessors of a typical microprocessor-based VCR system. Channel information from either timer IC_8 or remote IC_4 microprocessors is passed to tuner/PLL IC_1 via a four-bit parallel data bus.

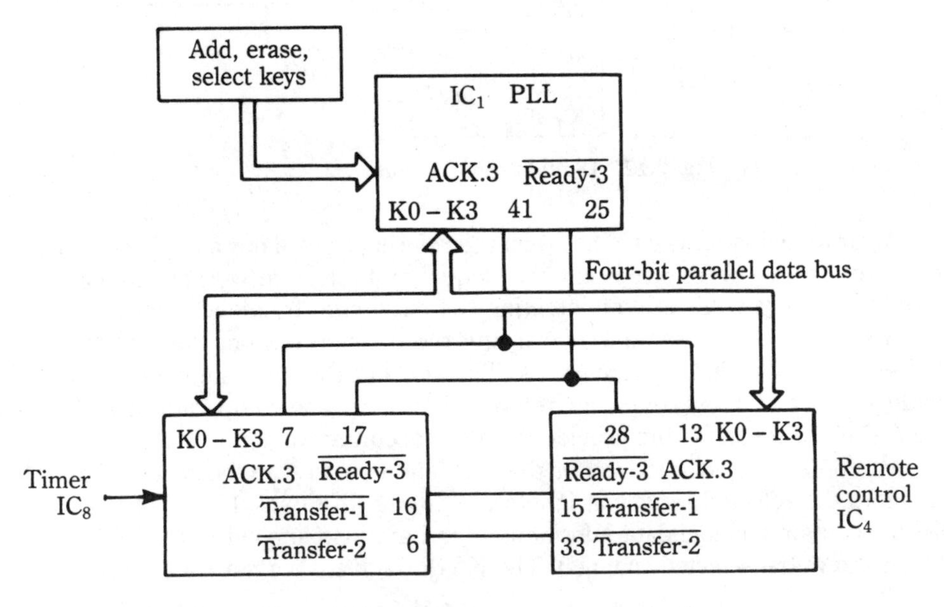

Fig. 9-13 Tuner PLL microprocessor communications.

Communications are passed between timer IC_8 and remote IC_4 microprocessors to establish which microprocessor is sending information. Upon receiving this communications signal, the other microprocessor goes into a standby or inhibit mode. The communications signal is sent via the $\overline{\text{transfer-1}}$ and $\overline{\text{transfer-2}}$ lines to the appropriate microprocessor.

As an example, if timer microprocessor IC_8 turns on in the timed-recording mode and sends channel-change information to IC_1, IC_8 pulls the $\overline{\text{transfer-1}}$ line low, informing remote-control microprocessor IC_4 that IC_8 is ready to send channel information and that IC_4 should go into standby.

The $\overline{\text{ready-3}}$ and acknowledge-3 (ACK.3) handshaking lines are also used in this control-data transfer between IC_8 or IC_4 and IC_1. The $\overline{\text{ready-3}}$ signal (pin 25 of IC_1) informs IC_1 that channel data bits are ready to be sent. When ready to receive data, IC_1 pulls the ACK.3 line high.

The following is a typical example of the communications sequence from timer microprocessor IC_8 to tuner/PLL microprocessor IC_1.

1. Timer IC_8 pulls $\overline{\text{ready-3}}$ and $\overline{\text{transfer-1}}$ lines low.
2. Remote IC_4 receives $\overline{\text{transfer-1}}$ signals and goes into standby.
3. Tuner/PLL IC_1 receives $\overline{\text{ready-3}}$ signals and pulls the ACK.3 line high.
4. Timer IC_8 receives ACK.3 signals, and outputs four-bit data for the tens digit, and returns $\overline{\text{ready-3}}$ and $\overline{\text{transfer-1}}$ lines high.
5. Tuner/PLL IC_1 receives data and returns ACK.3 line low. The sequence repeats to send ones-digit data.

9.8.2 Typical PS/digital tuning system

Figure 9.14 shows the circuits of a typical FS/digital tuning system with PLL and PSC or *pulse-swallow control.* (The PSC system uses a high-speed prescaler with *variable division* ratio, instead of the fixed division ratio prescaler in Fig. 9.12. The variable division ratio depends on the PSC signal from PLL IC_1. As the number of PSC pulses increase, the division ratio increases. A specific number of PSC pulses are produced by IC_1 in response to channel-selection commands.)

The multi-band tuner is controlled by circuits within PLL IC_1 which, in turn, receives commands from the remote-control circuits, as shown in Fig. 9.14. (IC_1 receives channel commands from the IR remote-control transmitter after the commands are decoded by the remote-control receiver, as described in Sec. 9.7.) IC_1 also monitors signals from the IF/demodulator circuits to know when a station is being received. These signals are the AFT up and down (pins 35/36) and station-detect signal (pin 34).

The detected video from the output of the IF/demodulator is passed to a sync amplifier and detector Q_7/Q_9. The detected sync signal (station detect) is passed to pin 9 of IC_3, amplified, and applied to the station-detect input at pin 34 of IC_1. When a station is tuned in properly, the sync is detected from the video signal and applied as a high to pin 9 of IC_3. This applies a high to pin 34 of IC_1, indicating to IC_1 that video with sync is present.

To maintain proper tuning, IC_1 monitors the AFT up and down signals (at pin

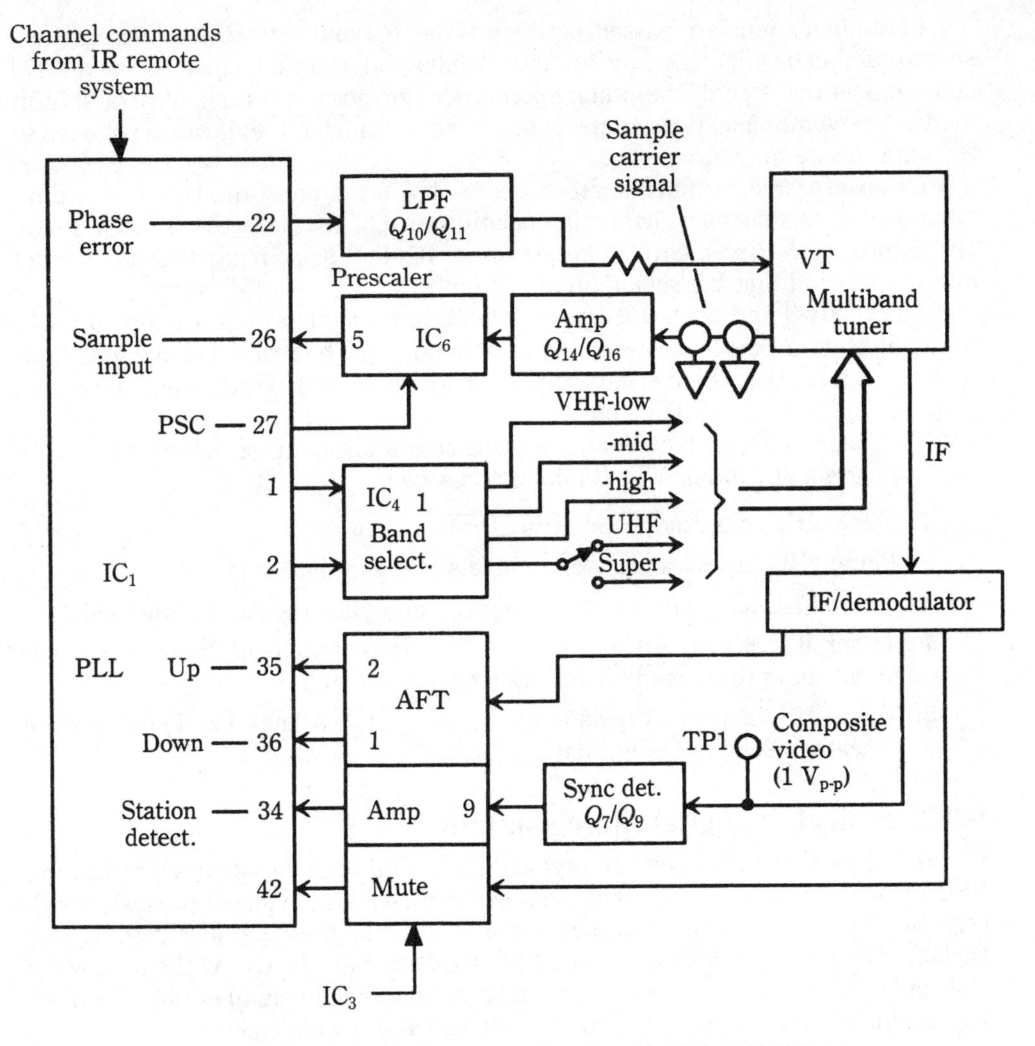

Fig. 9.14 FS/digital-tuning system with PLL and PSC.

35/36) from IC_3. The AFT circuit of IC_3 is a *window detector*, monitoring the AFT voltage and outputting a high at pins 1 or 2, depending on the magnitude and direction of the AFT voltage swings (should the tuner-oscillator frequency tend to drift).

When a channel is selected, band-switching information is supplied from IC_1 (at pins 1 and 2) to IC_4, which develops four band-switching outputs. (In this particular circuit, one of the four outputs is switched by a normal/cable switch to provide the five bands shown.)

The tuner oscillator passes a *sample carrier* signal to oscillator amplifier Q_{14}/Q_{16}. The amplified oscillator signal is then passed to prescaler IC_6. The amount of frequency division is determined by PSC pulses from IC_1. The frequency-divided output of IC_6 (pin 5) is then passed to the sample input of IC_1 (at pin 26). When a channel is selected, circuits within IC_1 produce the appropriate number of PSC pulses at pin 27. The PSC pulses are applied to IC_6 and produce the correct amount of frequency division.

The divided-down oscillator signal (sample output) at pin 26 of IC_1 is divided down again within IC_1 and compared to an internal 5-kHz reference signal. The phase error of these two signals appears at pin 22 of IC_1 and is applied to *low-pass filter* (LPF) Q_{10}/Q_{11}. The dc output from the LPF is applied to the tuner oscillator. This voltage sets the tuner oscillator as necessary to get the proper frequency for the selected channel.

9.8.3 Troubleshooting FS/digital tuning circuits

The most common symptoms for failure of any FS/digital tuning system are a combination of no stations received, failure to lock on channels, picture snowing, audio noise, and color dropping in and out. The author experienced all of these symptoms with a name-brand TV set. The cause was traced to poorly soldered terminals on both the tuner and PLL IC. However, the following troubleshooting steps led to the defective solder junctions.

The first troubleshooting step is to isolate the problem to the tuner and IF/demodulator, or to the PLL system. Start by checking for power to all ICs and components. For example, the tuner requires + 12 V. Once you are satisfied that power is available to all components, start the isolation process.

Select a channel and confirm that the band-switching signal for that particular channel appears at the tuner input and band-select output. For example, if channel 4 is available locally, select channel 4 and check that the VHF-low band-switching signal is present (high). This signal appears at pin 1 of IC_4 and at the tuner band-switch input. If the band-switching signal is not at pin 1 of IC_4, suspect IC_4 (or possibly IC_1).

Next, apply a *substitute tuner-control voltage* to the tuner oscillator, at the terminal marked VT. Then monitor the video signal at test point TP1 with a scope. (The TP1 signal is the usual composite-video output, with both video and sync information, and is about 1 V peak to peak.) Note that the substitute tuning-control voltage can be obtained from an external variable power source (0 to 30 V) or from a circuit such as shown in Fig. 9.15 (as recommended in the service literature).

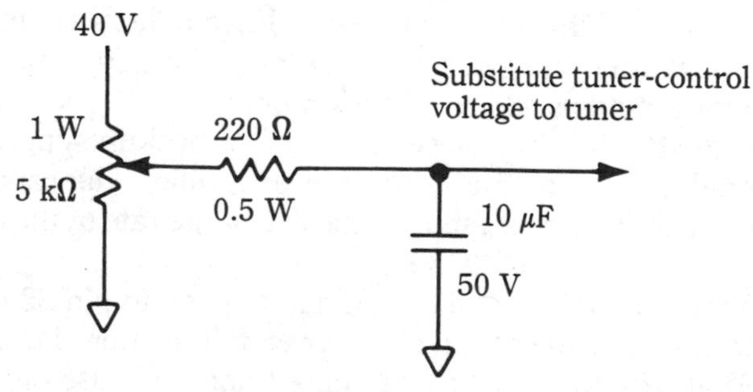

Fig. 9.15 Substitute tuning-voltage source for digital-tuning circuits.

If the video signal does not appear at TP1, suspect a problem in the tuner and IF/demodulator assembly. In most present-day TV/VCR equipment, this means replacing the complete assembly or possibly replacing individual tuner and IF/demodulator ICs.

If you can tune in stations using the substitute tuner-control voltage, the problem is most likely in the PLL components rather than in the tuner components.

If video appears at the test point with a substitute voltage, check for a high at the station-detect input of IC_1 (pin 34). If the station-detect input is missing, suspect IC_3 or the sync-detector circuits Q_7/Q_9.

If the station-detect signal at IC_{1-34} *is normal*, monitor the AFT up and down inputs to IC_1, pins 35 and 36, while changing the substitute tuning voltage. (Note that the substitute tuning voltage can come from any external source, but must match the normal voltage range at terminal VT of the tuner.)

If there are no logic changes at pins 35 and 36 as you tune through a station (as you vary the voltage at VT), suspect the window-detector circuit within IC_3 (or the AFT detector within the IF/demodulator).

If the inputs at pins 35 and 36 appear to be normal, check for a sample oscillator signal at pin 26 of IC_1. If the oscillator signal is missing, suspect IC_6, the oscillator amplifier Q_{14}/Q_{16}, or possibly the shielded cable from the tuner. (It is also possible that IC_6 is not receiving proper PSC pulses from pin 27 of IC_1, although you should still get sample signals from IC_6 to IC_{1-26}.)

If the input at pin 26 of IC_1 *appears to be normal* but the outputs at pins 1 and 2 are absent or abnormal, suspect IC_1.

9.9 Timer/keyboard/display

The timer function of a VCR permits the VCR to record program material at specific times on specific days while not attended by the user. Most modern VCRs use a microprocessor IC as the heart of the timer function.

In many VCRs, the timer IC also assumes many additional duties over and above that of automatic program recording. Some typical functions for the timer microprocessor include control of the VCR operating controls (the keyboard) and operating indicators (the display). The following paragraphs describe troubleshooting for such microprocessor-based circuits.

Figure 9.16 shows the timer/keyboard/display circuits for a typical VCR with microprocessor control. Timer microprocessor IC_8 monitors all input/output commands of the tuner/timer/charger, drives the multidigit front-panel display, performs timekeeping functions and record operations.

Microprocessor IC_8 receives power from a 5-V backup supply line. This line has a 0.22 F (farad, not microfarad) capacitor as a filter. The capacitor maintains power IC_8 in the event of a power failure. The discharge rate of the capacitor is over one hour.

During normal operation, the 60-Hz signal applied to pin 22 of IC_8 is used to maintain proper time. In the event of a power failure (for the memory backup period), an internal 32.8-kHz oscillator at pins 1 and 40 is used for timekeeping.

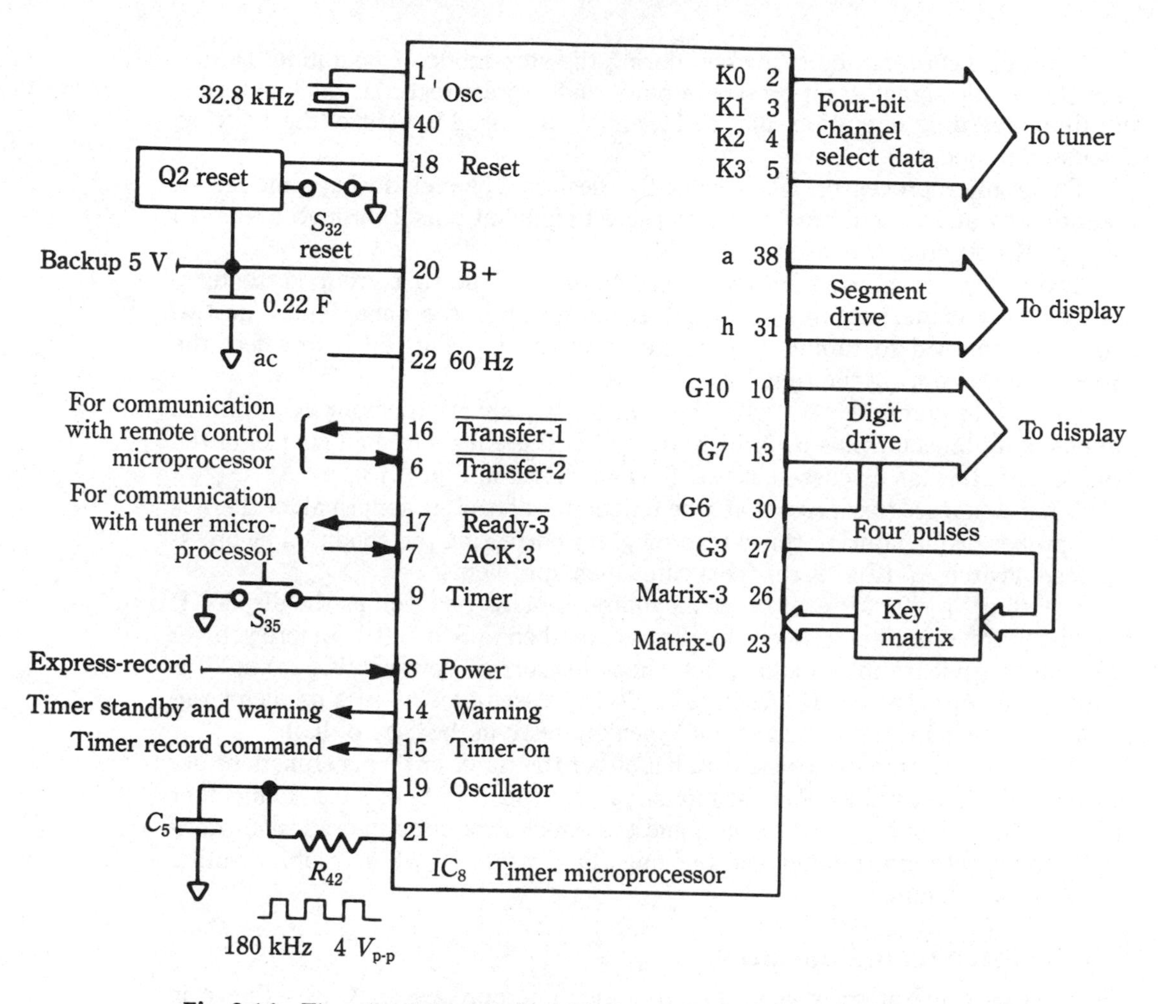

Fig. 9.16 Timer/keyboard/display circuits with microprocessor control.

The long-term accuracy of the 60-Hz signal is far better than the 32.8-kHz signal. However, during the one-hour battery-backup period, the 32.8-kHz signal is sufficiently accurate to maintain timing.

Microprocessor IC_8 generates eight scanning pulses, G3 through G10, to drive the display and key-matrix circuit. In this case, the key-matrix circuits consist of a series of switches connected to generate four possible outputs (matrix 0 through matrix 3 at pins 23 through 26).

The key matrix includes an express-record switch that turns on an express-record circuit within IC_8. The express-record circuit allows immediate recording for 30, 60, or 90 minutes or for 2-, 3-, or 4-hour periods. The express-record enable signal is applied to pin 8 of IC_8. (This enable signal is 5 V divided down from the 12-V power applied to the tuner.)

Express-record can be turned on during the stop mode of operation. During play, the enable signal is not present at pin 8, and express record is inhibited. During time-recording operation, pins 9, 14, and 15 are used to control the VCR, as discussed in Sec. 9.9.1.

Timer microprocessor IC_8 selects the desired channel during time-record operation by applying a four-bit channel-select signal at pins 2 through 5 (the K0 through K3 outputs discussed in Sec. 9.8).

Two groups of information are passed from IC_8. The first group is the tens-digit of the channel required, and the second group is the ones digit. The two groups are passed to tuning PLL microprocessor IC_1 (Fig. 9.12) to select the proper channel during the timed-record mode.

For IC_8 to communicate with the tuning PLL microprocessor IC_1, various handshaking lines (at pins 6, 7, 16, and 17 of IC_8) are used to gain control of the four-bit data bus (as discussed in Sec. 9.8 and shown in Fig. 9.13).

If you do not get timed recording (the function is absent or abnormal) or there is an improper display during timed recording, try correcting the condition by pressing reset switch S_{32} (Fig. 9.16). This cures many problems.

Pressing S_{32} should reset timer microprocessor IC_8, as well as the display. If not, check the timer and related circuits as described in Sec. 9.9.1. Before you get into timer-circuit troubleshooting, first check for correct power to IC_8 at pin 20, a proper reset signal at pin 18 (when reset S_{32} is pressed), a 32.8-kHz oscillator signal at pin 1 (which should be about 4 V peak-to-peak in the case of IC_8).

If all of these signals are present, but either the timer or display functions are not normal, then troubleshoot the circuits as described in Sec. 9.9.1. Remember that the timer, display, tape counter, and system-control circuits of most modern VCR circuits are interrelated, and this must be considered when troubleshooting any of these circuits.

9.9.1 Timed-record operation

Timed-record operation is controlled by timer microprocessor IC_8, as shown in Fig. 9.17.

After the proper record time is programmed into IC_8, and timer switch S_{35} is pressed, a low is applied to pin 9. At this time, one of these signals appears at pin 14. Dependent upon the previous state of the signal at pin 14, the signal can be steady-low, steady-high, or warning (1-Hz pulses).

If the timer operation is off, and the timer switch S_{35} is pressed, the signal at pin 14 goes steady-high (standby condition), or a 1-Hz square-wave (warning condition) is generated. The 1-Hz square-wave warning is present when there is a cassette tape within the VCR that has the record tab broken out, or when no cassette is present. (The record tab is discussed in Sec. 9.2.)

The signal at pin 14 drives Q_{101} to turn on the timer LED. If the warning signal is present, then Q_{101} is turned on and off at the 1-Hz rate, causing the timer LED to blink on and off. This alerts the user of a problem within the cassette (either no cassette or a cassette that is not to be recorded upon).

With a cassette installed (with record tab in place), the signal at pin 14 is high.

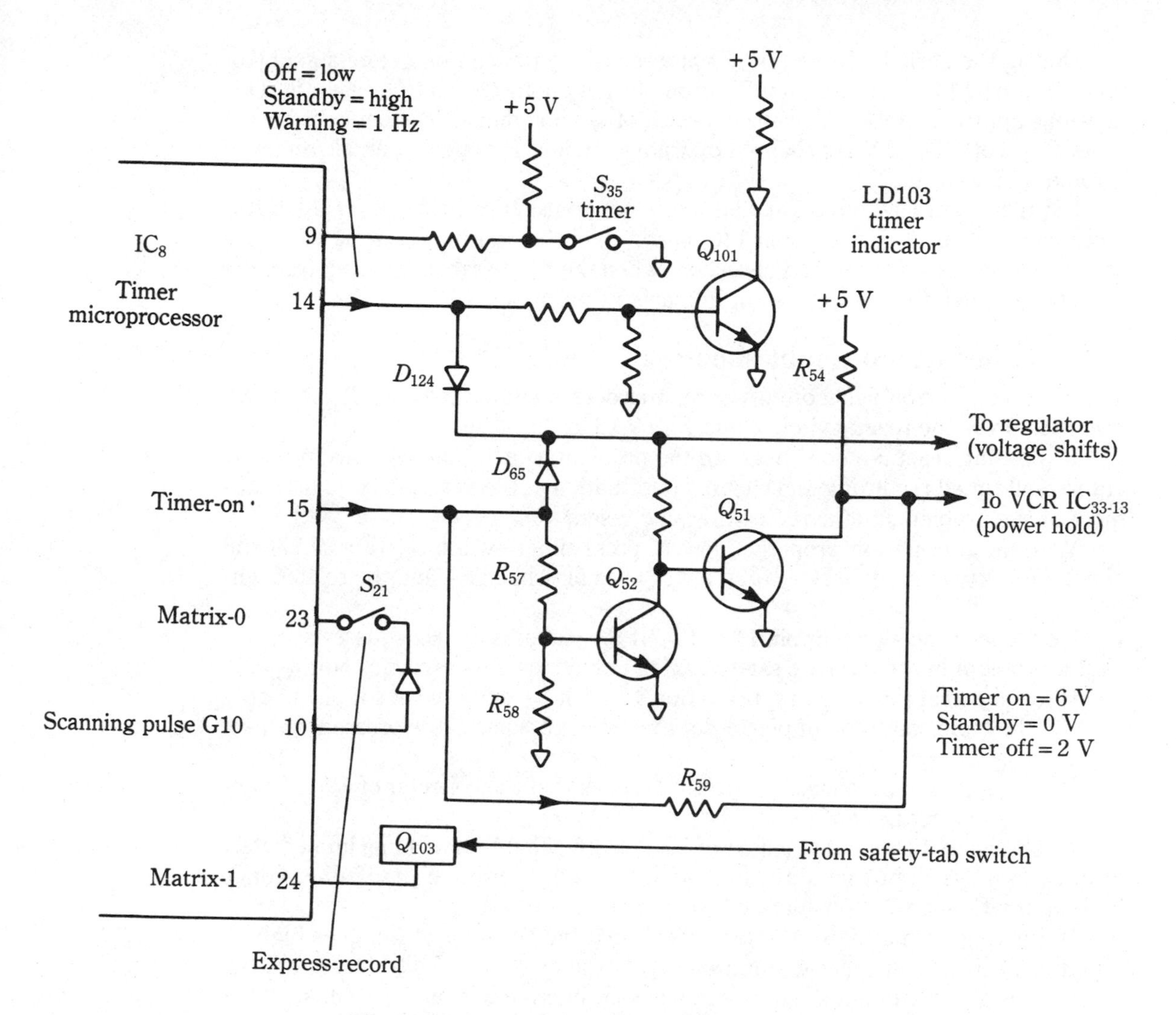

Fig. 9.17 Microprocessor timed-record operation.

This high is passed through D_{124} to bias the VCR control circuits, consisting of Q_{51} and Q_{52}. The standby high is applied to the collector of Q_{52} and the base of Q_{51}.

During the standby mode (after reset S_{32} is pressed, but before IC_8 starts the record mode), the signal at pin 15 of IC_8 is low. This low is applied to the base of Q_{52}, keeping Q_{52} off and Q_{51} on. With Q_{51} on, the collector of Q_{51} goes low. This low is applied to power-hold IC_{33} (which is part of system-control as described in Sec. 9.2).

With a cassette installed (with record tab in place), the signal at pin 14 is high. This high is passed through D_{124} to bias the VCR control circuits, consisting of Q_{51} and Q_{52}. The standby high is applied to the collector of Q_{52} and the base of Q_{51}.

During the period when S_{35} is not pressed (the off condition before standby or record), pins 14 and 15 are low. With pin 14 low, both Q_{51} and Q_{52} are off. This develops approximately 2 V at the collector of Q_{51} and signals system-control that timer IC_8 is off. The 2 V is produced by voltage divider R_{54} and R_{59}, and is returned through pin 15 of IC_8.

Express-record switch S_{21} applies a scanning pulse to matrix 0 at pin 23 of IC_8. The scanning pulse is also applied to matrix 1, pin 24, during the time when the safety-tab switch is actuated. This occurs when the safety tab is removed from the cassette or when the cassette is not in place, as discussed.

9.9.2 Timed-record troubleshooting

The first step in troubleshooting any microprocessor timer or timer/display function is to press the reset switch (S_{32} in Fig. 9.16 in this case).

If pressing reset S_{32} does not cure the problem, check that you have properly entered all timed recording and channel information at the key matrix. (This cures most of the problems not cured by pressing reset.)

With all information properly entered, press timer switch S_{35} (Fig. 9.17) and check for a low at pin 9 of IC_8. If the low is missing, check S_{35} and the related wiring.

Next check the signal at pin 14 of IC_8. If the signal is a 1-Hz squarewave, suspect a problem in the cassette safety-switch circuits or possibly transistor Q_{103}.

If the signal at pin 14 is low, press timer switch S_{35} once and see if pin 14 goes high. If the logic condition of pin 14 does not change when S_{35} is pressed, suspect IC_8.

If pin 14 goes high when S_{35} is pressed, check that the collector of Q_{51} is low or 0 V. If not, suspect Q_{51} and/or Q_{52}.

Next, check the signal at pin 15 of IC_8. Pin 15 should remain low immediately after S_{35} is pressed, but should go high after a standby period. If pin 15 does not go high at any time after S_{35} is pressed, suspect IC_8.

If pin 15 goes high after standby, check that the collector of Q_{51} goes high (to about 5 V). If not, suspect Q_{51} and/or Q_{52}. If the collector of Q_{51} goes high, but timed-record operation does not occur, suspect the system-control circuits.

9.10 Tape-counter and time-remaining circuits

Most modern VCRs have some form of electronic display to show the amount of time consumed (or remaining) for a particular cassette. Early-model VCRs use mechanical tape counters to show this information.

In some cases, modern VCRs use the same electronic display for both the tape-counting function and the record-time function. Likewise this same display can also be used to indicate various system-control functions.

Figure 9.18 shows the tape-counter and time-remaining circuits for a typical VCR with microprocessor control.

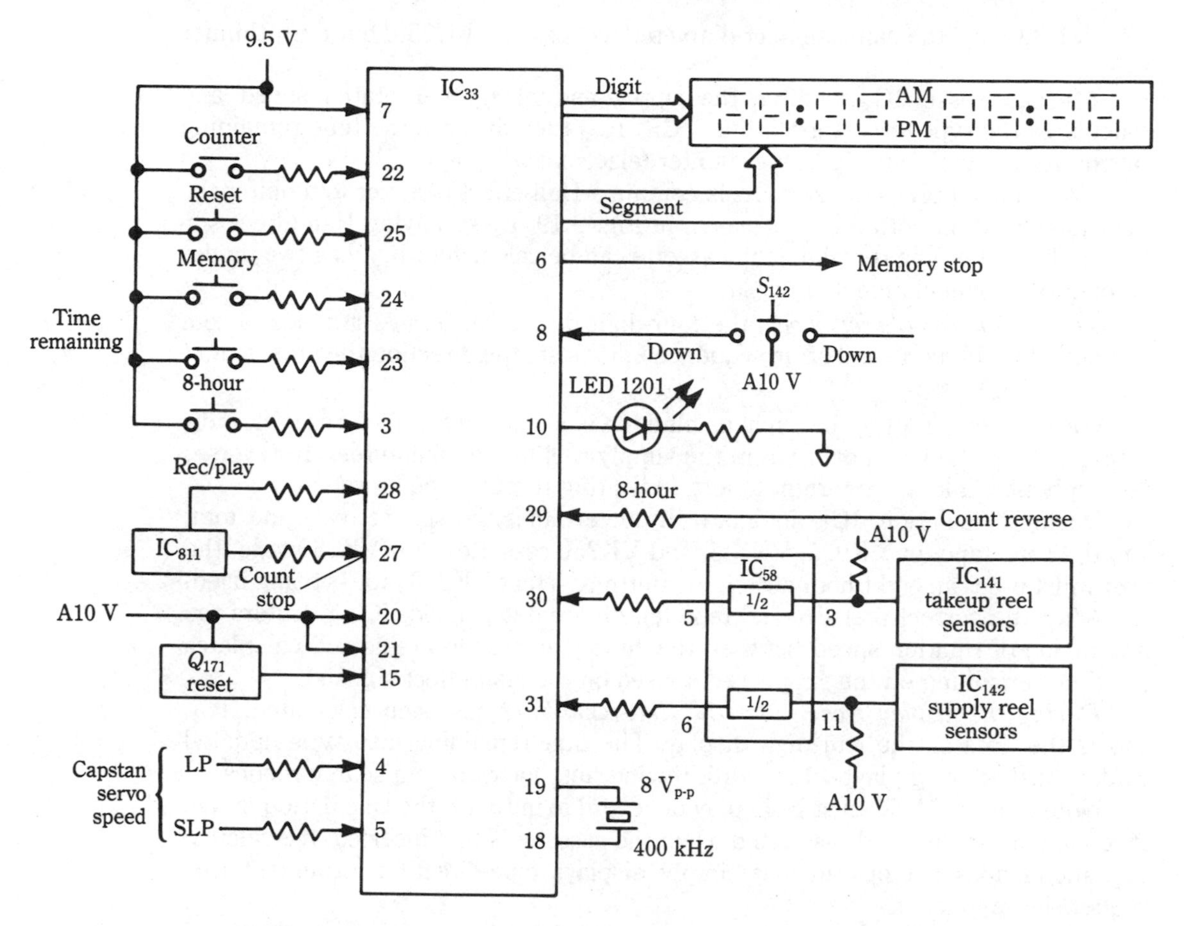

Fig. 9.18 Digital tape-counter and time-remaining circuits.

In this case, a single microprocessor IC_{33} is used for both the counter and time-remaining functions (but not for system-control or record-timer circuits). In this section, be concerned only with the tape-counter and time-remaining circuits, because both system-control and record-timer functions are covered in other sections of this chapter.

9.10.1 Microprocessor calculator and tape sensors

Because the VCR described here is an RCA unit, RCA cassette sizes are mentioned for reference. Four sizes of RCA cassette tapes are available for this unit: VK065, VK125, VK250, and VK330. Two characteristics distinguish the difference between these four types: *thickness* of the tape and *diameter* of the supply and takeup reels.

The VK065 and VK125 cassettes have normal-thickness tape and larger-diameter reels. The VK250 has normal-thickness tape but smaller-diameter reels.

The VK330 has the same smaller-diameter reels as the VK250, but even thinner tape.

Microprocessor IC_{33} monitors the supply- and takeup-reel rotation speed, and calculates the type of cassette in the VCR. IC_{33} then displays the time remaining on the front-panel four-digit tape counter (electronic display).

Both the supply and takeup reels contain a Hall-effect detector to monitor the rotation speed on both reels, as shown in Fig. 9.19. By supplying Hall-effect signals to IC_{33}, the time remaining on cassette can be calculated. IC_{33} has two modes of operation: calculation and display.

During the calculation mode, the four-digit display generates a series of four hyphens that blink on and off in sequence to indicate the direction the tape is moving.

For example, in play, the hyphens blink on in sequence from left to right, indicating that the tape is moving from the supply reel to the takeup reel. In reverse, the hyphens blink on from right to left, indicating reverse tape movement.

During calculation, IC_{33} automatically determines the size of reels and total length of the tape for VK065, VK125, and VK250 cassettes. For VK330 tape, the user must press an eight-hour switch to inform IC_{33} that VK330 tape is being used.

After the correct reel size and tape length are determined, IC_{33} monitors the proportion of rotation speed between the supply and takeup reels and calculates the time remaining on the supply reel (based on the Hall-effect signals).

During the display mode, after the time-remaining has been calculated, IC_{33} shows the time on the four-digit display. The time-remaining display is updated continuously during playback, record, rewind, fast-forward, and search modes.

Note that the VCR must be in play or record to initialize the calculation mode. If rewind or fast-forward is selected when the cassette is first inserted, the calculation mode does not operate but simply displays tape-direction movement (the sequencing hyphens).

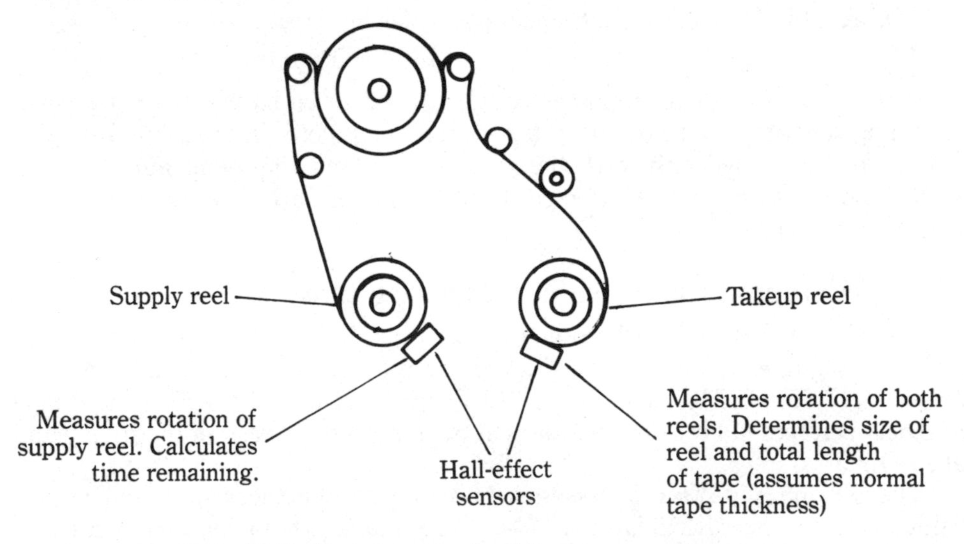

Fig. 9.19 Hall-effect detectors in digital tape-counting systems.

9.10.2 Tape-counter operation

As shown in Fig. 9.18 when the VCR is turned on, 9.5-V power is applied to pin 7 of IC_{33}. This enables the output drives to power the four-digit, seven-segment display.

When power is first applied to IC_{33} at pins 20 and 21, power is also applied to Q_{171}. Transistor Q_{171} generates a reset pulse that is applied to pin 15 of IC_{33}. This pulse resets all internal counters, and initializes IC_{33} to start operation of the time-remaining mode (Sec. 9.10.4). Note that IC_{33} can also be reset manually by the front-panel reset switch connected to pin 25.

Either tape-counter mode or time-remaining mode is selected by corresponding front-panel switches (counter pin 22 or time-remaining pin 23). IC_{33} returns to the mode that the VCR was in (counter or time-remaining) when power was turned off.

In the play mode, IC_{141} (the takeup reel Hall-effect sensor) generates pulses in proportion to the revolutions of the takeup reel. These pulses are divided by two within IC_{58} and applied to pin 30 of IC_{33}. In normal (forward) playback, the count-reverse input at pin 29 of IC_{33} goes low. This tells IC_{33} to count the reel pulses at pin 30 in the upward direction (incrementing). In reverse play, rewind, or any reverse tape mode, the count-reverse line at pin 29 is high, instructing IC_{33} to decrement the count.

Note that in the tape-counter mode, the supply-reel sensor signal at pin 31 of IC_{33} (from IC_{142} through IC_{58}) is not used.

9.10.3 Tape-counter troubleshooting

If a malfunction exists during the tape-counter mode of IC_{33}, but the time-remaining function is normal, check that pin 22 of IC_{33} goes high when the front-panel counter switch is pressed. If not, check the switch and related wiring. If normal, suspect IC_{33}.

If both the tape-counter and time-remaining modes are absent or abnormal, check for proper inputs to IC_{33}. If the display is off, look for power at pin 7. Then look for power and reset inputs at pins 20, 21, and 15. Also check for 400-kHz clock signals at pin 18. The clock signals should be about 8 V peak to peak. If any of the inputs to IC_{33} are missing, trace back to the input source.

If all inputs are present, check for normal operation of the counter and time-remaining switches by switching between the two functions. If either function is absent or abnormal, suspect the corresponding switch and/or IC_{33}.

If all inputs and switch functions appear to be good, then look for pulses at pin 30 of IC_{33} during playback. If the pulses are missing, suspect IC_{58} or IC_{141} (probably IC_{58}). If the pulses are present at pin 30, but there is no counter function, suspect IC_{33}.

9.10.4 Time-remaining operation

When IC_{33} is placed in the time-remaining mode by pressing the time-remaining switch (pin 23), four inputs are required for IC_{33} to calculate the correct time remaining: takeup-reel rotation at pin 30, supply-reel rotation at pin 31, capstan servo speed for LP (Sec. 9.11) at pin 4, and capstan servo speed for SLP at pin 5.

Without these signals, IC_{33} never leaves the calculation phase of the time-remaining function. With the signals, IC_{33} calculates the type of cassette and then changes from the calculation phase to the display phase of operation. The time-remaining is then displayed and updated in all modes (except special effects).

Note that when the VCR is in the special-effects modes of field-still (and being frame-advanced) or slow-motion, the time-remaining calculation becomes very erratic and inaccurate, so IC_{33} is inhibited from counting the reel pulses. This is done by a count-stop input at pin 27, supplied by a decoder within IC_{811}.

Note that if the VCR is first placed in rewind or fast forward, the calculation mode is not entered, and only the tape movement direction is displayed. The calculation phase of time-remaining operation is entered every time a cassette is loaded and the VCR is set to play. The cassette-down switch applies a high to pin 8 of IC_{33} to indicate that the cassette is loaded and the VCR is set to play. The cassette-down switch applies a high to pin 9 of IC_{33} to indicate that the cassette is loaded. IC_{811} applies a high to pin 28 to indicate that the VCR is in play or record mode.

The calculation mode takes about 20 to 60 seconds for the time-remaining calculation to conclude before the time-remaining is displayed. After the first record or play function is actuated (pins 8 and 28 high), IC_{33} keeps track of the time-remaining until the cassette is ejected. When ejection occurs, the cassette-down switch applies a low to pin 8, instructing IC_{33} to clear all registers and memory.

As discussed, IC_{33} cannot calculate the time-remaining properly when the thin eight-hour tape (such as RCA VK330) is used. In this case, the user presses the eight-hour switch (pin 3) to tell IC_{33} that a thin eight-hour tape is installed.

9.10.5 Time-remaining troubleshooting

If a malfunction exists in the time-remaining mode, but the counter mode operates normally, you can assume that the clock, power, and reset functions are good. Also, because the takeup-reel sensor input at pin 30 is used in both counter and time-remaining modes, the pulses at pin 30 are probably good.

The first place to check is at the supply-reel sensor input, pin 31 of IC_{33}. If these pulses are missing, suspect IC_{58} and/or IC_{142}. If pulses are present at both pins 30 and 31, suspect IC_{33}.

If a malfunction exists in both counter and time-remaining modes, check for all inputs to IC_{33} (as discussed in this section and in Sec. 9.10.3). If any of the inputs are missing, trace back to the source. If the inputs are present, suspect IC_{33}. Also check the display using the general guidelines described in Sec. 9.9.

9.11 Microprocessor-controlled servo circuits

Early-model VCRs generally use a single motor that drives both the head scanner and tape capstan through belts and gears. Modern VCRs use separate direct-drive (DD) motors for scanner and capstan. The direct-drive motors are electronically synchronized as to phase and frequency by microprocessor-based servo circuits. First, the next section reviews the basics of VCR control and synchronization.

9.11.1 Record and playback heads

Figure 9.20 shows the relationship of the recording and playback heads on a typical two-head VCR. The same principles apply to multi-head models. Instead of moving the tape at a high speed, the video heads are rotated to produce a high relative speed between the head and tape. This increases the frequency range necessary for video (4 to 5 MHz). The actual tape speed is in the range of 2 cm/s, and the

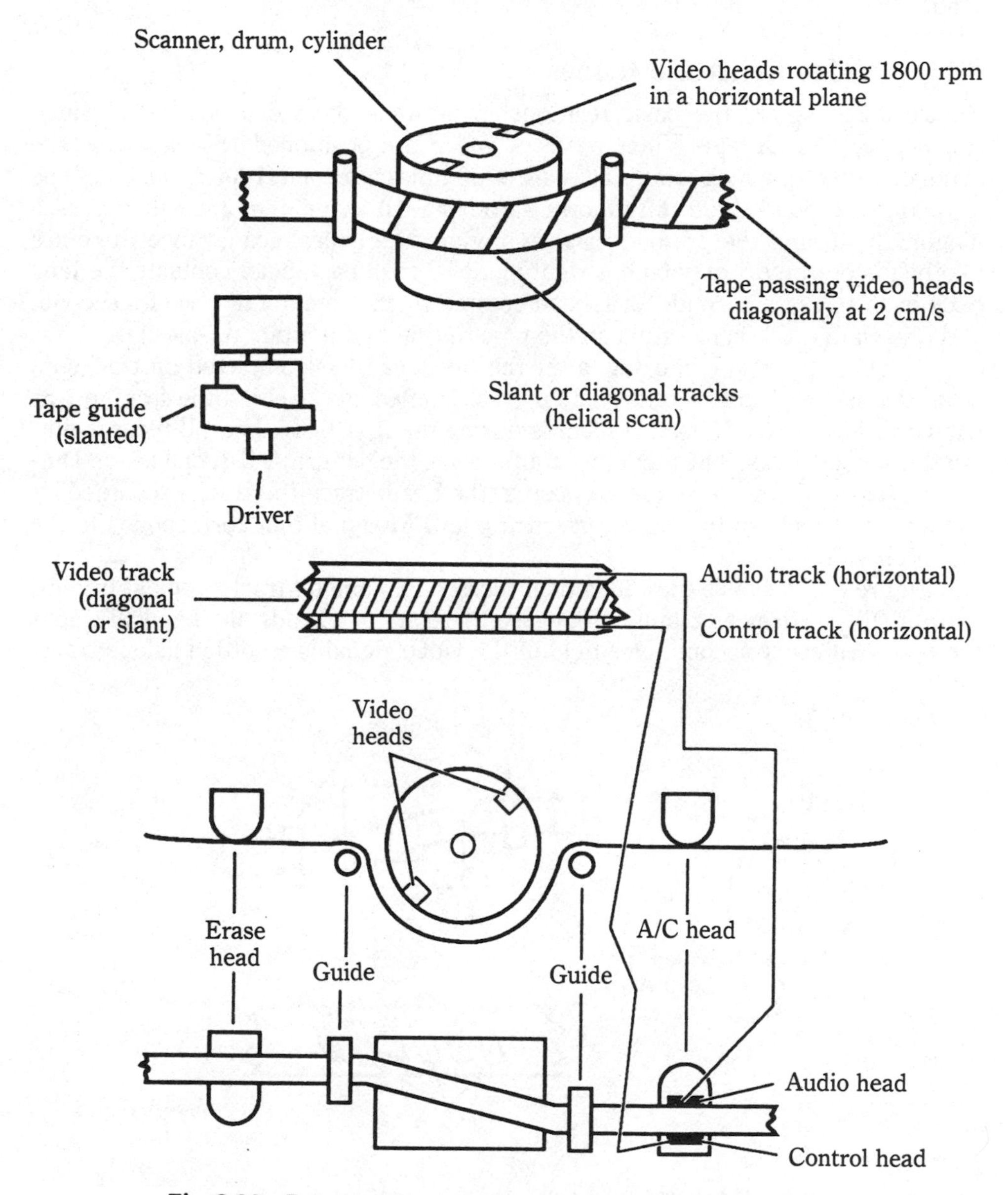

Fig. 9.20 Relationship of the recording and playback heads.

video heads are rotated at 1800 rpm. This results in a *relative speed* in the range of 5 to 7 m/s (225 to 275 in/s).

Note that the video heads rotate in a horizontal plane (on a drum, cylinder, or scanner, depending on the literature you read), while the tape passes the heads diagonally. This is known as *helical scan* and produces *slant tracks* or *diagonal tracks* for video recording. The audio head and control-track head (mounted one above the other) are stationary and separate from the video heads, as is the erase head.

9.11.2 Video fields and frames

Figure 9.21*a* shows the basic relationship between the video heads and video tracks recorded on tape. Video heads A and B are positioned 180° apart on the drum or cylinder, which rotates at a rate of 30 times a second (1800 rpm). The tape is wrapped around the drum to form an omega (Ω) shape. The tape then passes diagonally around the drum surface to produce the helical scan. Since there are two heads on the drum (which is rotating at 30 rps), each head contacts the tape once each 1/60 of a second. Each head completes one rotation in 1/30 of a second, and one slant track is recorded on the tape during half a rotation (1/60 s).

Because the tape is moving, after the first head has completed on tracks on tape, the second head records another track immediately behind the first one, as shown in Fig. 9.21*a*. If head A records during the first 1/60 s, head B records during the second 1/60 s. The recording continues in the pattern A-B-A and so on. During playback, the same sequence occurs (the heads trace the tracks recorded on the tape and pick up the signal, producing an FM signal that corresponds to the recorded video signal).

Figure 9.21*b* shows the theoretical relationship among tracks, fields, frames, and the TV vertical sync pulses. Because there are two heads, 60 diagonal tracks are recorded every second. One field of the video signal is recorded as one track,

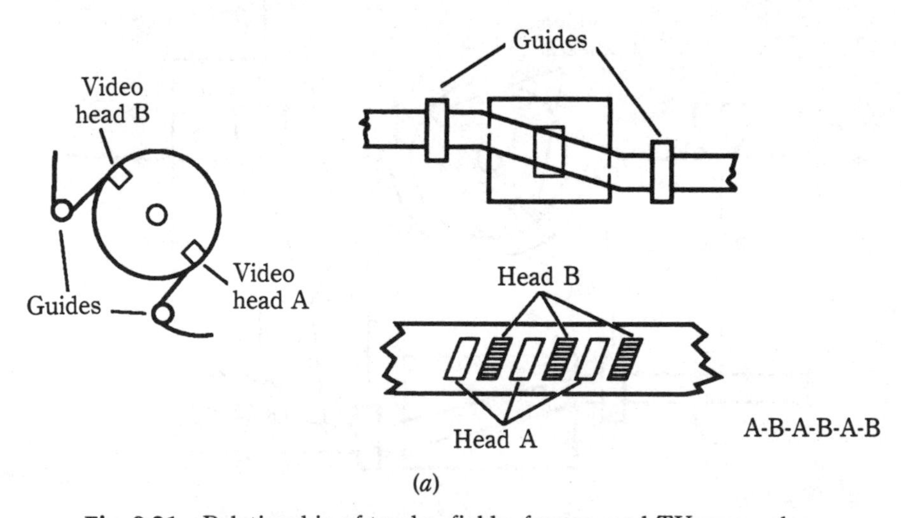

Fig. 9.21 Relationship of tracks, fields, frames, and TV sync pulses.

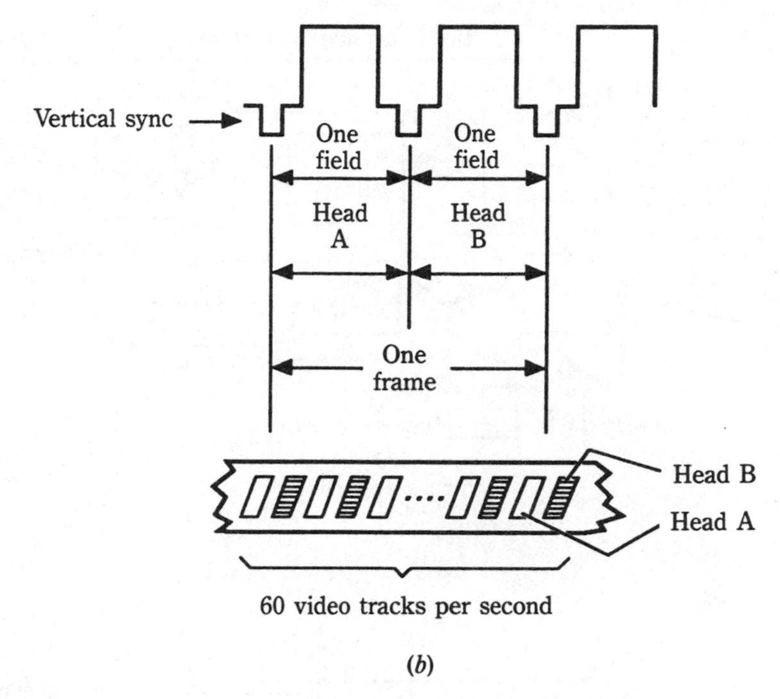

(*b*)

Fig. 9.21 Continued.

and two fields (adjacent tracks A and B) make up one frame. In actual practice, there is some overlap between the two tracks. For example, the video signal recorded by head A (just leaving the tape) is simultaneously applied to head B (just starting the track). During playback, this overlap is eliminated by electronic switching so that the output from the two heads appears as a continuous signal.

9.11.3 Basic VCR servo system

It is obvious that no matter how precisely the tracks are recorded, the picture cannot be reproduced if these tracks are not accurately traced by the rotating heads during playback. In addition to mechanical precision, VCRs use an automatic self-governing arrangement called the *servo system*.

Figure 9.22*a* shows operation of a basic servo system for a typical two-head VCR. The vertical-sync pulses of the TV-broadcast signal are used to synchronize the rotating heads with tape movement. The TV-sync pulses are converted to a 30-Hz control signal (often called the CTL signal). This CTL signal is recorded on the tape by a separate stationary control head.

One major function of the servo is to rotate the cylinder at precisely 30 Hz during record. Note that 30 Hz is one-half the vertical-sync frequency (60 Hz) of the input video signal. With a 30-Hz speed, the vertical-blanking period can be recorded at any desired point on each video track. In TV, the vertical blanking occurs at the bottom of the screen, where blanking does not interfere with the pic-

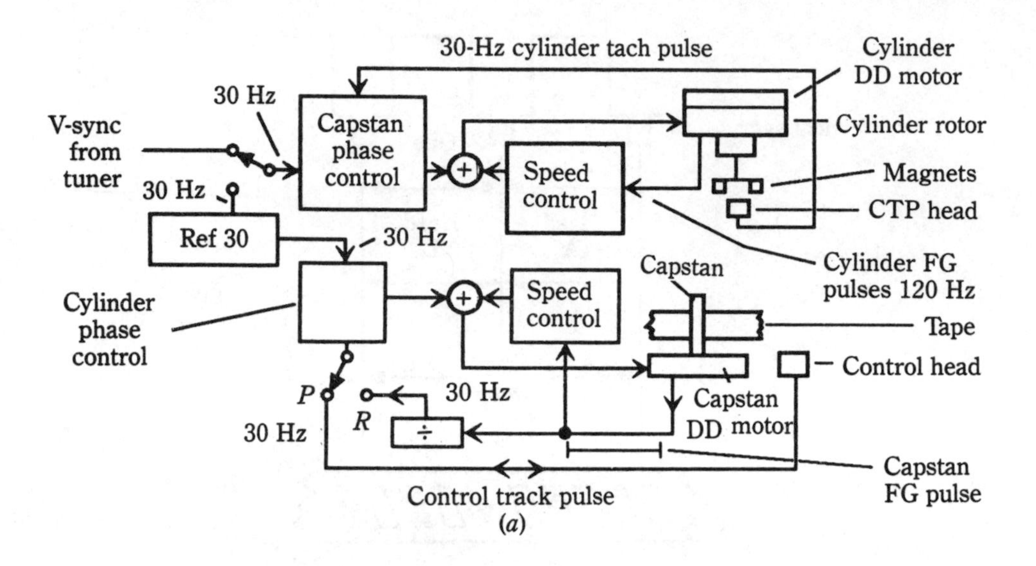

(*a*)

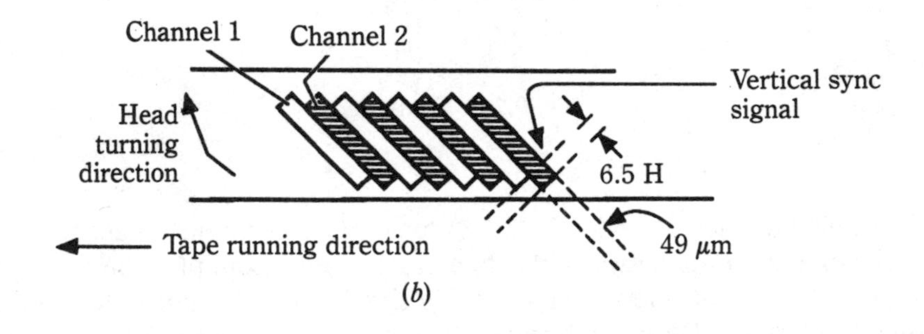

(*b*)

Pulse	Frequency	
Cylinder FG	120	
Capstan FG	SP (2 hour)	720
	LP (4 hour)	360
	EP (6 hour)	240
	Slow (slow motion)	120
	Quick (fast motion)	720
	Search	2160
Cylinder tach	30	
Ref 30	30	
Control track	30	

(*c*)

Fig. 9.22 Basic servo system with digital control.

ture. For this reason, the vertical sync is recorded at the bottom (or start) of each video track, as shown in Fig. 9.22*b*.

In the system of Fig. 9.22, there are two heads (channel 1 and channel 2), and each head traces on track for each field. Two adjacent tracks or fields make up one complete frame. To ensure that there are no blanks in the picture, the information recorded on tape overlaps at the changeover point (from one head to another). This changeover point also must occur at the bottom of the screen, where the changeover does not interfere with the picture.

To ensure proper changeover, the vertical-sync signal is recorded precisely in a position 6.5 H from the changeover time of the channel 1 and channel 2 tracks. (The term 1 H refers to the period of time required to produce one horizontal line on a TV screen, or about 63.5 μs, and is sometimes referred to as line period. 6.5 H is 6.5 times 1 H.)

The precise timing requires that the speed and phase of both the cylinder motor and capstan motor be controlled (because the cylinder motor determines the position of the tape). In older VCRs, the synchronization is done by driving both capstan and cylinder from a common motor through belts. In a modern VCR servo system, five separate (but interrelated) signals are used to get the precise timing as shown in Fig. 9.22*c*. The following paragraphs describe each of these signals, and in general terms, how the signals are used.

Cylinder FG pulses The cylinder FG pulses are developed by a generator in the video-head cylinder. In the system of Fig. 9.22, the generator consists of an eight-pole magnet installed in the cylinder rotor and a detection coil in the stator. When the cylinder rotates at 30 rps, the stator coil detects the moving magnetic fields and produces the cylinder FG pulses at a frequency of 120 Hz.

Capstan FG pulses The capstan FG pulses are developed by a generator in the tape capstan and are applied to the capstan speed-control circuits, as well as the capstan phase-control circuits, as well as the capstan phase-control circuits (through a divider) during record. In Fig. 9.22, the generator consists of a 240-pole magnet installed in the lower part of the capstan shaft and a detection coil in the stator. (In many modern VCRs, all of the servo-control pulses, cylinder, capstan, tach, reference, and control are developed by Hall-effect generators, instead of a magnet/coil detector.)

When the capstan rotates, the stator coil detects the moving magnetic fields and produces the capstan FG pulses. The frequency of the capstan FG pulses depends on the capstan speed (which also controls tape speed). The table in Fig. 9.22*c* shows some typical capstan FG pulse frequencies for various playing times and tape speeds. Note that not all VCRs have the same six play modes or tape speeds shown in Fig. 9.22.

Cylinder tach pulse The cylinder tach pulses (CTPs) are developed by another generator in the cylinder and are applied to the cylinder phase-control circuits. In some VCRs, the generator consists of a pair of magnets installed symmetrically in a disk in the lower part of the cylinder shaft, and a stationary pickup head. Hall-effect generators are used in other VCRs.

With the system of Fig. 9.22, the CTP pickup head detects the moving magnetic fields when the cylinder rotates. The pulse frequency is a constant 30 Hz. In effect, the tach pulse indicates video-head channel switching and is used as a comparison signal in the cylinder phase-control circuits during both record and playback.

REF30 pulse The reference signal for the phase-control system of both the capstan and cylinder motors is taken from a crystal oscillator with a frequency of 32.765 kHz. A frequency of 30 Hz is obtained when the crystal oscillator signal is divided. The REF30 pulse is used for the cylinder phase control only during playback. During record, the cylinder phase control receives V-sync pulses from the tuner.

Control track pulse The 30-Hz control-track pulses are the broadcast V-sync pulses recorded on tape during record. At playback, the pulses are picked up by the control head and applied to the capstan phase-control circuit.

9.11.4 Cylinder-servo operation

Figure 9.23 shows the overall operation for the cylinder-servo system of a typical VHS VCR. Figure 9.24 is the diagram for the phase and speed loops of the cylinder servo. Note that the phase-loop circuit is contained in IC_1, while the speed-loop portion is in IC_2. The outputs of the phase and speed loops are added together to determine the drive voltage to the cylinder motor-drive IC (Fig. 9.22). The phase-loop portion of the cylinder servo is a pulse-width modulation (PWM) system.

During playback, the reference signal for the phase loop is taken from a 3.58-MHz signal at pin 25 of IC_1. This is divided down to 30 Hz. The comparison signal is derived from the 30-Hz cylinder input at IC_{1-14}. This signal is delayed by two *multivibrators* (MV).

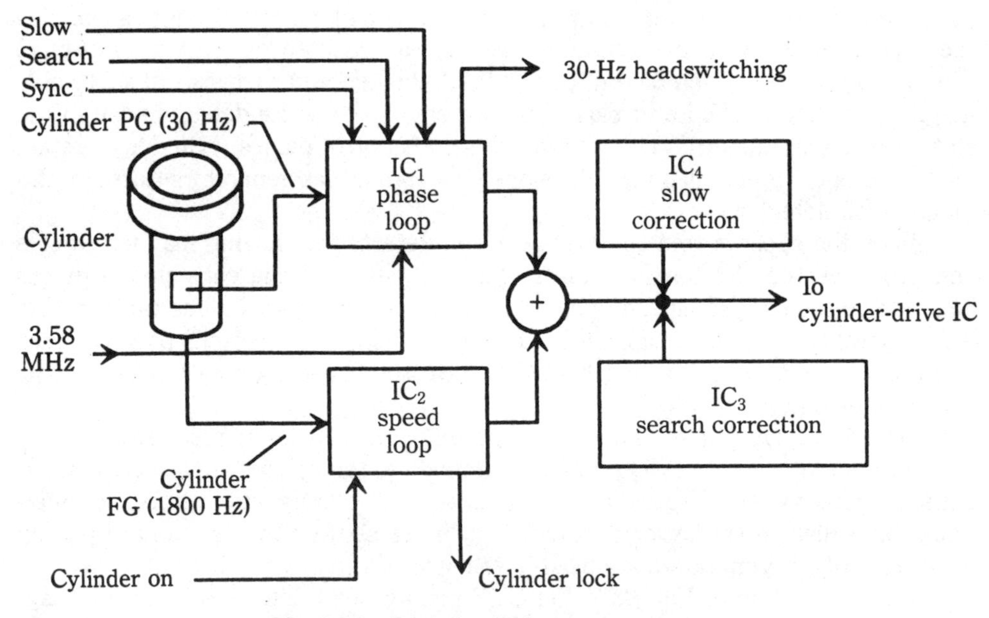

Fig. 9.23 Digital cylinder-servo system.

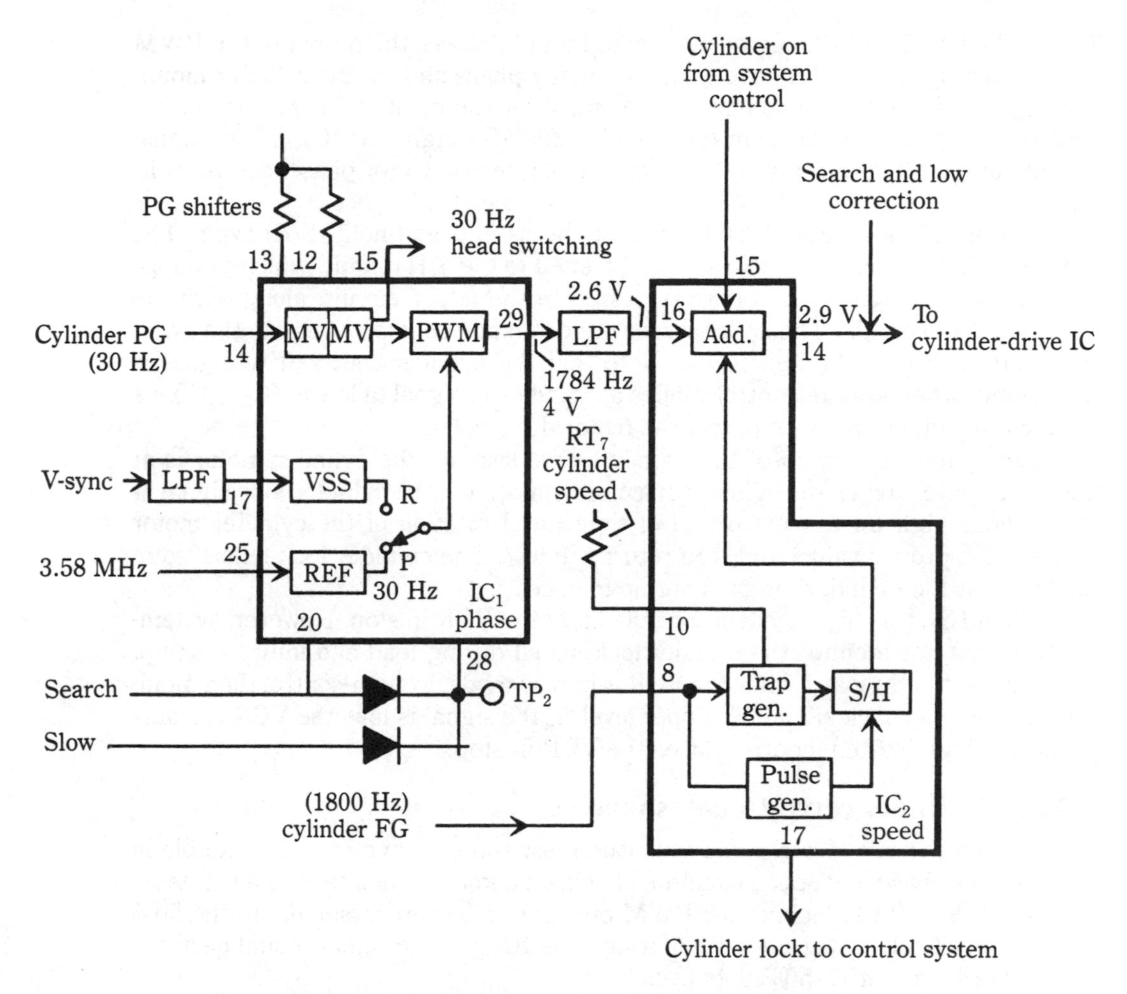

Fig. 9.24 Microprocessor circuits for digital cylinder-servo.

The time constant of the two MVs is controlled by the two shift adjustments at pins 12 and 13 of IC_1 (often called the PG shift adjustments). One output from the MVs is the 30-Hz head-switching signal. The other output from the MVs is applied to the PWM circuit, which also receives the 30-Hz record or playback reference signal. The two signals are compared in the PWM circuit to reference signal. The two signals are compared in the PWM circuit to determine the correction necessary for proper phase.

During record, the reference signal is the divided-down vertical sync signal at IC_{1-17} (instead of the 3.58-MHz signal). The output of the PWM at IC_{1-29} is a 1784-Hz signal, which is passed through an LPF to remove the ac component. The resultant dc output (about 2.6 V) from the LPF is applied to pin 16 of the speed-loop IC_2.

During search and slow modes, the system-control circuits apply a high at

IC_{1-28}. This high, which can be monitored at TP2, places the output of the PWM system in a fixed 50% duty cycle, inhibiting any phase shift of the cylinder motor.

IC_2 is essentially the coarse speed control for the cylinder-servo system. IC_2 receives a primary input from the cylinder 1800-Hz signal at IC_{2-8}. The signal applied to pin 16 of IC_2 from IC_1 is a form of vernier speed (or phase) control voltage.

The speed-loop circuits in IC_2 are of the sample-and-hold (S/H) type. The 1800-Hz signal at IC_{2-8} is processed and passed to the S/H circuit, which develops a correction voltage. This voltage is passed to the adder circuit, along with the phase-control voltage. The adder passes the combined control voltage to the cylinder motor-drive IC through IC_{2-14}. Note that the cylinder motor-drive signal is active only when system control applies a cylinder-on signal (a low at IC_{2-15}). This enables the adder circuit to output the motor-drive voltage.

During initial start-up of the cylinder-servo system, the cylinder motor is at zero rotation. So when the cylinder-on command occurs, the cylinder-lock signal at IC_{2-17} goes high for a short period of time (until rotation of the cylinder motor reaches the proper value) and then returns to low. The cylinder lock tells system control that the cylinder motor is not up to speed.

When IC_{2-17} is high, system-control places the VCR in stop. However, system-control does not monitor the cylinder-lock signal during load and initial start-up. After proper loading and when the VCR is in play mode, system-control then monitors the cylinder-lock signal for proper level. If the signal is low, the VCR remains in play. If high, system-control places the VCR in stop.

9.11.5 Cylinder-servo troubleshooting

If the picture is out of horizontal sync (the most common symptom for trouble in the cylinder phase and speed circuits), play back a known-good tape or a test tape. Connect 5 V to TP2, locking the PWM output of microprocessor IC_1 to the 50% duty cycle. Check for the presence of a signal at IC_{1-29}. The signal should be about 4 V at 1784 Hz, with a 50% duty cycle.

If the signal is absent or abnormal, suspect the LPF. If the signal at IC_{2-16} is normal, check for a low at IC_{2-15}.

If IC_{2-15} is high, indicating that the cylinder is locked, check the system-control circuits. If IC_{2-15} is low and there are about 2.6 V at IC_{2-16}, try monitoring the dc voltage at IC_{2-14} while adjusting the cylinder-speed control RT7. Adjust RT7 for about 2.9 V at IC_{2-14}, and check that the TV picture is in horizontal sync.

Note that if the phase-control microprocessor IC_1 is disabled by 5 V at pin 28, there might be some noise bars floating through the picture, even if the picture is in sync. Remove the 5 V, and recheck for noise bars and horizontal sync. If you cannot get proper horizontal sync by adjustment of RT7, with all inputs to IC_2 correct, suspect IC_2.

9.11.6 Capstan-servo operation

Figure 9.25 shows the overall operation for the capstan-servo system of a typical VHS VCR. Note that the capstan servo is functionally similar, but not identical, to

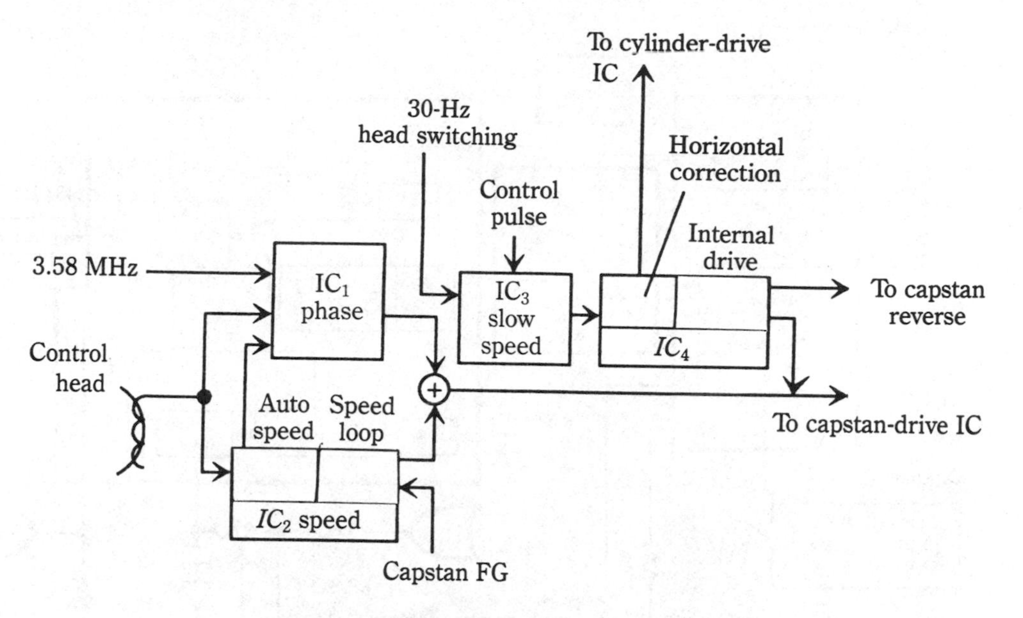

Fig. 9.25 Digital capstan-servo system.

the cylinder-servo described in Sec. 9.11.4. Both use PWM for speed and phase control. The use of PWM is typical for most modern VCRs (although some VCRs use sample-and-hold circuits in the microprocessor). The servos of most modern VCRs use three-phase motor drives (direct-drive or DD) with Hall-effect control.

As shown in Fig. 9.25, the phase-loop portion of the capstan servo is in IC_1, and the speed-loop portion is in IC_2. During playback, the auto-speed select circuits in microprocessor IC_2 monitor the control-track signal (picked up by the control head) and determine the proper speed for the capstan motor. The correction voltage from the phase loop and speed loop are added together and passed to the capstan motor-drive IC.

During slow-motion and special-effects modes, correction signals developed from IC_3 and IC_4 are passed to the capstan motor-drive IC along with the speed-loop correction voltage. During slow-motion operation, the phase-loop portion of the servo (IC_1) is inhibited, and phase correction is done by IC_3.

9.11.7 Capstan phase-loop operation

As shown in Fig. 9.26, the capstan phase loop uses a 3.58-MHz signal at $IC_{1\text{-}25}$ as a reference for the PWM comparator. During playback, the comparison signal is derived from the control-track signal recorded on tape. During record, the comparison signal is taken from the capstan signal (the frequency of which depends on playing time).

The control-track signal and the capstan signal are passed through the programmable divider in microprocessor IC_3, which does not divide the signal in normal play or record modes, but does divide in the search modes. This division is necessary since capstan speed is increased but phase-lock must be maintained.

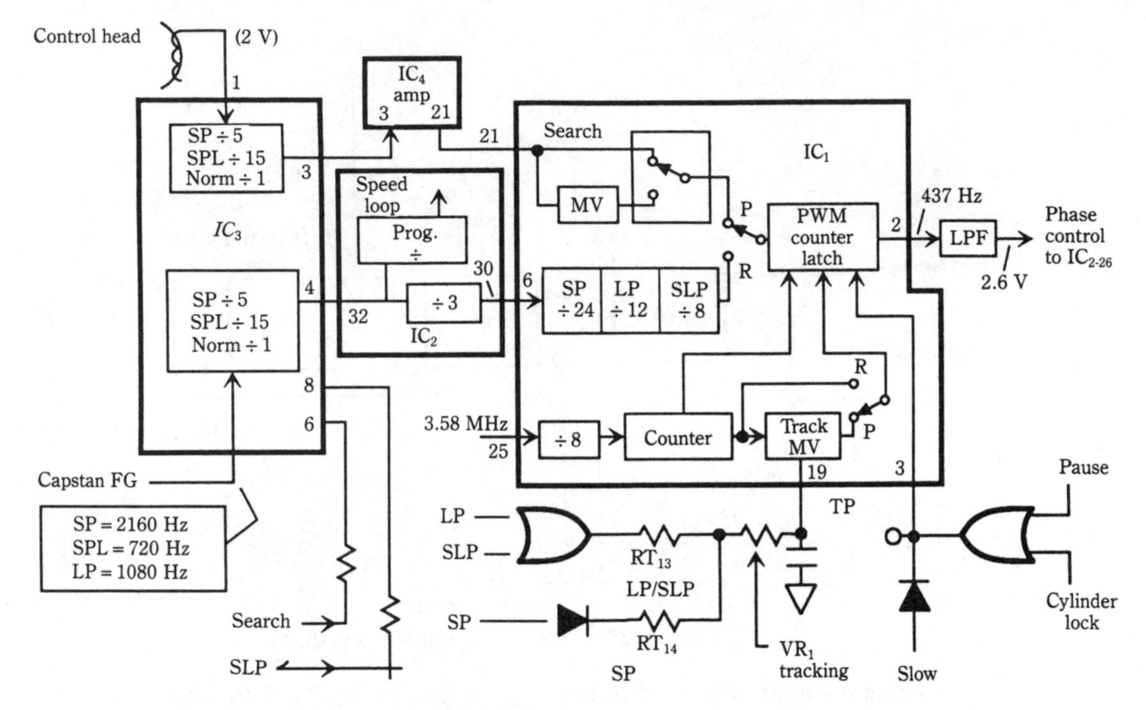

Fig. 9.26 Microprocessor circuits (phase loop) for digital capstan-servo.

To accommodate the increased frequency of the control-track and capstan signals (SP is increased by 5; SLP by 15), the signals are divided down to normal frequencies so that both the phase and speed loops can control the capstan motor. The control-track signal is passed from IC_{3-3} to IC_{4-3}, where the signal is amplified and applied to the phase-loop IC_{1-21}. The capstan signal at IC_{3-4} is passed through a divider within microprocessor IC_2, and applied to IC_{1-6}.

During playback, the control-track signal at IC_{1-21} is applied to the input of the PWM counter-latch circuit. During record, the capstan signal is divided down to 30 Hz and applied to PWM system. The capstan-servo system adjusts capstan speed (and thus tape speed), thereby maintaining the divided-down capstan signal at 30 Hz.

To get proper tracking, the 3.58-MHz reference signal at IC_{1-25} is divided down and delayed by an MV. The time constant of the tracking MV is determined by the resistance and capacitance at IC_{2-29}. The front-panel tracking control VR1 and preset controls RT_{13} and RT_{14}, determine the amount of delay to the reference signal to get proper phase of the control-track signal, minimizing noise in the video signal.

The output of the PWM system is a 437-Hz signal appearing at IC_{1-2}. The duty cycle or *pulse width* of the PWM output signal is proportional to the phase error of the system. The output is passed to an external LPF, which removes the ac component and passes the dc component to the speed loop. The dc component of the PWM signal is about 2.6 V.

During special-effects, slow-motion, pause, or cylinder-lock conditions, a high is applied to IC_{1-3}. This disables the PWM system and generates a fixed 50% duty cycle for the 437-Hz signal at IC_{1-2}. In effect, the capstan phase-control is removed from the circuit.

9.11.8 Capstan phase-loop troubleshooting

The most common symptom of problems in the capstan phase-loop circuits is a *noise bar* (or bars) floating through the picture but with the picture properly synchronized. (If the capstan speed-loop is malfunctioning, you get excessive audio *wow* and *flutter*, along with picture instability, out of sync, etc.) If you suspect problems in the capstan phase-loop circuits, set the VCR to pause, and check the 437-Hz PWM signal at IC_{1-2} for a 50% duty cycle. If absent or abnormal, suspect IC_1.

If the PWM signal is good, return the VCR to play, and play back a known-good tape or a test tape. Check for the presence of the control-track signal at IC_{3-1}. If the control-track signal is absent or abnormal, suspect a defective control-track head, connector cable, or possible tape drive-path problem (tape not moving past the control head properly).

If the control-track signal is good, check for a signal at IC_{3-3}. If it is absent or abnormal, suspect IC_3. Then check for a signal at IC_{4-21} and IC_{1-21}. If the signal is absent or abnormal, suspect IC_4. If the signal at IC_{1-21} is good, check the 34.58-MHz reference signal at IC_{1-25}.

If both the comparison signal (IC_{1-21}) and reference signal (IC_{1-25}) are normal but the capstan phase loop appears to have no control of the capstan (noise bars, etc.), check that IC_{1-3} is low (normal phase-loop operation). If IC_{1-3} is low, suspect IC_1. If IC_{1-3} is high (in normal play), check the pause, slow, and cylinder-lock lines from system control.

9.11.9 Capstan speed-loop operation

As shown in Fig. 9.27, the capstan speed-loop develops a correction voltage resulting from a difference in frequency between the capstan signal and the divided-down 3.58-MHz signal. The capstan signal is passed through the programmable divider in IC_3 to IC_{2-32}. IC_3 divides the capstan signal by 5 in the SP search mode and by 15 in the SPL search mode. The capstan signal is not divided by IC_3 in normal playback. The capstan signal is further divided with microprocessor IC_2 and passed to a digital counter, which is part of the PWM system of IC_2.

The 3.58-MHz reference signal is first divided by 8 in IC_1 and passed to IC_{2-35} as a 477-kHz signal. The divided-down signal is then processed and applied to a counter within IC_2 for comparison with the divided-down capstan signal. The PWM circuit develops a dc output voltage corresponding to the difference in frequency between the capstan signal and the 3.58-MHz reference.

Capstan speed control RT_8 at IC_{2-27} adjusts the level of the dc correction voltage for the speed loop. The correction voltage is combined with the capstan phase-control signal by an adder in IC_2. The combined speed and phase-loop voltages are applied to a variable amplifier. The amplifier gain is determined by the auto-speed-select circuit in IC_2. The output at IC_{2-28} is passed to the capstan motor-drive IC.

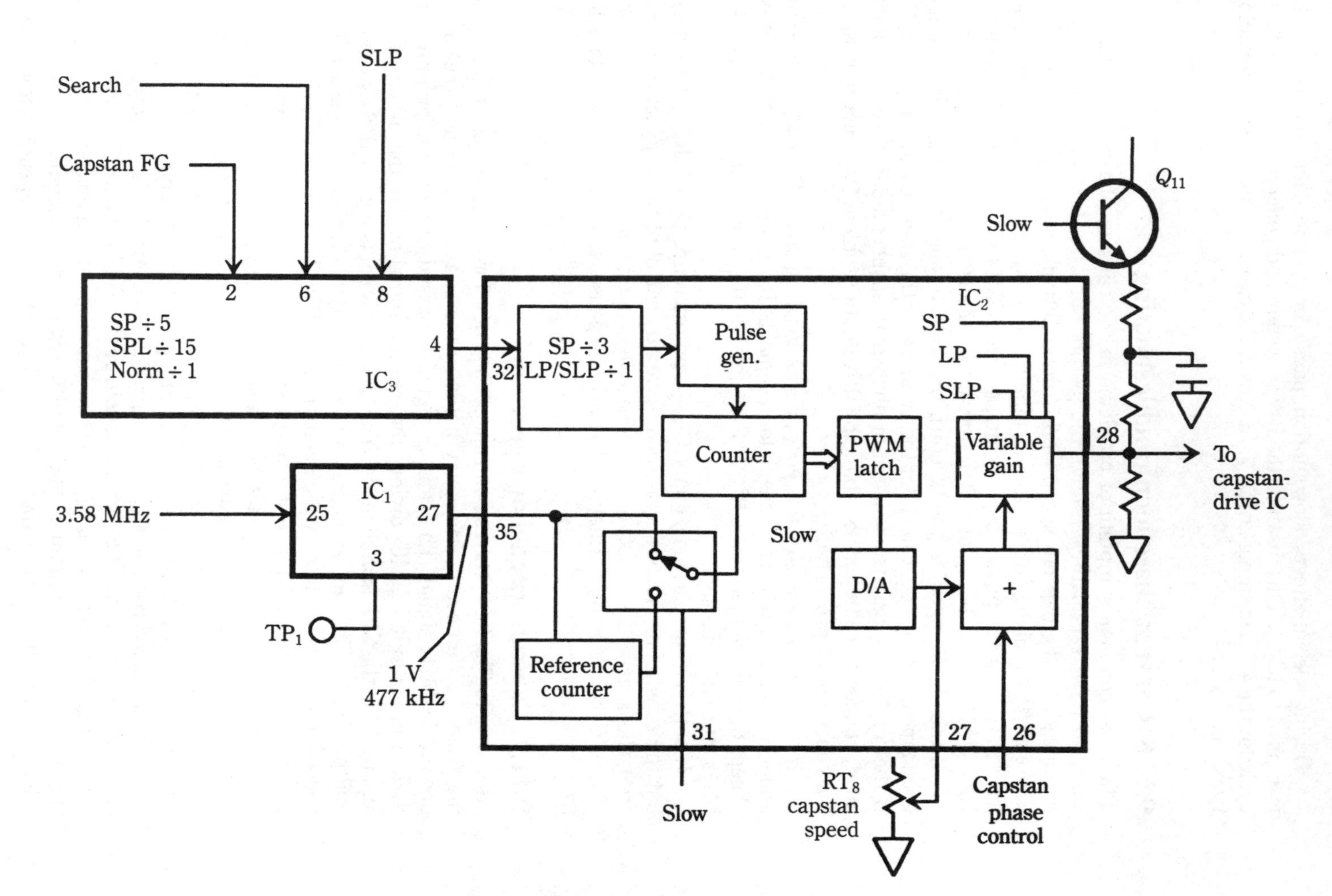

Fig. 9.27 Microprocessor circuits (speed loop) for digital capstan-servo.

During slow motion, the capstan servo speed loop is disabled by a high at IC_{2-31}. The output drive voltage is also pulled up to about 3.5 V by turning on Q_{11}. In still operation, the output drive voltage is pulled to ground by IC_4.

9.11.10 Capstan speed-loop troubleshooting

The most common symptoms of problems in the capstan speed loop are excessive audio wow and flutter, picture instability (out of sync), or both. The first step is to disable the phase-loop circuit by connecting TP1 (IC_{1-3}) to power (usually about 5 V). This disables the phase-loop operation and sets the output of the PWM system to 50 percent. Check for about 2.6 V at IC_{2-26}. If absent or abnormal, suspect a problem with the 50% duty cycle square-wave output from the phase loop or in the LPF between phase-loop and speed-loop (Fig. 9.26).

If the input from the phase loop at IC_{2-26} is normal, monitor the capstan free-run waveform while slowing rotating capstan speed control RT_8. Adjust RT_8 to get the proper sampling pulse and trapezoid lockup. (The service literature describes the exact procedures. Follow them.)

If the capstan speed loop can be locked in (picture stable, minimum or no wow and flutter) by adjustment, with the phase-loop disabled, suspect the phase-loop components (Fig. 9.26). If you cannot correct capstan speed-loop problems by adjustment of RT_8, look for the correct inputs to the capstan speed-loop at IC_{2-32} and IC_{2-35} (Fig. 9.27).

If the input signals appear normal, suspect IC_2. If the signals are absent or abnormal, look for a 3.58-MHz reference at IC_{1-25} and for a capstan FG signal at IC_{3-2}. If either signal is absent or abnormal, trace the 3.58-MHz reference line and/or capstan FG line. If inputs at IC_1 are good, suspect IC_1 and IC_3.

10
Troubleshooting digital TV/video circuits

This chapter is devoted to troubleshooting digital equipment where the bulk of the processing circuits are contained within a few ICs. The step-by-step examples in this chapter represent typical digital video circuits such as found in digital TV, video-display terminals, and CDV (Laserdisc) special effects.

10.1 Introduction to digital TV

The digital-TV circuits described here receive conventional TV signals and produce corresponding pictures and sound (although both video and sound are generally superior to those of conventional TV).

The key to digital TV is in A/D and D/A conversion (Sec. 3.3). In the simplest of terms, the analog signals (composite TV video and sound) at the output of a TV tuner are converted to the digital equivalent by an A/D process somewhat similar to that found with compact discs. The resulting digital signals are then processed to produce the corresponding video and audio in digital form.

When the processing is complete, the digital signals are restored to analog form by D/A converters and applied to the picture tube and audio circuits. The A/D conversion, digital processing, and D/A conversion all take place in digital ICs.

10.1.1 Advantages of digital TV

The obvious advantage of digital processing is in the quality of the signal reproduced. As in the case of CDs, the tuner signals to be processed are sampled (at high frequency), the sampled signals are stored digitally, and after processing, the restored signals are retrieved.

Another advantage is that a digital TV set can easily be adapted to the three basic TV systems because the sampling clock for the A/D converter is phase-locked to the broadcast color-burst frequency. Simply by changing the frequency of the clock, the system can accommodate NTSC (3.58 MHz) or PAL (4.43 MHz) color-burst systems. The same digital TV can also be used for SECAM in black and white. However, the system must be modified for color SECAM (because the PAL and SECAM color techniques are different).

For those not familiar with the three TV broadcast systems, the NTSC (National Television Standards Committee) system (525 lines) is used exclusively in North America and widely in Latin America and Japan; PAL (phase alternation by line, 625 lines) is used in the United Kingdom, most of Western Europe, Australia, New Zealand, and South Africa; SECAM (sequential color with memory, or in French, *sequential couleur avec memoire*, 625 lines) is used in France and Eastern Europe.

Because the TV picture can be stored (in digital form), a digital TV can reduce flicker caused in interlaced scanning and can increase the *apparent resolution* of the picture. Digital TV eliminates interlace by storing all 525 lines (or 625 lines) and displaying the complete picture on the screen all at once instead of having only half the scan lines on the screen for each field of video, as is the case with conventional TV.

In addition to these obvious advantages, digital TV is generally superior because sync is checked on each horizontal line with PLL circuits, and there are a minimum number of capacitors, inductors, and RC (resistive-capacitive) circuits to break down or to distort video signals.

10.1.2 Special effects and tricks

In addition to superior performance, the big selling feature of digital TV is the ability to do tricks or display special effects. The following is a brief summary of these special effects. Not all digital TVs can do all of the tricks described here. All of the effects are programmed into digital ICs.

The *mosaic* and *paint* effects are available on some digital-TV systems. With mosaic, the screen is a picture composed of small blocks. The paint effect is similar to an oil painting and is called *posterization*.

The *freeze mode* freezes the current picture being viewed. Either field or frame reproduction can be selected to minimize jitter. Simultaneously, a real-time picture can be set into one corner of the picture tube screen if desired.

The *preview mode* displays the still picture of nine (or more) channels in sequence, arranged in three rows and three columns on the screen. At predetermined times, the display is changed so that all channels programmed into the tuning-system memory are scanned.

The *picture-in-picture mode* inserts a miniature, real-time picture from an external video input (possibly another channel) in the corner of the screen. The main picture and insert picture can be exchanged at any time by pressing a button.

The *strobe mode* displays eight (or more) time-sequenced still pictures at once, while showing the real-time picture in the lower corner of the screen. The *editing mode* allows the user to change the still pictures displayed in the strobe mode.

10.1.3 The basic five-chip digital TV

Figure 10.1 shows the so-called basic five-chip digital TV system in block form. If you look closely, you will find seven chips or ICs. However, the clock chip is generally not counted, and there are typically two identical audio-processor chips (for stereo operation).

Each of these chips is discussed in the following sections. However, remember two points when studying the five-chip system. First, not all digital TVs have all five chips. Second, although analog signals are fed and analog signals come out, all the signal processing is done digitally in the five-chip system. This is not necessarily true for some digital TVs that do not use the five-chip system.

10.2 Typical digital TV circuits

This section describes the video and related circuits found in a Zenith Digital System 3 TV. The circuits described here use the full five-chip configuration, and have a general circuit arrangement as shown in Fig. 10.1.

10.2.1 Modular configuration

The circuits for the Digital System 3 are contained on six modules. Five of these modules include conventional circuits and/or perform TV-set operations that are used on other (nondigital) Zenith sets. The five conventional modules are not described here, except for the input-output relationship to the digital circuits.

All circuits that are unique to the digital functions are contained on one module, designated as the digital main module 9-535, and generally referred to as the *9-535* or *main* module.

10.2.2 Digital main module

Figure 10.2 shows the IC layout of the 9-535 module. From a troubleshooting standpoint, there are several important features to remember concerning the digital main module.

First, with one exception, all of the ICs on the 9-535 (both digital and nondigital) plug in. The exception is the ADC IC_{1404}, which is permanently installed on the module. As with any plug-in device, if all other troubleshooting steps fail, you can replace the ICs one at a time until the problem is cleared (or you can replace the entire module at unbelievable expense to the customer).

All the voltages and signals to and from the digital ICs can be measured at this module. If the voltages and/or signals are incorrect, you can trace back to the source from this module. That approach is used here. Before you get into the circuit details, see how the Zenith digital ICs compare with the ICs shown in Fig. 10.1.

10.2.3 CCU IC_{6001}

The central control unit (CCU) IC_{6001} is an eight-bit microprocessor containing a 6.5-kilobyte ROM and a 120-byte RAM. IC_{6001} has direct electrical connection to nine of the 14 ICs on the main module. The program executed by the ROM section of IC_{6001} affects operation of four other digital ICs on the 9-535.

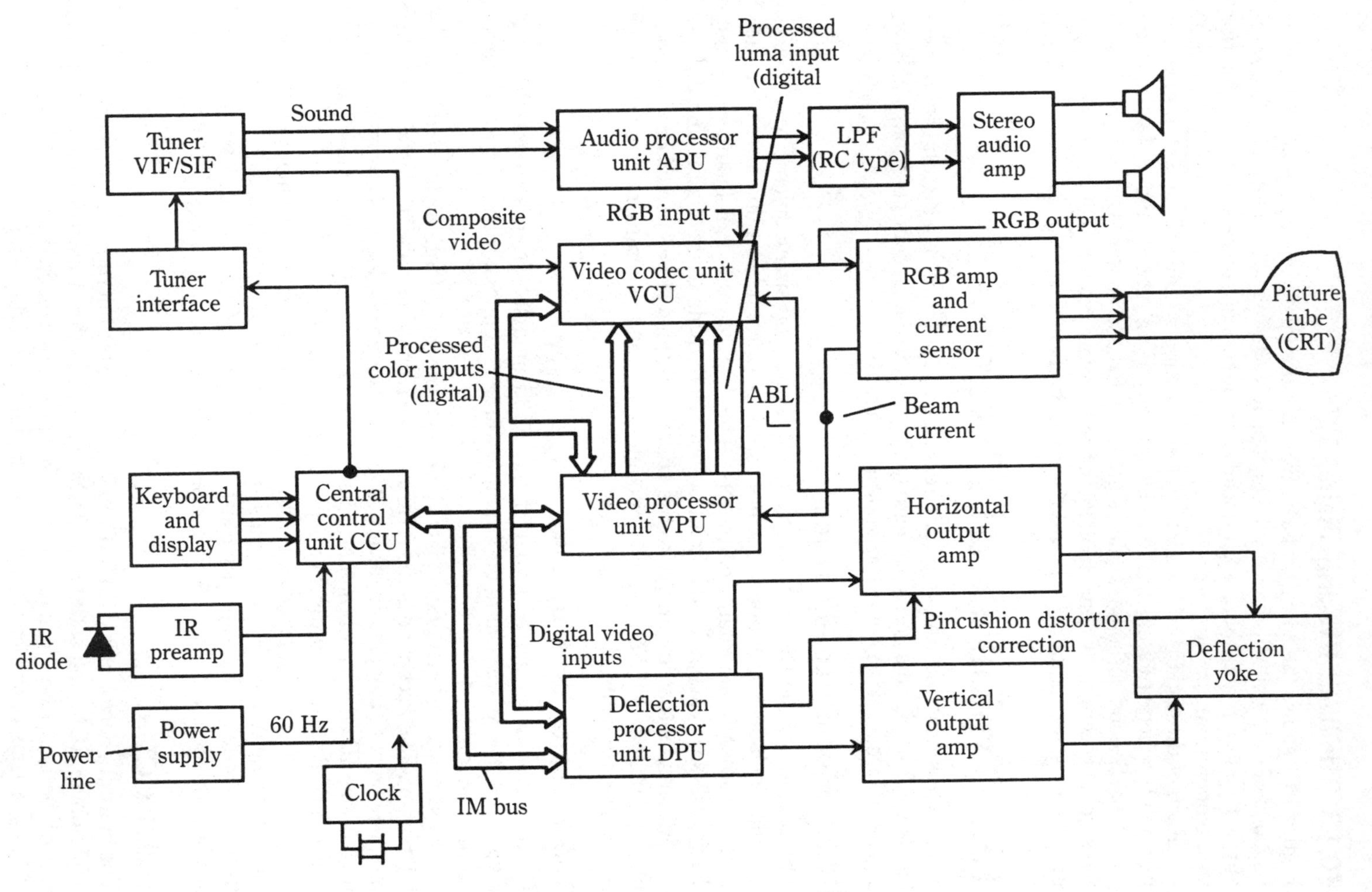

Fig. 10.1 Basic five-chip digital TV.

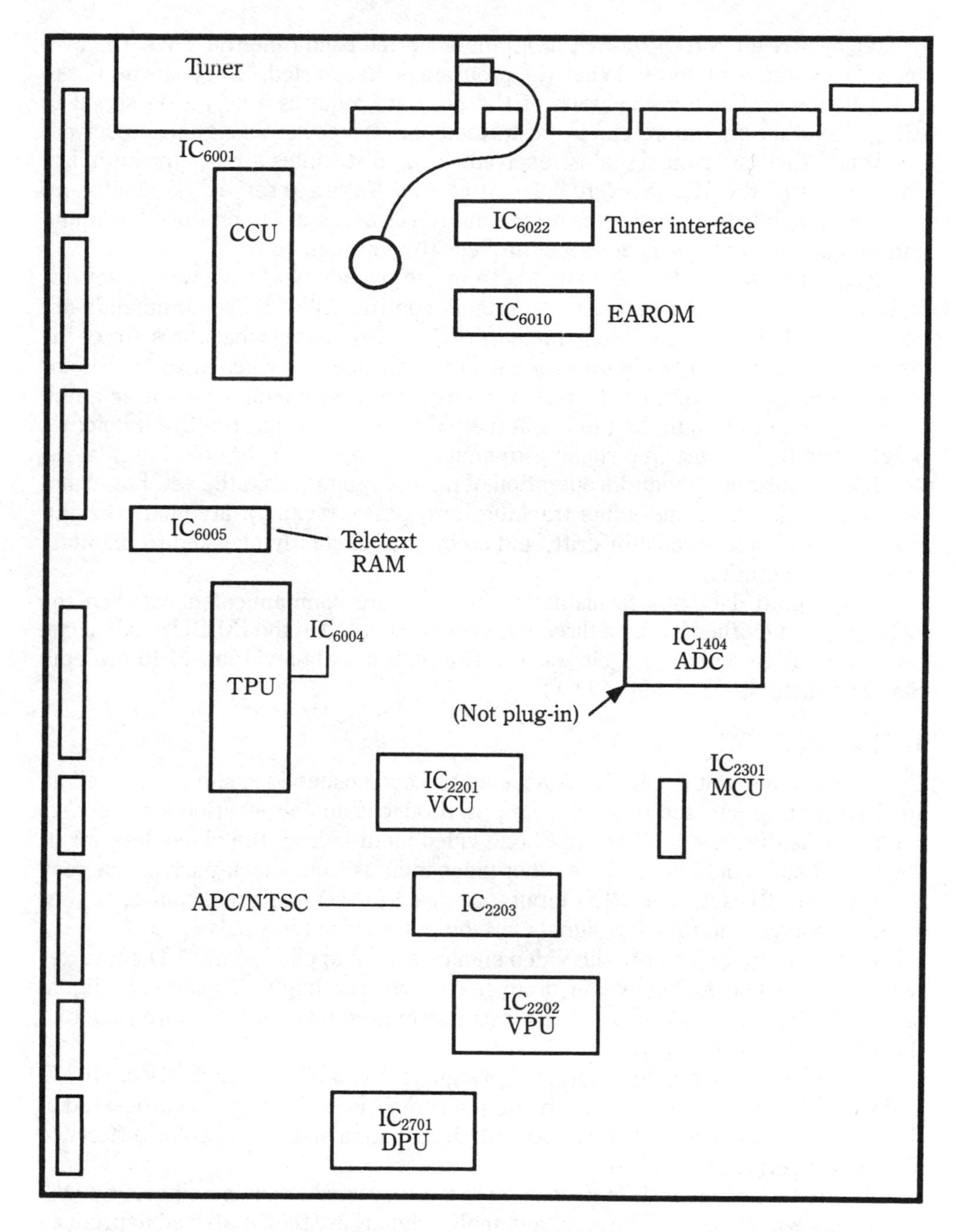

Fig. 10.2 Digital main-module layout.

IC_{6001} is reset and the stored program is started each time the TV set is connected to a source of power. When the program is first started, IC_{6001} sets all ICs to the wait mode. Further execution of the program requires a power-on signal to IC_{6001}. No other operations can be performed until the power-on signal is received.

When the power-on signal is received, IC_{6001} distributes a reset (or initialization) signal to the ICs (Sec. 10.3.4). After the ICs are reset, IC_{6001} sends and receives data bits to and from the remaining ICs as necessary to produce the sound and video. This process is described in Sec. 10.4 through 10.6.

IC_{6001} also responds to commands from the outside world (viewer or service technician) through the keyboard or remote control. All of these commands are coded digital signals (data bits) applied to IC_{6001}. No matter what the source, the commands cause changes in operation (channel changes, changes in volume, color level, black level, picture level, etc.). Similarly, the service technician (or an ambitious tinkerer) can input data to IC_{6001} that (1) centers the picture, (2) sets picture height, and (3) sets the free-running frequency of the master (or color) oscillator.

IC_{6001} constantly monitors operation of many circuits within the set. For example, automatic black and white tracking (gray-scale tracking), automatic brightness limiting, color-oscillator drift, and so on are constantly checked to maintain proper set operation.

The digital data bits to maintain operation are communicated between the CCU IC_{6001} and other ICs by a three-wire connection called the IM BUS. IC_{6001} can both send and receive data bits on the IM BUS (an abbreviation of Intermetall Semiconductors, a division of ITT).

10.2.4 VCU IC_{2201}

The video codec unit (VCU) IC_{2201} accepts two composite-video signals, as well as an RGB (red, green, and blue) video input. (Codec is an abbreviation for encoder-decoder.) As discussed in Sec. 10.6, one video input is from the video detector in the IF and audio module, and the other video input is from a jack-pack on the rear panel of the TV set. The RGB inputs are used for (1) teletext displays, (2) on-screen displays, and (3) RGB signals for computer video terminals.

Either of the two composite-video signals can be applied to an A/D converter within IC_{2201}. The A/D converter produces a corresponding video signal in digital form. The RGB inputs are not *digitized* (the term used by Zenith) but are routed to the RGB outputs of IC_{2201}.

The VCU IC_{2201} applies digital video signals to two ICs: the DPU (Sec. 10.2.7) and the VPU (Sec. 10.2.5). After the digital composite-video signal is processed in these two ICs and the APC (Sec. 10.2.6), digital-luma and digital-color-difference signals are returned to IC_{2201}.

The luma and color-difference signals are converted to analog signals by D/A converters within IC_{2201}. The resultant analog signals are then matrixed to produce the RGB output from VCU IC_{2201}.

An *automatic brightness-limiting* (ABL) signal, developed by sensing the average total picture-current on the horizontal output module, is applied to IC_{2201}. Black-and-white operation of the set is determined by this ABL signal, and an automatic black-and-white tracking signal applied to the VPU.

10.2.5 VPU IC_{2202}

The video processing unit (VPU) IC_{2202} accepts a digital composite-video signal. A digital bandpass filter within IC_{2201} separates the luma and chroma information.

The digital-luma signal is output from IC_{2202} and applied to the APC (Sec. 10.2.6). (On this particular set, a binary-code converter converts the seven-lead output into an eight-lead format.)

The digital-chroma or color-difference signals R-Y and B-Y are applied to the APC. The color-difference signals are *time-shared* (or multiplexed) with other digital information that determines picture operation.

The other information is obtained by applying three monitor or sensing signals to IC_{2202}. These signals are: (1) signals from the output module that indicate the beam current for each picture-tube gun, (2) a signal from a photo sensor that senses the ambient light level around the TV set, and (3) an input signal that represents the cut-off current for each picture-tube gun. The three monitor signals are time shared with the chroma signals.

IC_{2202} contains a phase comparator used to synchronize the 14.3-MHz oscillator in MCU IC_{2301} (Sec. 10.2.8). The phase comparator receives the RF-burst portion of the chroma signal within IC_{2202} and the master-clock signal from IC_{2301}. The control loop for synchronization is completed by applying a digital correction signal back to MCU IC_{2301}. The correction signal is gated into the MCU by a clock signal.

10.2.6 APC and NTSC IC_{2203}

The *automatic picture control* (APC) IC_{2203} is also known as the NTSC IC. IC_{2203} is not part of the basic five-chip configuration but is an IC specifically developed to process NTSC video signals. Zenith includes APC and NTSC IC_{2203} to improve operation.

The outputs from IC_{2203} are corrected color and luma signals returned to VCU IC_{2201}. These output signals are corrected automatically (by IC_{2203}) for brightness, contrast, color separation, and flesh tone.

10.2.7 DPU IC_{2701}

The *deflection processor unit* (DPU) IC_{2701} contains circuits that process the digital composite-video signal into correctly timed and phased horizontal- and vertical-drive signals.

A standard signal detector (SSD) circuit within IC_{2701} recognizes the composite-video signal as either a color or a black-and-white signal. In either case, both the horizontal and vertical drive signals are locked to the 14.3-MHz clock from MCU IC_{2301}.

IC_{2701} contains counters, dividers, and phase comparators to develop proper horizontal- and vertical-drive signals. For example, to get the proper horizontal frequency when a color signal is present, a programmable counter is set to divide the clock signal by 910. To phase the same horizontal signal properly, the counter output is phase-compared to the flyback phase.

To get proper vertical-drive signals, the horizontal signal is divided by 525 twice. To phase the same vertical signal properly, the counter output is phase-compared to the integrated vertical-pulse signal (obtained from a sync comparator within IC_{2701}).

DPU IC_{2701} also develops four keying and blanking signals. These signals are applied to the VCU, APC, and VPU to process the digital video signal.

10.2.8 MCU IC_{2301}

The master clock unit (MCU) IC_{2301} contains a VCO that generates the main timing signals for all ICs. The 14.3-MHz output of IC_{2301} is locked to the RF-burst portion of the received color signal by a phase comparator in the VPU.

The digital-correction voltage (PLL) from the phase comparator is applied to IC_{2301}, which contains a D/A converter that converts the digital-correction voltage to an analog signal used by the VCO.

The MCU IC_{2301} output is applied to seven of the ICs on the 9-535 module. The clock output is used to (1) synchronize A/D and D/A converters for both video and audio signals, (2) time horizontal- and vertical-sweep signals, and (3) time movement of the data from one circuit to another.

10.2.9 TPU IC_{6004} and RAM IC_{6005}

The teletext processor unit (TPU) IC_{6004} and related RAM IC_{6005} are not part of the basic five-chip system (Fig. 10.1), so we do not describe them here. Both ICs are part of the Zenith Teletext System, which is not currently in use.

10.2.10 EAROM IC_{6010}

The electrically alterable ROM (EAROM) IC_{6010} is a nonvolatile, reprogrammable IC capable of storing 128 eight-bit words (and is not part of the basic five-chip system). As in the case with other EAROMs (Chapter 4), the words are retained when power is removed. The words can be changed when a programming voltage (about 20 V) and a write signal are applied to IC_{6010}. When a read signal is applied, the words can be transferred out of the EAROM.

The information stored in IC_{6010} consists of such factors as favorite channels, color-oscillator frequency, vertical height, and so on. IC_{6010} is connected only to the CCU. All information to and from the EAROM is passed through the CCU. Data bits are transferred between the EAROM and CCU via the IM bus.

10.2.11 Tuner control, or interface IC_{6022}

The tuner control, or interface IC_{6022}, operates with the CCU (and two transistors) to develop the frequency-control voltage for the tuner mounted on the 9-535 module. The digital-tuning system uses frequency synthesis (FS) with PLL control, such as described in Sec. 9.8.

10.3 Basic digital TV troubleshooting

Before getting into troubleshooting details for the Zenith Digital System 3, the next section reviews the basic troubleshooting approach for any digital TV (and for most digital equipment). Remember that this chapter is concerned with locating troubles caused by the digital ICs, not with troubleshooting the remaining circuits (which are the same as those found in other nondigital TVs).

10.3.1 Plug-in IC replacement

Because all the Zenith digital ICs plug in (except the ADC), it is practical to try correcting the problem by replacement. (This is not practical where ICs are wired in.) By replacing each IC in turn with a known-good IC and checking to see if the trouble symptoms are eliminated, you can cure about 90% of the problems in a digital TV (except for those problems caused by improper adjustment, which are covered in the service literature).

Of course, it is helpful if you start by replacing the ICs most likely to cause the problem. For example, in the Zenith set, if the problem is bad video, with good audio, start by replacing the VCU, VPU, APC, and NTSC. If the problem appears to be in deflection, either horizontal or vertical, start with the DPU. If the problem cannot be easily localized, start with the CCU.

Obviously, to make a logical choice for replacement of ICs, you must be able to group those ICs that perform a specific function. You do just that for the Zenith set in the troubleshooting discussions of Sec. 10.4 through 10.6. For now, consider the terrible possibility that all of the digital ICs have been replaced with known-good ICs, all adjustments have been made (in accordance with the service literature), but the problem remains.

10.3.2 Modular replacement

At this point, if a known-good digital main module is available, you can try substitution. If this cures the problem, you can then trace the problem on the device module. During replacement, you might find that the problem is one of poor electrical connection (such as loose connectors, dirty contacts, etc.).

Unfortunately, modular replacement might not always be practical. You might not have a replacement module readily available, or you might have only one shop-standard module (that you will not surrender at any price). A shop-standard module is recommended if you plan to service a particular model of digital TV. Of course, you might not want to invest in dozens of known-good modules for a variety of digital TVs (if such a variety ever comes to pass).

If you do choose modular replacement, take special care when reinstalling the shielding systems. Digital signals are quite high in amplitude and frequency. This can cause interference in nearby electronic equipment or in the picture being received. Such problems are most evident in areas where signal levels are weak, and where antennas are used (instead of cable). To eliminate problems, make certain that all shield covers are reinstalled and that locking clips are tight. All ground connections should be carefully resoldered (poor solder connections cause more problems than any other trouble source).

10.3.3 Power and ground connections

The first step in tracing problems on a known or suspected defective module (when the problem is not corrected by substitution or known-good ICs and adjustment) is to check all power and ground connections to the ICs. Make certain to check all power and ground connections to each IC because many ICs have more than one power and one ground connection. For example, additional grounds are required for the MCU (in the Zenith set) because the PAL and SECAM clock oscillators must be disabled (so they won't create interference). Also, look for any special power inputs. For example, IC_{6001} requires a 60-Hz sync signal (from the power line through the power-supply module).

10.3.4 Reset signals

With all of the power and ground connections confirmed, check that all the ICs are properly reset. The reset connections are shown in Fig. 10.3*a*.

CCU IC_{6001} is reset when the power cord is first connected and standby power is applied. The remaining ICs in the digital system are reset (at a 4-MHz rate) from IC_{6001} (at pin 39) when a power-on signal is applied to IC_{6001}.

One simple way to check the reset function is to check for reset pulses at the appropriate pins. For example, pin 4 of IC_{6001} should be low (near zero) when standby power is first applied (because of the drop across R_{6041}) and then rise (to about 4.5 V) when C_{6013} charges through CR_{6001}.

The low applied to the IC_{6001} causes the circuits within IC_{6001} to reset. After reset, when pin 4 rises to 4.5 V, IC_{6001} remains ready to perform the CCU functions, unless the voltage at pin 4 drops to a low for a prolonged period and produces a reset.

If the IC_{6001} reset is not as described, suspect C_{6013}, CR_{6001}, and R_{6041}. However, remember that IC_{6001} is not reset by temporary drops in the standby power (because of the charge on C_{6013}) but should be reset once each time the standby power is removed and reapplied.

As in the case of any digital device, if the reset line is open or shorted to ground or to power (5 or 12 V), the ICs are not reset (or remain stuck in reset) no matter what control signals are applied. This brings the entire digital operation to a halt. So if you find a reset line that is always high, always low, or apparently connected to nothing (floating), check the line carefully.

10.3.5 Clock signals

The clock signals for the digital ICs are shown in Fig. 10.3*b*. Note that there are two clocks in the Zenith set. The CCU IC_{6001} has a 4-MHz clock at pin 1. This clock signal can be monitored at the C_{6031} test point. The other clock signal (14.3 MHz) is taken from the MCU (IC_{2301}). As discussed, this clock is locked to the RF-burst by a PLL phase comparator in VPU IC_{2202}. The PLL connections are at pins 5 and 6 of IC_{2301}. The 14.3-MHz master clock (for NTSC) is connected to the remaining ICs as shown in Fig. 10.3*b*.

It is possible to measure the presence of a clock signal with a scope or digital-logic probe. However, a frequency counter provides the most accurate measurement. Obviously, if any of the ICs do not receive the clock signal, the IC cannot function. On the other hand, if the clock is off frequency, all of the ICs might appear to have a clock signal, but IC function can be impaired. (Note that crystal-controlled clocks do not usually drift off frequency but can go into some overtone frequency, typically a third overtone.)

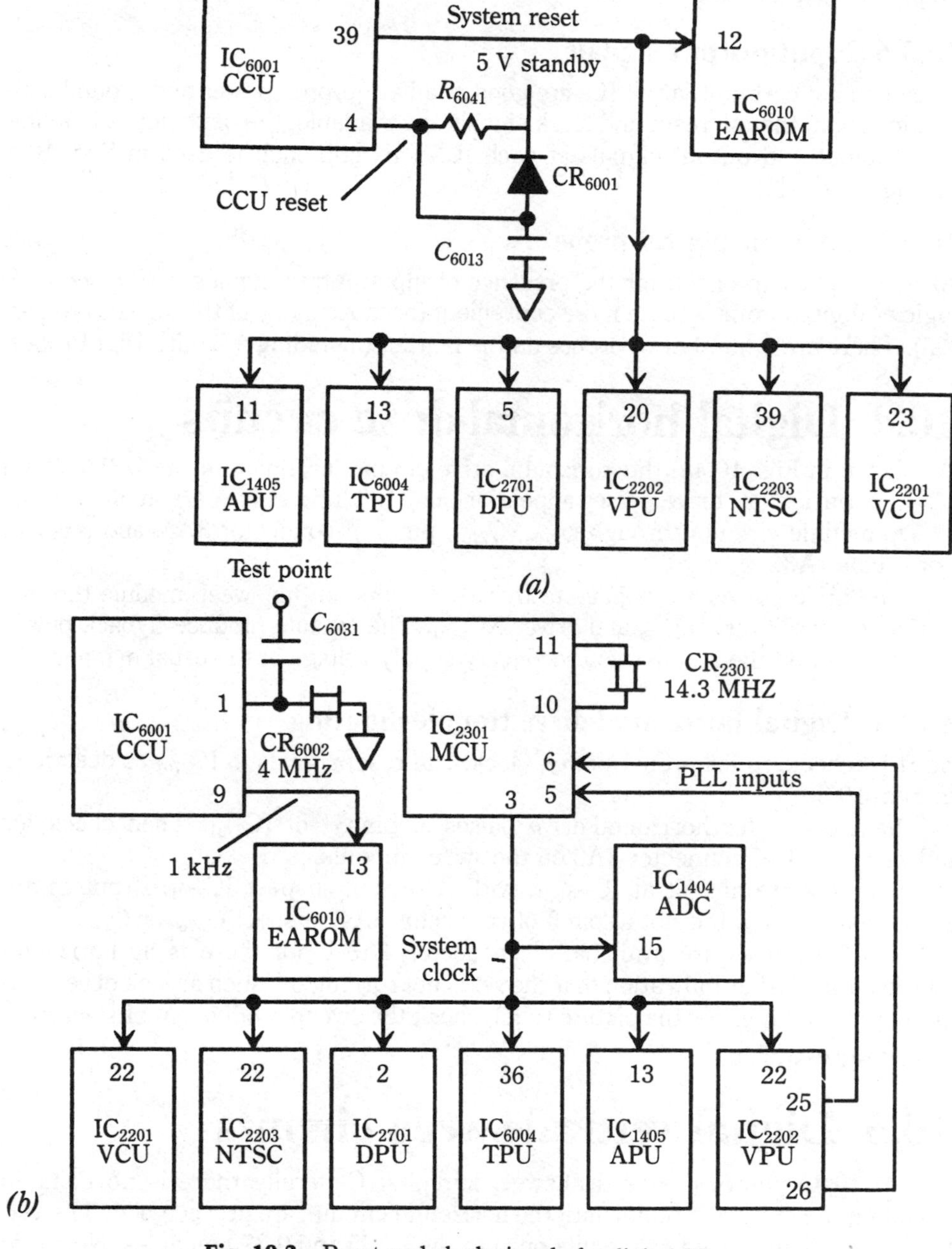

Fig. 10.3 Reset and clock signals for digital ICs.

If the 4-MHz clock is absent or abnormal, suspect CR_{6002} and IC_{6001}. If the 14.3-MHz clock is absent or abnormal, suspect CR_{2301}, IC_{2301}, and IC_{2202}. Also check for pulse activity on the PLL lines between IC_{2301} and IC_{2202}. Generally, the presence of pulse activity on these lines shows that the PLL circuits are probably good.

10.3.6 Input/output signals

Once you are certain that all ICs are good and have proper power and ground connections, and that all reset and clock signals are available, the next step is to monitor all input and output signals at each IC. This approach is used in Sec. 10.4 through Sec. 10.6.

10.3.7 Logic or digital probe

Although you can check for the presence of input/output signals with a scope, a logic or digital probe is often more convenient (because many of the signals are digital). There are a number of probes on the market (including a Zenith Digi-Probe).

10.4 Digital horizontal-drive circuits

As shown in Fig. 10.4*a*, the horizontal-drive circuits originate in the DPU. When IC_{2701} is turned on, drive pulses appear at pin 31. These pulses are applied to the sweep-module circuits through Q_{2700}, Q_{2701}, pin 5 of connector 3A4, and pin 1 of connector 4A3.

The drive pulses are applied to flyback circuits on the sweep module through horizontal predriver XQ_{3202} and driver XQ_{3201}. The circuits produce flyback pulses for the picture tube and develop secondary supply voltage in the usual manner.

10.4.1 Digital horizontal-drive troubleshooting

Start by checking for proper power, clock, and reset signals to IC_{2701} as described in Sec. 10.3.

Next check for horizontal-drive pulses at pin 31 of IC_{2701}. Then check for pulses at pin 1 of connector 4A3 on the sweep module.

If pulses are absent at $IC_{2701\text{-}31}$, with IC_{2701} on, suspect IC_{2701}. If pulses are present at $IC_{2701\text{-}31}$ but not at pin 5 of connector 3A4, suspect Q_{2700} or Q_{2701}.

If drive pulses are available at connector 4A3-1 but there is no horizontal sweep (or any other indication that the set is not turning on, such as lack of secondary supply voltages for the picture tube), check the sweep-module circuits, starting with XQ_{3202}.

10.5 Digital vertical-sweep circuits

Figure 10.4*b* shows the vertical-sweep circuits. Generally there is no point in checking the vertical circuits until the horizontal circuits are proved good. The vertical sweep is produced by circuits on both the 9-535 and 9-370 modules. Note that the 9-370 module provides both power-supply and vertical signal functions (in addition to other functions not related to our digital ICs).

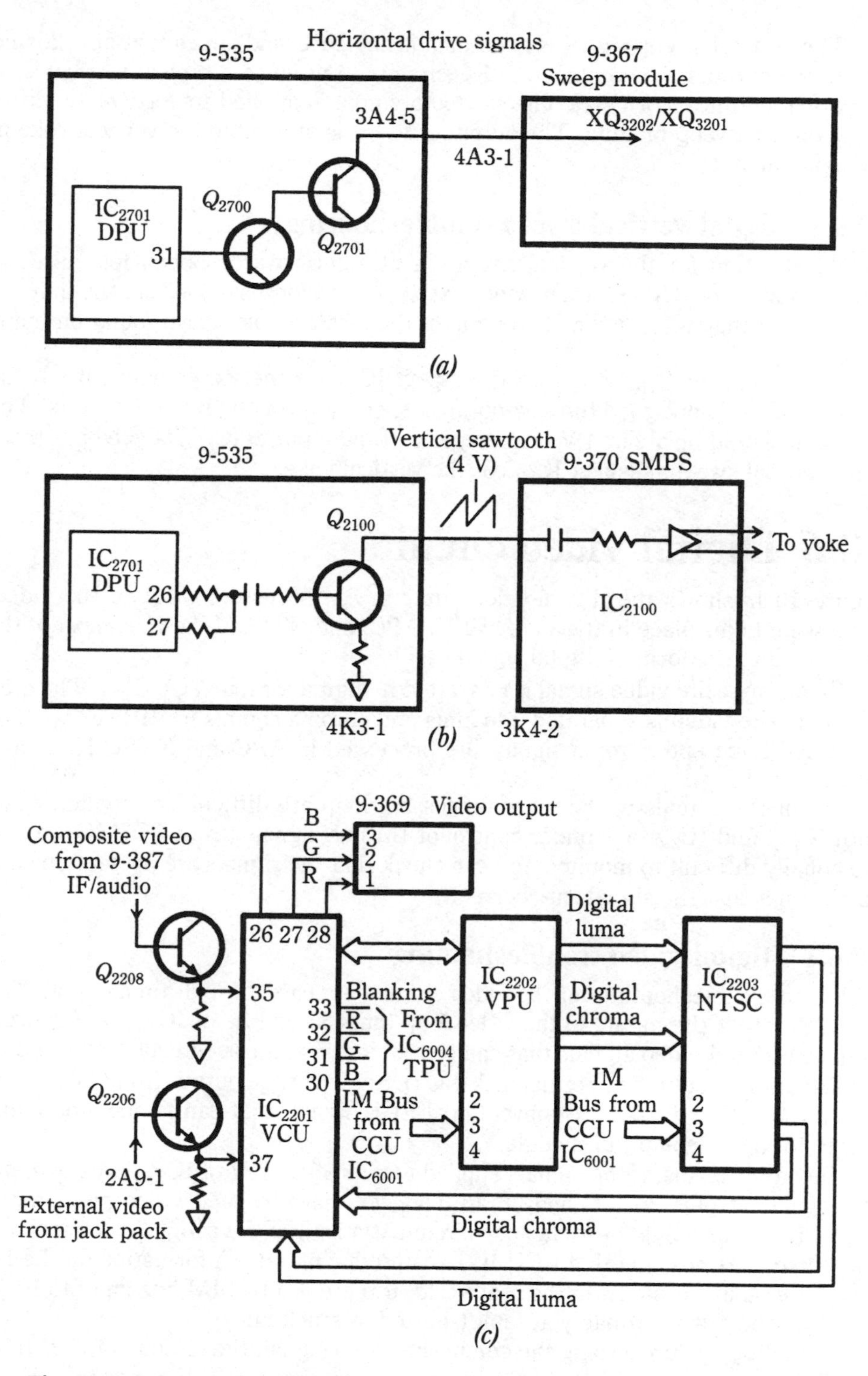

Fig. 10.4 Horizontal drive, vertical sweep, and video processing for digital TV.

The vertical-sweep signal is generated in the DPU and appears at pins 26 and 27. These outputs are combined in RC circuits to produce a typical vertical sawtooth at the emitter of Q_{2100}. The sawtooth signal is applied to vertical amplifier IC_{2100} on the sweep module. The output of IC_{2100} is applied to the vertical yoke in the usual manner.

10.5.1 Digital vertical-sweep troubleshooting

Troubleshooting for the vertical circuits is straightforward. Check for pulses at pins 26 and 27 of IC_{2701}, for a sawtooth at Q_{2100}, and for a sawtooth at the input of IC_{2100}. Note that IC_{2100} is an IC version of the vertical-yoke drive found on many sets.

If the input to Q_{2100} is not good, suspect IC_{2701} or the RC components. If the input to IC_{2100} is not good (no sawtooth), suspect Q_{2100} or the related circuits. The sawtooth should be about 4 V peak to peak. If the input to IC_{2100} is good but there is no vertical sweep, suspect IC_{2100} or the vertical yoke.

10.6 Digital video circuits

Figure 10.4*c* shows the digital-video circuits. As discussed, most of the video-processing takes place in the VCU, VPU, APC, and NTSC. Likewise, most of the processing in the form of digital signals.

The composite video signal is converted to digital form in $VCUC_{2201}$. The digital composite video is separated into luma and chroma signals in VPU IC_{2202}. The separated luma and chroma signals are processed in APC and NTSC IC_{2203} and returned to IC_{2201}.

All of the signals in the video-processing loop are difficult to monitor. Also, both IC_{2202} and IC_{2203} are under control of IM bus signals (from CCU IC_{6001}) that are equally difficult to monitor. You can check that the signals are present on each line but not that the signals are correct.

10.6.1 Digital video troubleshooting

Although troubleshooting for the video circuits appears difficult (in a digital TV), remember that the inputs to the video loop (pins 35 and 37 of IC_{2201}) are conventional baseband video signals that can be monitored and traced back to the source (IF and audio module or external jack-pack). Similarly, the output from the loop (at pins 26, 27, 28 of IC_{2201}) are conventional RGB signals that can be monitored and traced to the video-output module.

In simple terms, if the signals applied to pins 35 and 37 of IC_{2201} are good, but the signals at pins 26, 27, and 28 are bad, you have isolated video problems at IC_{2201}, IC_{2202}, or IC_{2203}, or to the loop circuits (typically PC wiring and connectors). Of course, it is possible that CCU IC_{6001} is producing false information on the IM bus, but this is not likely. In the real world, if there is a bad IM bus input to IC_{2202} or IC_{2203}, the bus is probably at fault (shorted or stuck lines).

In addition to processing the composite video signal, the circuits of Fig. 10.4*c* are also the point at which viewer commands are processed. For example, if the tint button is pressed, commands are applied to the CCU through an IR remote-control system (Sec. 9.7). These commands are converted to digital signals on the IM bus (by the CCU) and are applied to IC_{2202} and IC_{2203} (at pins 2, 3, and 4).

10.7 Digital on-screen display (OSD) circuits

Many digital devices are provided with some form of OSD (on-screen display) circuit. The on-screen channel display of TVs, and the on-screen programming display of VCRs are typical examples. The OSD circuits use a *character generator* IC to produce the desired numbers and letters. The character generators are similar (in some cases identical) to the character generators for video terminals described in Chapter 4 (Fig. 4.16).

The character generator receives sync signals from the same source as the picture tube (so that the characters appear at a given position on the screen). Some OSD circuits include positioning controls. The characters to be displayed are determined by a microprocessor which, in turn, receives commands from push-button controls. In some cases, the commands are hard-wired into the microprocessor.

10.7.1 Typical OSD circuit

Figure 10.5 shows the date and time OSD circuits for the *electronic viewfinder* (EVF) of a camcorder. These OSD circuits provide date and time information that appears on the EVF and that can be recorded on tape if desired. Figure 10.5 also

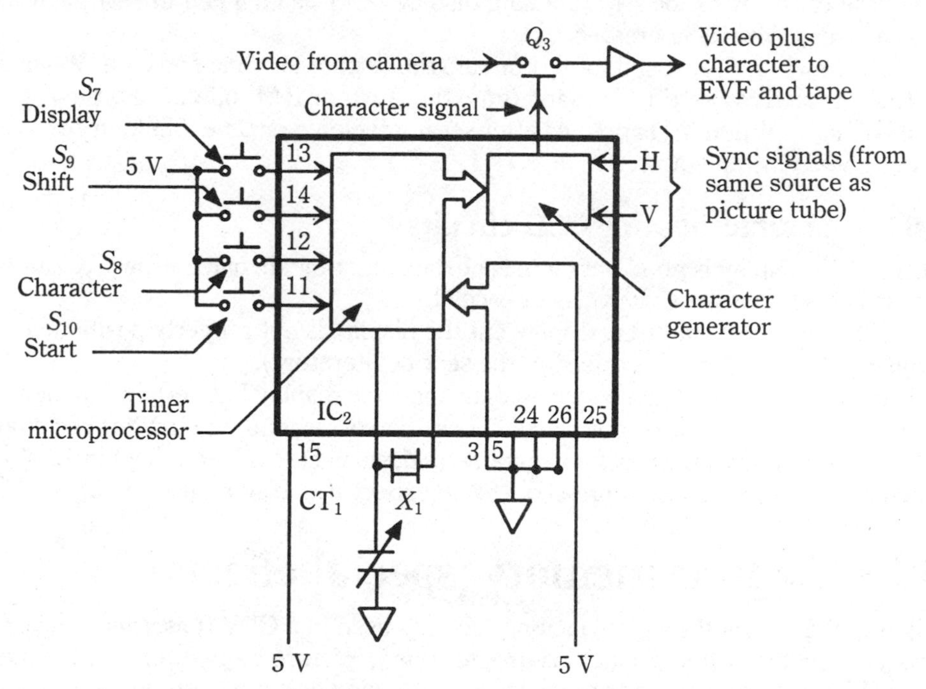

Fig. 10.5 Typical OSD on-screen display circuit.

shows the hard-wire and push-button command connections to produce various combinations of date and time.

Note that most of the OSD circuits are contained within IC_2. When 5-V power is applied to IC_2 at pin 15, crystal X_1 generates a 32.786-kHz clock pulse, and the timer-microprocessor in IC_2 starts operation. CT1 provides a means of adjusting the clock frequency (to properly position the display).

The character generator in IC_2 receives data and time information from the timer and generates corresponding character signals synchronized with the H- and V-signals from the sync generator (the same source of H- and V-sync applied to the base of Q_3. When the character signal is high, Q_3 turns on, and the character is added to the video passing from the camera to the EVF and tape.

The display mode for date and time is selected by the logic at pins 3, 5, 24, 25, and 26 of IC_2. In the configuration shown, the date display is set to the month, day, year format because pin 25 is high (5 V), and the time is set to the 12:15 A.M., 11:15 P.M. format because pins 3, 5, 24, and 26 are low (ground or 0 V).

The date and time setting and display switches S_7 through S_{10} determine the display generated by the character generator. when display switch S_7 is pressed, the character display appears in the EVF screen. The display mode is changed to date, time, data and time, and display-off repeatedly, every time S_7 is pressed.

When shift switch S_9 is pressed, the display blinks. The blinking display is changed to A.M. (P.M.), hour, minute, year, month, and day repeatedly, every time S_9 is pressed.

When character switch S_8 is pressed, the blinking number is incremented. In the case of the A.M. or P.M. blinking display, A.M. is changed to P.M., and vice versa, every time S_8 is pressed.

Start switch S_{10} is used to start or stop the setting for time and date. When S_{10} is first pressed for setting date and time, the A.M. or P.M. blinks, and the setting can be made. When S_{10} is pressed after setting the date or time, blinking stops and date/time counting starts.

10.7.2 Troubleshooting OSD circuits

If the EVF display is good but there is no date and time display, suspect IC_2 and/or Q_3. It is also possible that X_1 is not oscillating.

If there is date and time display but the display is not properly positioned, try adjustment of CT_1 (as described in the service literature).

If there is a properly positioned date and time display but certain functions of the display are absent or abnormal, check the corresponding circuits. For example, if the blinking number is not incremented, suspect S_8. Check that pin 12 of IC_2 goes high (5 V) when S_8 is pressed. If so, suspect IC_2. If not, suspect S_8.

10.8 Digital memory (special effects)

Figure 10.6 shows the digital memory circuits used in a CDV (Laserdisc) player to produce special effects (such as freeze-frame, picture-in-picture, picture step, etc.). Although the circuits shown are for CDV, similar circuits are found in digital TV and in many present-day VCRs.

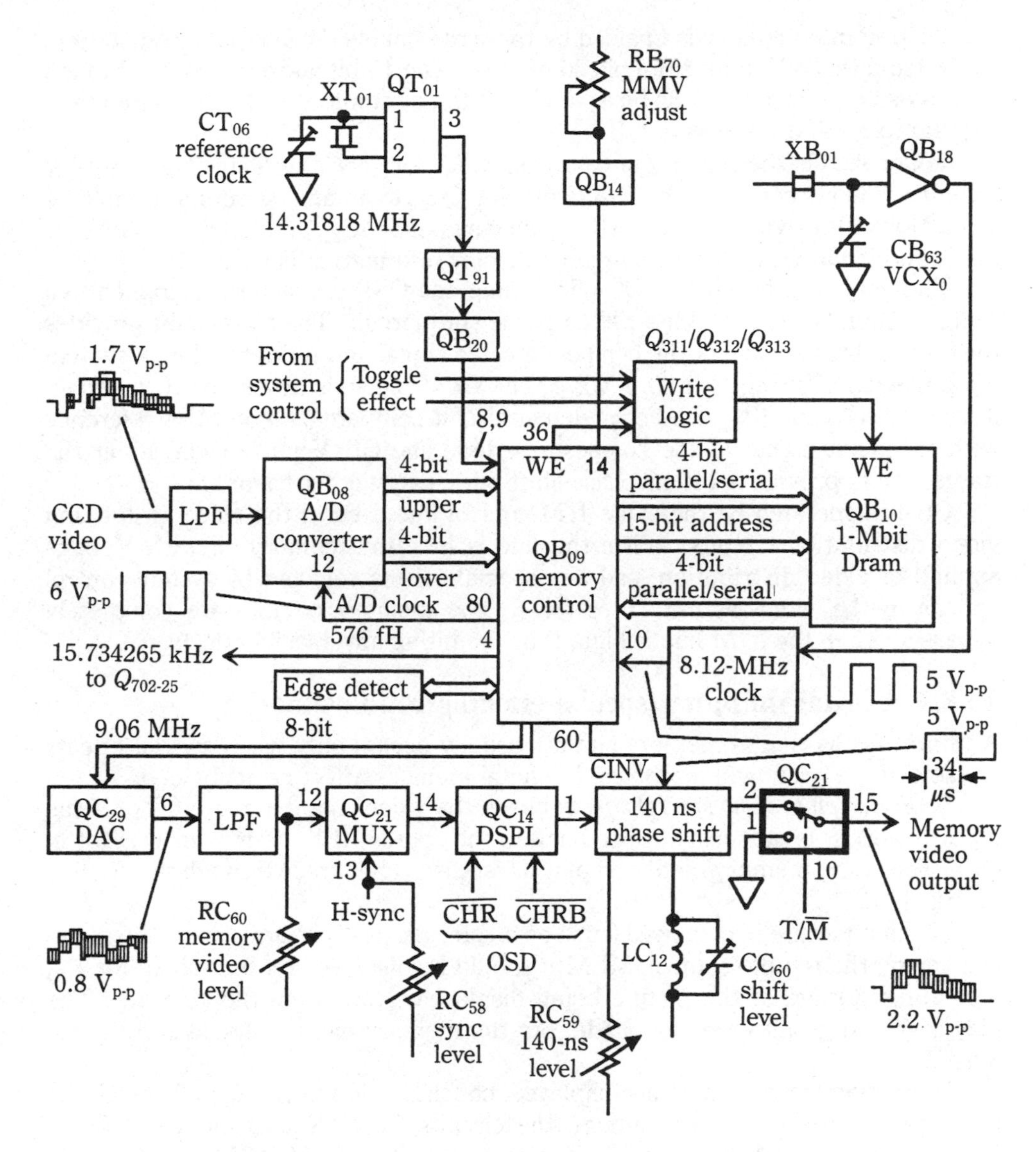

Fig. 10.6 Digital-memory (special effects) circuits.

Video is applied to the digital-memory circuits at A/D converter QB_{08} through an LPF. The video is converted to eight-bit data (four-bit upper and four-bit lower) by QB_{08} at a rate of 576 samples per horizontal line. The A/D clock input at QB_{08-12} is developed by memory control QB_{09}. The A/D clock is at a frequency of 9.06 MHz (which is 576 times the horizontal rate, or 576 fH).

Memory control QB_{09} controls both the writing and reading operations to and from the memory IC QB_{10}. The 18.12-MHz (1152 fH) input clock at pin 10 of QB_{09} is required to write each sample into memory through the four-bit parallel-serial bus.

Writing into memory is enabled by the write-enable (WE) inputs from system control and the WE input from pin 36 of QB_{09}. The 15-bit address bus A0 through A14 is used to write the four-bit by eight-bit (four samples, or 32 bits) data into a four-sample RAM address in QB_{10}.

QB_{09} controls the reading of data from QB_{10} for special effects. The eight-bit memory video is converted by D/A converter QC_{29} to an analog output at pin 6 of QC_{29}. Horizontal sync is added to the analog signal by IC_{21}, and the composite output is applied to QC_{14} (where on-screen display information is inserted).

The output from pin 1 of QC_{14} (including any OSD characters) is applied to a switch within QC_{21d} through a 140-ns phase-shift circuit. The phase shift provides the correct burst-phase of the composite video signal and is operated by a switching pulse (CINV) from pin 60 of QB_{09}. The CINV pulse is developed by an edge-detect circuit and QB_{09}. The edge-detect circuit compares a 3.58-MHz reference with the burst signal of the composite video. The CINV pulse switches at the frame rate to provide a 140-ns phase-shift burst every other frame.

QC_{21} is operated by the same T/M control line used in the video switch and video distribution circuits. When the line is low, the memory circuit output is applied to video distribution, and any special effects selected by system-control appear in the video at pin 15 of QC_{21}. The memory circuits are completely bypassed when the T/M line is high, thus inhibiting any special effects.

10.8.1 Digital memory (special effects) troubleshooting

A failure in the digital-memory circuits usually occurs only when special effects are selected. Rarely will a problem in digital memory affect normal video.

If the symptom is no special effects, check the video signal input to QB_{08}. If the input is absent or abnormal, trace back to the source. If there is a good input to QB_{08}, select the memory mode and play a test disc. Then trace through the circuits of Fig. 10.6.

When a picture is displayed (such as a color-bar test pattern), press the memory key on the remote control. MEM.P should be displayed on the screen for 1 s, indicating storage of the picture being displayed. Now press the stop key. The player should go into the stop mode, but the memory picture should still be displayed.

If the memory picture is not displayed, check the output from pin 6 of QC_{29}. If the signal is present, trace through the circuits from QC_{29} to pin 15 of QC_{21}. Remember that the 140-ns phase-shift circuits require a 5-V CINV signal from pin 60 of QB_{09} to produce the correct output.

If the signal at pin 6 of QC_{29} is absent to abnormal, check the eight-bit bus input to QC_{29}. Again, it is not practical to interpret the eight-bit code from QB_{09} to QC_{29}. However, all of the bus lines should show pulse activity (and all should show the same resistance to ground, with power off) when checked as described in Sec. 8.6.

If one bus line shows no activity, that line is suspect. If all lines show no activity when the memory mode is selected, suspect QB_{08}, QB_{09}, QB_{10}, and the related circuits (edge detect, clock, etc.). Also try performing all of the memory adjustments described in service literature.

If the symptom is poor special effects (tearing, loss of sync, distorted video, etc.) but there is good normal video, check all of the circuits in Fig. 10.6. Pay particular attention to all of the inputs from other circuits. For example, if the horizontal sync is missing during memory, check for proper H-sync signals at pin 13 of QC_{21}, and for proper adjustment of sync level RC_{58}.

Index

D

U

V

W

About the author

For more than 39 years, **John D. Lenk** has been a self-employed consulting technical writer specializing in practical troubleshooting guides. A long-time writer of international best-sellers in the electronics field, he is the author of more than 70 books on electronics which together have sold more than 1 million copies and have been translated into eight languages. Mr. Lenk's guides regularly become classics in their fields and his most recent books include *Complete Guide to Modern VCR Troubleshooting and Repair*, *Digital Television*, and *Practical Solid-State Troubleshooting*, which sold over 100,000 copies.